PUSH BUTTON AGRICULTURE

Robotics, Drones, Satellite-Guided
Soil and Crop Management

PUSH BUTTON AGRICULTURE

Robotics, Drones, Satellite-Guided Soil and Crop Management

K. R. Krishna

AAP | APPLE
ACADEMIC
PRESS

Apple Academic Press Inc. | Apple Academic Press Inc.
3333 Mistwell Crescent | 9 Spinnaker Way
Oakville, ON L6L 0A2 | Waretown, NJ 08758
Canada | USA

©2016 by Apple Academic Press, Inc.

Exclusive worldwide distribution by CRC Press, a member of Taylor & Francis Group

No claim to original U.S. Government works

Printed in the United States of America on acid-free paper

International Standard Book Number-13: 978-1-77188-304-7 (Hardcover)

International Standard Book Number-13: 978-1-77188-305-4 (eBook)

Library and Archives Canada Cataloguing in Publication

Krishna, K. R. (Kowligi R.), author
Push button agriculture : robotics, drones, satellite-guided soil and crop management /
K.R. Krishna.

Includes bibliographical references and index.
Issued in print and electronic formats.
ISBN 978-1-77188-304-7 (hardcover).--ISBN 978-1-77188-305-4 (pdf)
1. Agricultural innovations. 2. Agricultural productivity. 3. Crops. 4. Soil fertility. I. Title.

S494.5.I5K75 2016   338.1'6   C2016-902255-2   C2016-902256-0

Library of Congress Cataloging-in-Publication Data

Names: Krishna, K. R. (Kowligi R.), author.
Title: Push button agriculture : robotics, drones, satellite-guided soil and crop management
/ K. R. Krishna.
Description: Waretown, NJ : Apple Academic Press, 2016. | Includes bibliographical references and index.
Identifiers: LCCN 2016016120 (print) | LCCN 2016022094 (ebook) | ISBN 9781771883047 (hardcover : alk. paper) | ISBN 9781771883054 ()
Subjects: LCSH: Agricultural innovations. | Robotics. | Aerial photography in agriculture. | Agriculture--Remote sensing. | Drone aircraft.
Classification: LCC S494.5.I5 K74 2016 (print) | LCC S494.5.I5 (ebook) | DDC 338.1/6--dc23
LC record available at https://lccn.loc.gov/2016016120

CONTENTS

LIST OF ABBREVIATIONS

ACPA	Asian Conference on Precision Agriculture
ALI	advanced land imager
AMSRE	advanced microwave scanning radiometer
APAL	Apple and Pear Australia Limited
APM	ardupilot mega
ASTER	advanced space-borne thermal emission and reflection radiometer
AT	autonomous tractors
BMP	best management practice
CAI	cellulose absorption index
CC	canopy conductance
CEC	cation exchange capacity
CNES	Centre National d'Etudes Spatials
CSM	crop surface models
CWSI	crop water stress index
DSM	digital surface models
EC	electrical conductivity
EMI	electromagnetic induction
EOLSSS	Encyclopedia of Life Support Systems
EOSAT	Earth Observation Satellite Company
FAA	Federal Aviation Administration
GIS	geographical information system
GNSS	Global Navigation Satellite Systems
GRDC	Grain Research Development Centre
GSD	ground sample distance
HALE	high altitude long endurance
HD	high definition
ICCRC2012	International Conference on Control, Robotics and Cybernetics
ICRoM	Robotics and Mechatronics
IROS	Intelligent Robots and Systems

LAI	leaf area index
MEMS	micro-electronic mechanical systems
MSS	multi-spectral scanners
MYP	maximum yield practice
NASA	National Aeronautical Space Agency
NDVI	normalized difference vegetation index
NIR	near infra red
OLI	operational land imager
PBA	push button agriculture
RFID	radio frequency identification
RS	remote sensing
SALUS	system approach for land-use sustainability
SIC	Satellite Imaging Corporation
SMAP	soil moisture active passive satellite
SNMS	soil nitrate mapping systems
SWIR	short wave infrared sensors
TIRS	thermal infrared sensor
TLS	terrestrial laser scanning
TM	thematic mapper
UAS	unmanned aerial systems
UAV	unmanned aerial vehicle
UNCSAM	United Nations Centre for Sustainable Agricultural Mechanization
VRI	variable rate irrigation
WSC	water status of crops

ACKNOWLEDGMENTS

During the course of the compilation of this book several of my friends, colleagues, professors at farm universities, farmers, agricultural industries, particularly those dealing with farm vehicles and implements, agricultural robots, unmanned aerial vehicles (drones), and precision farming, have offered or permitted use of published material and photographs. I wish to acknowledge their encouragement. Following is the list of officers and agricultural industries who have been generous:

- Dr. Ann Anderson, Autonomous Tractor Corporation Inc., Fargo, North Dakota, USA
- Dr. Arnaud Frachdiere, Director General, Vitirover Inc., La Gare, Saint de Emilion, France
- J. Burton, CEO, AgRobotics Inc., 1818 North Taylor Street, Little Rock, Arkansas 72207, USA
- Christophe Millot, Wall-Ye Inc. Macon, France
- Daniel Danford, CASEIH Inc. Oakes Road, Sturtevant, Wisconsin, USA
- Dr. David Dorhout, Head, Dorhout, RD LLC, Waltham, Massachusetts, MA, USA
- Dr. David Nelson, Nelson Farms, Fort Dodge, Iowa, USA
- Davon Libby, DigitalGlobe Inc., 1601, Dry Creek Drive, Longmont, Colorado, USA
- Dr. Dean Carstens, Proprietor, Twin Diamond Industries LLC, 1306 K Road, Minden, Nebraska, NE-68959, USA
- Mr. Donald Effren, President, AutoCopter Corporation, Charlotte, North Carolina, USA
- Dr. Eric Lund, Veris Technology, Salina, Kansas, USA
- Dr. Hans Peter Thamm, Geo-Technics, Linz, Germany
- Dr. Jeannette Allen and James Irons, Landsat Program, Hydro-spheric and Bio-spheric Sciences, National Aeronautics and Space Agency, Goddard Space Flight Centre, US Mail code 610.9 Org. Code 618, USA
- Dr. John Kawola, CEO, Harvest Automation Inc., 85, Rangeway Road, Building 3, Suite 210, Billerica, Massachusetts, MA 01862, USA
- Dr. John Read and Dr. John Taylor, Chief Officers, Ag Armour Inc., 120 Eastgate Dr., Washington, Illinois 61571, USA

- Dr. Joseph Barnard, Barnard Microsystems, 44–54 Coleridge Road, London N8 ED, United Kingdom
- Dr. Kent Cavender-Bares, Chief Executive Officer, RowBot Systems, Carnegie Robotics LLC, Pittsburgh, Pennsylvania, USA
- Ms. Lia Reich, Director, Communications, Precision Hawk Inc., Noblesville, Indiana, USA
- Carrie Laboski, Department of Soils, University of Wisconsin, Madison, WI, USA (for cover picture of maize crop)
- Mr. Mathew Wade, SenseFly, a Parrot Company, Switzerland
- Mrs. Paula-Landoll-Smith, 1900 North Street, Landoll Corporation, Marysville, Kansas, KS 66508, USA
- Dr. Philip Garford, Garford Farm Machinery Inc., Peterborough, England, United Kingdom
- Dr. Rebecca and Anthony Montag, Montag Manufacturing, Inc., 3816–461St. Avenue Emmetsburg, IA 50536, Iowa, USA
- Dr. Rory Paul, CEO, Volt Aerial Robotics, Chesterfield, Missouri, USA
- Dr. R. Scott Johnson of Des Moines, representing Kinze Manufacturing Inc., 2172 M Avenue, Williamsburg, Iowa 52361, USA
- Mrs. Silvia McLachlan, Agricultural Division, Trimble Navigation Systems, Sunnyvale, CA 94085, California, USA
- Ms. Theresa Eveson and Steven Vale, Farmer's Guardian, Preston, United Kingdom
- Dr. Tony E Grift, Associate Professor-644, 1304 W Pennsylvania Av, Urbana Champaign, Illinois 61801, USA
- Dr. Tony Koselka, Vision Robotics Corporation, 11722 Sorrento Valley Road, San Diego, California, CA 921211, USA

I wish to thank Dr. Uma Krishna, Sharath Kowligi, and Roopashree Kowligi, and offer my best wishes to Tara Kowligi.

PREFACE

Agriculture is one of the greatest of inventions by human beings. Agriculture is a gift by *Homo sapiens* of early Neolithic settlements. Agriculture weaned humans away from the hunting and gathering way of food procurement. Agriculture, at the same time, meant hard toil and drudgery in open fields. Human drudgery and hard labor in crop fields crept in rather stealthily. Human drudgery as a constraint was felt but still went unnoticed through several millennia. Farm toil of varied intensities/skills was required to match yield aspirations. Agriculture, even today, demands hard toil for lengthy stretches. Farmers in many regions accept incessant labor to mend crops almost as a natural consequence. Farmers, in fact, co-exist in farms along with crops and bestow a lot of time and accept physical hardships to produce crops. Of course, we have reduced agricultural drudgery to a great extent, but yet much needs to be accomplished.

Since 10,000–12,000 B.C., agriculture has witnessed a series of great inventions and improvements to the way crops are raised and utilized. Agriculture is a highly varied occupation. Agriculture experienced a conspicuous effect due to the invention of the chisel plow and its usage. To a certain extent, the plow reduced farmers' drudgery of digging soil in fields. The plough helped in sowing seeds systematically and in required densities. Ploughs dragged by human beings, or draft animals, have helped to generate food grains. Similarly, as time lapsed, inventions such as the cotton gin, automotive tractors, pivot irrigators, combine harvesters, and a variety of grain processers have all reduced physical toil and improved food production efficiency. We should note that each significant invention must have made humans elated. Since food grain generation became that much easier, it reduced physical exertion and became physiologically more compatible, and leisurely and yet accurate. We now know that none of these inventions and gadgetry introduced was an elixir that might have imagined or dreamt. Yet, we can realize that for several centuries, farmers have bestowed most of their time on crops. In many regions, farming is a major occupation, even today, and a fraction of the human population involved is high, at 70% or even more. Farm workers are traceable in great numbers in many agrarian regions of the world, and their gainful employment depends on crops and financial resources from farms. The introduction of farm machinery has progressively reduced human involvement in crop production and food grain processing. We still have no idea about the extent to which human labor and toil could be removed during crop production. A few countries have only 3–5% of the human population engaged as farm workers.

"Push button agriculture" is a concept of great significance to global food grain generation systems. It is a kind of agrarian revolution that has begun earnestly for the benefit of farmers and human population worldwide. Push button agriculture is forecast to dominate the global crop production strategies in the very near future. Soon, we may trace push button (or touch screen) systems in all agrarian belts. Push button agriculture is actually a phrase that connotes that farmers are destined to utilize machines, computers, vehicles, and other gadgets that are easily maneuvered by using electronic push button systems. Such autonomous, pre-programed machines are set to alleviate human drudgery. At the same time, they offer greater accuracy to all crop production procedures. They are intended to enhance crop production efficiency per unit area, with less drudgery. Push button agriculture, as described in this volume, is a conglomerate of three rapidly developing technologies, namely robotics, drones, and satellite-aided crop production. Push button agriculture is forecasted and billed to be one of the most conspicuous effects on global agriculture in the near future.

Push button agriculture, as defined and described in this volume, involves recent and futuristic techniques. It involves high technology robotics with machine vision, sensors, electronic controls, and satellite guidance. Robots capable of a series of agricultural tasks that are simple to highly complex are described in Chapter 2. Robotics is already conspicuous, and it is destined to dominate the crop fields in near future. Autonomous tractors with GPS-RTK guidance systems are now a kind of rage among farmers/companies in North America. Small robots are vying for their fields in many of the agrarian regions. Small robots of really great versatility and ability for different tasks are being researched and developed. Robots that perform in plantations are gaining popularity.

Computers and excellently matching software have the ability to make robots handle almost all possible activities during food grain generation, and on plantations.

Drones have arrived at agricultural farms as a great boon, in terms of reducing drudgery during scouting, monitoring, spraying, and applying nutrients to crop fields. Drones are again highly versatile in their ability to perform tasks while airborne. Drones replace a sizeable amount of human drudgery and reduce the cost of labor. Drones conduct certain tasks with great ease and at remarkably high speeds. They cover large farms in a matter of minutes, which is perhaps unthinkable if human skilled labor is enlisted. At present, drones are being introduced into crop fields worldwide. Drones offer digital maps/data that is unmatchable. Drones collect data with great accuracy and rapidity; hence they are destined to dominate agrarian regions. Chapter 3 offers greater details on types of drones and their role within the larger concept that we now know as push button agriculture.

Satellites are in vogue in global crop production zones for the last four or five decades. We already know great deal about satellites and their utility in providing imagery of natural resources, vast agrarian expanses, small farms, and their ability to guide farm vehicles. Satellite imagery has been a boon to farmers with regard to early warning of pestilence, disease affliction, and disasters such as floods, erosion, drought, etc. Satellite-guided farming, particularly periodic monitoring and for controlling farm vehicles' movement and performance using GPS coordinates, is rapidly gaining ground in all agrarian regions. Examples of agricultural satellites and their role during crop production are discussed in greater detail in Chapter 4.

Now, in the context of this book, "push button agriculture" is devised as a concept, for the first time, to suggest that we can produce crops by amalgamating, inter-phasing and coordinating robots, drones, and satellites with great care. We are on the verge of great strides in electronically linking several robots, semi-autonomous/robotic vehicles with drones and satellites. Swarms of robots capable of accomplishing seeding, spraying, and harvesting in a matter of short time and with exceedingly high accuracy are being touted by agricultural engineers.

Push button agriculture is not a fantasy. It is not a concept from the realm of science fiction. Instead, it is a reality; indeed a revolution that has begun earnestly in agrarian regions of world. It is believed that as we develop further and master push button techniques and control farm operations with greater versatility and accuracy, we will have caused a benefical effect on global farming and food generation systems. Push button agriculture would lead us to food generation using nil or fewer farm labourers — a situation imagined never before. It would allow us to improvise on accuracy, as well as on grain/fruit production efficiency. Most importantly, push button techniques reduce the use of harmful chemicals, fertilizers, and water. They avoid exacerbating climate change effects, greenhouse gas emission, and ground water contamination. In due course, push button techniques that suit agrarian belts of different continents should be in place for use by farmers—it is a matter of research efforts. Push button agriculture is actually forecasted to cause a kind of revolution with regard to food generation systems adopted by human beings. Food grains and, to a certain extent, fruits, textiles, and fuel could be generated without human drudgery. Push button agriculture should be a great gift and rescue act by agriculture engineers to support the global farming community that would be burdened to produce food grains for over 9 billion human beings by 2050.

Dr. Krishna Kowligi
Bangalore, India
January, 2016

ABOUT THE AUTHOR

Dr. K. R. Krishna

Formerly Visiting Professor and Research Scholar at the Soil and Water Science Department at the University of Florida, Gainesville, USA; Independent Researcher and Author

K. R. Krishna, PhD, is retired from the International Crops Research Institute for the Semi-Arid Tropics (ICRISAT) in India. He has several years of experience in crops research. He has been a Cereals Scientist in India and a Visiting Professor and Research Scholar at the University of Florida, Gainesville, USA.

Dr. Krishna is a member of several professional organizations, including the International Society for Precision Agriculture, the American Society of Agronomy, the Soil Science Society of America, the Ecological Society of America, and the Indian Society of Agronomy. He received his PhD in Agriculture from the University of Agricultural Sciences, Bangalore, India.

GENERAL INTRODUCTION

Push button agriculture is about an imminent revolution in the way food-generating systems of the world are mended, maintained, harvested and the resultant grains/fruits or other products are handled. The theme of this book focuses on minimizing or perhaps totally eliminating human drudgery in the farms, fields and plantations, plus improvement of food grain production efficiency. The techniques adopted aim at a high level of autonomous gadgets and vehicles. Robotics, drones and satellite mediation of crop production strategies form the crux of the volume. Push button agriculture is explained as an amalgamation of robots, drones, and satellite guidance for the benefit of the farming community worldwide. The aim is to bring in electronic controls, autonomous machinery, and excellent computer-aided analysis and decision support systems to regulate soils, crops, and environment. Farm monitoring, supply of inputs such as fertilizers, water, herbicides, pesticides and agronomic procedures are all decided and regulated by computer programs that take due care about climate change effects while enhancing crop productivity. This concept 'push button agriculture' has been touted to help us in enhancing food grain/forage generation to meet the needs of the forecasted 9 billion plus human population and larger farm animal population by 2050. The most striking advantage of robotics/drones is the ability to reduce human drudgery in crop fields to negligible levels, if adopted with due care and greater intensity. It emphasizes the role of agriculture engineering during food generation.

Push button agriculture is not a fantasy, not part of any kind of science fiction, but it is a reality. It is a concept that has taken roots but yet it is rudimentary. Some aspects of push button agriculture are gaining in popularity in intensive cropping belts of North America and Europe.

There are five chapters that describe our efforts towards development of robots, drones and satellite techniques applicable to farming. Chapter 1 provides an introduction that describes, briefly, the historical developments and inventions that led to reduction in human drudgery in farms during crop production. It underlines the advantages gained by humans from mechanization, electrification and more recently computer decision support during crop production. Chapter 2 deals extensively with agricultural robots and their utility during soil and crop management. Chapter 3 has detailed information and discussions on agricultural drones. It highlights the various advantages of drones over other methods of scouting, monitoring, supplying fertilizers, spraying herbicides and pesticides to crops. There are elaborate discussions with illustrations about satellite-guided crop production systems in the Chapter 4. Summary and future course of push button systems are summarized and evaluated in Chapter 5.

CHAPTER 1

PUSH BUTTON AGRICULTURE: AN INTRODUCTION

CONTENTS

1.1 BACKGROUND

Historical records suggest that regular agricultural practices and crop production took roots in the Asia-Minor region, more popularly known as 'Fertile Crescent' (Godwin, 1965; Kipple and Ornelias, 2000; Simmonds, 1970). Agricultural crop production, it seems, was independently invented, devised and modified in several other regions on earth, either simultaneously or at different times during past 10,000 years (Fuller, 2005; Asouti, 2006; Feldmann, 1970; Krishna and Morrison, 2010; Krishna, 2002; 2012). Seeding, in other words, dibbling dry seeds into moist earth helped human species reduce dependence on hunting and collection of roots, tubers and fruits naturally available to him. However, with lapse of time, his dependence on crop production and farming, in general, increased and he got himself weaned from hunting and gatherer's behavior. There were many aspects staying hidden or unnoticed with this great invention, for example, agriculture, and his ingenuity to establish large farming stretches to feed higher population. A few of the conspicuous constraints were dastardly 'agricultural drudgery,' hard labor, difficult situations created by environmental vagaries, general scarcity of resources such as seeds, water, fertility; periodic occurrence of natural disasters, diseases, unsettling wars, etc. *Agricultural drudgery is among the most dastardly aspects that kept co-evolving through the ages along with crops, production trends and farming communities.* Diversification and expansion of crop production trends, in fact, required equally diverse and intense forms of drudgery, greater investment of human labor, his general acumen and capital, if any. Drudgery or hard labor in crop fields became well entrenched and got accepted as inevitable fate. The need for round the clock surveillance and specialized mending of crops at various stages from seeding to post harvest meant farmers should stay close to farms/crops and undergo hard toil and patiently too, if they ought to derive benefits from food crops. Hence, during past, all through the agricultural age, farmers co-existed with crops. Such co-existence is yet another inevitable consequence, in addition to drudgery at any time as and when demanded.

Human effort to reduce agricultural drudgery has been among the most conspicuous and prioritized items. No doubt, reduction in drudgery in fields and during crop processing has occurred, in spurts at times and rather slowly at others. There has been a consistent effort to invent, select and improve crop genotypes, introduce new crops, devise new methods and develop new farm vehicles and contraptions of use in the farms. The aim has been to accomplish agricultural activities, with greater ease, accuracy and efficiency. Several of the early inventions reduced drudgery in comparison to hard labor that was required prior to it. For example, dibbling seeds at each spot was replaced by plows that opened furrows, so that seeding could be done using coulters. The oxen or horse drawn plow meant greater ease, accuracy and efficiency. These were among early inventions or improvements in soil and crop management that reduced human drudgery during agricultural crop production. Animal draft was utilized by farmers for a considerably long stretch of time to drag plows, right until steam/IC engines fitted tractors took over soil tillage activity.

1.2 AGRICULTURAL MECHANIZATION AND ELECTRIFICATION

Within the realm of agricultural crop production, human ingenuity has been directed consistently to remove drudgery, labor requirements, improve crop productivity and economize on energy and capital needs. The improvements in agricultural operations have not been uniform all through the past 10–12 millennia. In fact, there have been spurts of great inventions or none at all for a stretch of time. Peterson (2012) suggests that introduction of mechanization, electric/fuel power and technology into agriculture has helped in making aspects such as tilling, seeding, inter culture, fertilizer distribution, plant protection, harvesting and grain processing very efficient. The progress has been most marked during past 200 years compared to the previous 5000 years and it was much slower prior to it, when crop production was still rudimentary in terms of technology and efficiency.

Horse drawn mechanical harvester was in vogue in North America since mid-19th century. It was devised by Cyrus McCormick in 1831. The invention and improvement of mechanical harvesters helped to reduce drudgery and requirement of human labor, while harvesting cereal grains/forage (The Great Idea Finder, 2014). During the past century and a half, tractors/harvesters have been consistently improved regarding the farm activities they perform, efficiency and extent of human drudgery they remove. It is said that development of plow with steel shares, mowers, seed drills and threshers, etc. were all efficient in reducing human labor needs and energy. Yet, it is the development of tractors with ability to sow and harvest large acres that became important. Initially, steam powered tractors/harvesters and later petrol dependent engines became highly acceptable during 1920s till 1950s. They replaced animal power for traction. Most farmers in the Great Plains were equipped with tractors/harvesters (Hurt, 1991; O'Dell, 2014). It is said that new machines reduced farmer's physical labor. The mechanical harvesters helped in collecting grains from 4 acres per day, if sickle wielding human labor could harvest grains from just one acre per day in 1830s (Ohio History Central, 2014). Further, we ought to realize that during past, from 1830s to 1930s, one farmer produced food grains for only 8 human beings. However, with rapid mechanization and electrification of farms, farm work was accomplished with greater ease, at a rapid pace and efficiently. A single farmer produced food grains for 128 human beings during late 1900s. With advent of Drones, Robots and GPS guided Precision Techniques this ratio is expected to be still better in future.

Reports by National Academy of Engineering of USA (2014) suggest that mechanization of farms in USA during early 1900s reduced human labor needs perceptibly. At the turn of 19th century in 1900, about 38% of total labor force in USA was constituted by farm labor. Mechanization and electrification reduced share of farm labor to just 3% by 2000. There were series of inventions and development of farm vehicles, implements and related contraptions that actually reduced animal draft and human labor. Many of these engineering marvels added accuracy and made life easy for farmers during crop production. Following is a time line of improvements effected in farm operation through engineering:

a) 1902 – Farm tractors based on internal combustion engines were produced;

b) 1904 – Crawler tractors were introduced;

c) 1917 – Fordson tractors capable of plowing 10–12 acres per day were manufactured;

d) 1918 – A horse drawn Corn silage harvester;

e) 1921 – Aerial pesticide sprayers were used;

f) 1922 – Power take off and several modifications, implements, hitches and methods to draw tractor power to drive other gadgets in farms was devised, for example to pump ground water;

g) 1932 – Diesel engines and rubber wheels are used that avoids hard-pan formation to a great extent;

h) 1935 – Rural electrification, push button water lifting pumps, grain processors, flour producers;

i) 1938 – Combine harvesters, also called 'Headers';

j) 1943 – Mechanical cotton spindle pickers were utilized in farms;

k) 1948 – Development of Centre-pivot irrigation systems;

l) 1954 – Development of Combine harvesters that harvest corn cobs, collect grains and forage;

m) 1966 – First multi-row gyral air seeder used in USA;

n) 1970 – Electronic monitoring of crop growth, nutritional status, disease/pestilence;

o) 1994 – GPS guided tractors and combine harvesters, seeding based on GPS signals;

p) 2000 – Satellite guided, seeding, irrigation, pesticide application, regular monitoring of farms, crop growth and grain formation, preparation of yield maps and soil fertility variation, application of fertilizers and irrigation based on aerial photos and GPS guidance.

[*Source*: National Academy of Engineering of USA, 2014; New Holland Agriculture, 2012; Note: The list pertains to mechanization in USA.]

These timelines relate to efforts to mechanize, electrify, and add accuracy to machines and reduce human labor needs during farming, mostly in North America. Machines like 'cotton gin' and 'pickers or strippers' offered great advantages in avoiding drudgery or even slavery in cotton fields. These are glaring examples of how engineering could remove human drudgery during crop production. Tractors, harvesters and electrification of farms took place either early during mid-19th century or at a period later in many other agrarian regions of the world. For example, Ford Company in USA produced a prototype gasoline based tractor called 'automobile plow.' In Europe, prototype gasoline-based tractors were in operation in early 1900. Regular petrol based tractors and reapers were in vogue in good number by early 1930s. The FIAT company in Italy produced Tractor model 702 and later a model 'La Piccola' in fairly large numbers between 1920 and 1950s (New Holland Agriculture, 2012). In Germany, for example, combine harvesters were used in large numbers in farms by 1934 (Freye, 2014). Similarly, in the Russian plains, mechanization and electrification took roots during early 1900s. In Asian countries such as India, Pakistan and China, mechanization began around 1940s, but its momentum picked up during 1970s in India and Pakistan. The Hindustan Machine Tools company in Pinjore, Punjab, India released tractors in large number, beginning from 1953 (Indian Mirror, 2013). The first GPS guided Tractors were produced by 'New Holland Agriculture' company during 2011 (New Holland FIAT, 2012). It was called 'Sky watch GPS/GPRS tractor' (New Holland Agriculture, 2012). In India, electrification and mechanization enormously reduced agricultural drudgery and labor needs during crop production. It helped in removing human labor and animal draft to a large extent. Yet, in remote corners of rural belt of Asian Nations, animal traction is in vogue. Mechanization and electrification of farm activity in the African continent is highly dependent on region, its geographic characteristics, weather pattern, crops grown and profitability of farm enterprises. In the Sahel, subsistence farming with animal traction is in vogue. In this region, human labor is used liberally to conduct seeding, inter-culture, fertilizer supply, harvest and even processing. However, in South Africa, for example, mechanization and electrification of cereal production zones is prominent. In some cases, timing is important. For instance, in the Russian Steppes, introduction of farm machinery and their acceptance depended on their appropriate timing and farmers' economic conditions (Pomfret, 2002).

During the past century, agricultural output was improved perceptibly to feed an enlarged population. Also, progressively, crop production efficiency was improved through mechanization. Farm labor and drudgery was decreased with introduction of tractors, weeders, harvesters, etc. Farm operations became easier. It is said that during the next few decades, as human population and demand for food increases further, the TFP (total factor productivity) needs to be enhanced further. According Reid (2011) mechanization is one of the factors that provided better TFP. Basic resources such as land, water and human labor need to be managed efficiently. Again, mechanization, electrification and electronic systems that bestow greater accuracy and snag free farm activity are essential.

Introduction of electricity into farms and use of electric power was a major impetus to farmers worldwide to aim at better/higher crop produce, efficient management of farm operations and above all farmers profited with greater ease and reduced involvement of human labor. Electric pump sets reduced human and animal energy involvement in irrigation, in processing crop produce, mainly activities such as winnowing, seed separation, de-husking, powdering, etc. Farmers could literally press a few buttons, to start lift irrigation, or operate other instruments such as sprayers, dusters, choppers, etc. and derive the desired results. Way back, about 5–6 decades ago in 1951, when electric energy mediated farm operations were beginning to encroach farmers in developed world, a wide range of electrically controlled machines were devised. With regard to massive electrification of farm activity that occurred during early to mid-1900s, Bartlett (1951) had stated that 'Push Button Agriculture' has come here to stay. This inference was based on the observation that almost every farm activity was being more frequently powered by electricity. He believed that agricultural revolution aided by electrical gadgetry; electrically operated land and crop management machines would allow farmers a degree of leisure and convenience. Since electrically powered instruments were devoid of fatigue and erratic efficiency parameters, he believed that farm productivity could be enhanced measurably. About it, he even thought that electric power could overcome drought through efficient and continued lift irrigation when normal animal or human energy based activities would fail. Food requirements could be answered better through 'Push Button Agriculture' aided by electric power. Over all, Bartlett (1951) forecasted that electricity would solve core problems encountered during crop production. Invention and development of a range of stand by electric power systems helped in uninterrupted farm operations (Worley, 2012). They were highly valuable to the spread of electricity supported Push button controlled gadgets and farm machinery. In fact, it is generally opined that electricity changed the nature of farm work in North America. During 1941, in Alberta, it seems mere 5% of farmers utilized electric power driven push button machinery. Currently, almost every aspect of farming is electric or fuel powered (The Western Producer, 2014). Push button systems are in vogue in plenty. Yet, there are many aspects of farming that could involve drudgery, fatigue and continuous work in farms. For example, fertilizer supply to a large farm of 10,000 ha does involve long stretch of drudgery and continuous attention. However, during recent years, invention of satellite guided farm machinery, automated soil analysis and variable-rate applicators that are controlled by computers does allow even a higher degree of push button controlled farm activity (Brace, 2006; Krishna, 2012; Srinivasan, 2006; Stafford, 2005).

During mid-1900, there were many agricultural companies that professed rapid mechanization, electrification and adoption of Push Button Farming methods. They developed equipment's that were swift and efficient in accomplishing farm operation. For example, Clay Equipment's Inc., a company in Iowa had adopted 'Push Button Farming' as its motto (Waterloo Cedar Falls Courier, 2014).

During the period 1940–1970, Corn Belt in USA was mechanized, plus electrically operated machines and other gadgets were introduced to perform a great number of farm operations. Soil and crop management procedures especially tillage, fertilizer supply; irrigation and pesticide application were refined, made accurate and efficient. In addition to grain and forage production, many of the dairy farm activities were automated using electric gadgets. Push button systems controlled milking, collection, pasteurization and storage (Anderson, 2008). During this period, it seems modernization of erstwhile large farms was indeed costly and required massive government subsidies and liberal loans to farm companies.

Let us consider a few more descriptions about electricity powered 'Push Button Farms' that were in vogue during first quarter of 20th century in Europe. It is said that from a distance a Push button farm, for example, one that used electrically operated automatic machines, harvesters and processors looks much like the vintage farm that is excessively dependent on animal and human energy. The secret of the efficiency is however attributable to extensive power lines (wires), motors, and other machines. Plants in Push button farms grew better due to electric heaters, poultry yielded better and lighting improved farmers own efficiency since he could extend his work time at farm house (British Pathe, 2014). Since, early 1900s, farms everywhere in Europe and other continents have been progressively converted to automatic machines, tractors and electrically driven gadgets. Several aspects of farm, including agronomic procedures in the field became easy and got regulated by fuel/electric-powered vehicles.

In Australia, conversion of traditional horse drawn plows, lift irrigation, processing, etc. to 'Push Button' type farms occurred during second half of 20th century. Here, most machines became electrically controlled and operated at the push of a button. The up-gradation of machines from horse drawn or animal powered systems to automotive IC engine dependent or electric systems were gradual, but steady. During past decade, Combine harvesters also called as 'Headers' in Australian Farm communities are among the best examples for conversion of a traditional farm into 'Push Button Farm.' These 'Push Button Headers' employed in Australia are capable of harvesting, separating grains, processing, and developing crop yield map based on GPS guided instrumentation. Panels with electronic controls in the combine harvester allow farmers to make decisions at the touch of screen (not even a push of the button). Computer programs allow farmers to get a glimpse of soil fertility and crop productivity variations. Harvesting is highly efficient and rapid; in addition the farm drudgery of yester years is almost erased. Human labor requirement is enormously reduced (The Combine Forum, 2014). It is said the computer programs that deal with weather, terrain, crop, harvesting and processing are being improved at a rapid pace, and in future, such Push Button headers may actually perform more efficiently.

Irrigation, if any, was conducted manually by developing proper land-forms such as furrows and ridges, flat beds for sheet irrigation, raised beds and furrows, etc. Later, pipes were laid in order to spread the water as per requirements. Then, came the sprinklers, traveling sprinklers and drip irrigation. Electrically operated motors and lift irrigation allowed farmers to irrigate their crops at the Push of Button. Water requirements could be calculated using various computer programs that decide water needs based on a range of soil, crop and weather related parameters. During recent years, Push Button irrigation' has taken a further leap in accuracy and efficiency. Soil moisture monitors, GPS guided vehicles equipped to read water need based on computer models are employed to distribute water accurately and uniformly. For example, during past few years in Queensland, a few farmers have been adopting traveling sprinklers that apply water using 'ICalc Computer Model (software)' at variable rates. Soil moisture maps and GPS signals received guide a set of electronically controlled water applicators (Hunt, 2014).

Reid (2011) points out that during 19th century, several inventions related to crop production was adopted in America and Europe. Many of these inventions utilized animal draft and or human labor efficiently. A major turning point occurred when draft animals were replaced by automotive tractors. The Combine harvesters allowed rapid accomplishment of harvesting, separation of grains and cleaning (Reid, 2011; National Academy of Engineering of USA, 2000). Further, Reid (2011) argues that by late 20th century, electronically controlled hydraulics, change of implements, power and GPS guided variable-rate application of farm inputs such as fertilizers, water, and pesticides have enormously improved efficiency and accuracy of farm operation. Therefore, farm labor requirement reduced further. Introduction of electronics, GPS guided farm vehicles and variable rate applicators have also added advantages. They avoid excessive use of farm resource, add accuracy and flexibility in farm judgments, reduce loss of fertilizers and soil moisture from fields, etc. Precision farming techniques, for example, avoid GHG emissions and contamination of ground water since they avoid accumulation of soil nutrients. It is strongly forecasted, that in future, automation, inter/intra machine communications through electronics, computers that impart better coordination, introduction of drones/unmanned aerial vehicles, and robots will allow us to achieve much higher TFP. It reduces stress on farm labor, and helps to accomplish difficult activities at the push of a button.

Several activities related to harvest and processing of grains were combined in a single machine and it resulted in building of 'Combine harvesters.' Currently, we have literally 'Push Button' operated electronically controlled combine harvesters capable of series of activities related to harvest, separation of forage, processing of seeds, and developing grain/forage yield maps. The efficiency of combine harvesters in removing drudgery, taking over complicated yield rate calculation, developing productivity and grain yield maps has been improved enormously in the GPS guided combine harvesters. Forecasts suggest that global agricultural sector is almost at the doorstep of revolution involving use of robots in large scale. Initially, agriculture experienced drastic changes as plows, seed drills, tractors with hoes and discs were introduced. Now, robots capable of variety of farm activities could induce development of 'Push Button Agriculture,' with perhaps least involvement of human labor.

1.3   DEFINITIONS AND EXPLANATIONS

'Push Button' is defined as a simple switch mechanism that controls some aspect or total functioning of a machine. Push Buttons are typically hard material, like plastic. They are spring loaded and work at the touch or pressing of a human finger (Wikipedia, 2014). Push buttons are used to initiate or start a variety of gadgets/contraptions and their functions in our daily life. In the context of this volume, 'push button' connotes the state of agricultural operations, the ease with which farm machinery could be started/initiated and worked out in crop fields. With regard to agriculture, therefore, 'push button' is defined as an electrical switch operated by pressing, which actually closes or opens a circuit that consequently has effects such as initiating a farm machine, starting a tractor fitted with discs, or coulters, or GPS guided harvest systems or combine harvester, etc., a light bulb, starting an irrigating lift pump, or sprinkler systems. Clearly, such a facility allows the farmers to leisurely operate and work out in farms with least drudgery or slavish tendencies. Currently, push button systems in farms may also include, initiating GPS guided variable-rate applicators that dispense, water or fertilizers (granules or liquid) or pesticides. They are based on computer aided decision-support systems that consider soil fertility or grain/forage yield maps, and of course yield goals. A farmer with 'push button' systems, powered by electricity and controlled using computer guided systems is prone to use less human labor and avoid drudgery. In addition, it might offer greater accuracy in farm operation. It is true and interesting to note that almost every aspect of farm operations could be started using push buttons. In fact, push button systems are found right at the front gate of farm, where in, the gate/barricade opens only when proper electronic codes and buttons are pushed. There are automatic farm gate kits that identify farm vehicles and open. Remote controlled farm gate openers are also in vogue. Sophistication is indeed rampant (Automatic Solutions, 2014). Overall 'Push Button Agriculture' means a type of farming where in almost all processes are highly streamlined and conducted by sophisticated machines. Glaringly, it avoids human drudgery and cumbersome manually operated systems. Farmers, in general, have strived hard to get into a situation wherein entire farm is under push button control systems. They have dreamt and/or achieved partially such systems at various points of time in history. Each system possessed a degree of sophistication greater than what they are actually capable of at that time. Let us consider a couple of examples in the following paragraphs.

Farmers in North America had embarked on rapid mechanization, automation and electrification of almost all activities necessary for crop production, maintenance of pastures and dairy cattle. They had envisaged a series of push button controlled farm vehicles and gadgets (Time Magazine, 1959). Agricultural student community and farmers alike, around Manhattan, Kansas State, USA it seems dreamt of 'Push Button Farming' as a clear possibility that should appear on earth well by 2000 A.D. (Novak, 1958). They forecasted that farmers could be leisurely watching their gadgets and instrumentation and just push the button to induce various crop production activities (Novak, 1958). During mid-1950s, the push button farming meant control of weather, growth pattern and harvest of crops staying in lounges. It has not occurred as yet in its complete sense but a few aspects of farm labor activity and drudgery has of course been conquered though mechanization, electrification and more recently through electronic and GPS controls.

In 1968, forecasts and imaginations about 'Push Button Farming' included, as essential, a set of highly automated activity utilizing well programed tractors and other machines in the field and for post-harvest activities. For example, Navarre (1968) explains that a farm of 1700 ac in a village called Dundee, Michigan, USA could utilize a combine harvester/ tractor that could harvest grains and simultaneously dibble seeds in the harvested locations. Of course, harvesting and seeding was to be controlled through panels full of push buttons and flickering lights. In comparison to suggested 'Push Button Farm' a farm close to Dundee farms, as routine adopted series of tractors to seed, inter-culture the seedlings, apply fertilizers and harvest grains in extended and disjointed operations. One of the coveted traits/facilities of 'Push Button Farming' was air conditioned driver's seat and cockpit that was filled with television screens, blinking lights conveying the progress of each of the activity such as harvest of crop, hay and grain separation, grain loading, and re-seeding. Such 'Push Button Farms' were to be profitable and efficient in terms of labor and resource use (Navarre, 1968). Any improvement in handling farm machinery and implements meant a step towards 'Push Button Technology.' For example, an electrically controlled hitching system, or a fertilizer dispenser system was considered highly useful. The 'Push Button' systems were extended into post-harvest storage and processes. For example, in USA, pushbutton operated vertical silos were most convenient for large quantities of cereal grains (Atwater and Atwater, 2014).

No doubt, many of these inventions such as tractors, harvesters, processors, etc., have led us to a kind of push button technology, yet human involvement is necessary to achieve and accomplish many of the tasks in the field. Several activities like methodical soil sampling, chemical analysis, preparation of soil fertility maps, seeding and gap filling, and periodic monitoring of entire farm was tedious, difficult and at times not feasible. The concept of Push Button Agriculture' as discussed in this book, however, relates to a situation currently being developed and that which occurs in future more commonly and deals with techniques, that reduce human surveillance and labor in farms. They are related to drones, robots capable of series of farm activities and satellite guided automatic farm vehicles that accomplish tasks with great accuracy.

During the course of agricultural history and evolution of farming *per se*, introduction of various agronomic procedures, implements and energy sources seems to have driven the farming community to imagine that drudgery would decrease and crop production efficiency would enhance. Sometimes such guesses have run wild and meant great many advantages to farmers. Situations such as introduction of a crop species that successfully provided good harvest, a new potent herbicide, pesticide/insecticides, fertilizer formulation, irrigation systems (e.g., drip irrigation), mechanized and automated tractor and combine harvester, each one has made farmers to believe that they have hit an elixir that answers crop production related problems. More recently and in the context of this book, introduction of drones, robots and satellite guided farm vehicles have made farmers to think about Push Button Agriculture. In many cases, they believed that it leads them to a kind of automated push button controlled farm activity.

Hydroponics involves culture of crops in water without the usual soil phase. Essential nutrients are channeled through flowing water or via stagnant water that is changed periodically to regulate nutrient concentration, their ratios and pH. Often, Hydroponic systems are designed to be contained in green houses, with sophisticated electrical controls to regulate the motors, water flow, water recycling, lighting (photosynthetic irradiance), ambient temperature and relative humidity. Hydroponics, as such are highly amenable for 'Push Button Farming' techniques. There are innumerable hydroponic companies producing a range of crops such as flowers, vegetables and fruits (not trees). Hydroponic companies dealing with individual crops and employing sophisticated electronic controls to regulate crop physiological aspects are found in USA, Europe and other regions. For example, Lanna Oriental is a Push Button Bell Pepper production Hydroponic based company found near Bangkok, in Thailand (Sutharoj, 2014). However, the concept of Push Button Farming considered in this book deals more with field crop production, and involves use of large tractors, combines, GPS guided equipments, etc., and is fairly different from hydroponic systems.

Aeroponic systems too are amenable to Push Button Technology, where in crops grown in green houses could be provided with nutrients and water using automatic machines such as aerosol creators, nutrient solutions and electric bulbs for photosynthetic radiation. Crops are grown using nutrient enriched aerosol. Plant roots are kept constantly in touch with nutrient aerosol.

In addition to crop production, dairy farms too went through a period of mechanization and adoption of push button technology. For example, right in 1879 milking machines based on vacuum devise was introduced into dairy farms in North America. The milking machine consisted of rubber tubes connected to udder and milk secretion was induced and later collected using partial vacuums created. During recent years, most of the dairy farms are automated and electric/electronic controlled milking devices, milk fat separators, packing machines that function based on push button techniques or previously programed (computer controlled) have taken over many functions. Aspects such as human labor, drudgery and tedious surveillance of farm animals have been mostly taken over by machines, electronic controls and GPS guidance.

Why do we need 'Push Button Farming'? There are a few explanations possible. There are several reports that suggest that human population may reach 9 billion by 2050 and this means we have to produce commensurately higher quantity of food grains and other products needed. About 3 billion tons food grains are required. Food grain production could be enhanced by adopting several methods such as expanding cropping zones, wherever possible, intensifying the crop production trends, finding substitutes to food grains (e.g., fruits) etc. Whatever the alternative, we have to achieve the goal with least disturbance to environmental parameters and with best efficiency possible with regard to use of natural resources. *Human involvement as field workers/laborers too needs to be carefully controlled. Excessive drudgery has to be avoided.* There are many regions within the food generating prairies that employ sizably large number of farm labor crossing 30–60% mark of the total population in that location (see Krishna, 2015). Some of the options available are introduction of better farming technique, mechanization and electrification that adds to efficiency of crop production systems; use of robots; drones (unmanned aerial vehicles) and precision farming techniques. In the context of this volume, the phrase 'Push Button Technology' is used to include *Robots* to accomplish as many agronomic procedures with least human labor consumption. Then, it envisages use of *Drones* that help farmers with scouting, detection of soil fertility variations, water deficits, disease/pest, crop growth pattern in general and grain yield variations. Push Button technology described here also includes the entire set of techniques used under the concept of *Precision Farming*. Precision Farming techniques such as preparing soil fertility/yield maps of large farms, use of management blocks, computer models to judge fertilizer/moisture needs based on maps derived from satellites or UAV; use of variable –rate applicators controlled by computer based decision support systems, and periodic surveillance using drones seems necessary. The above three modern aspects of farms, for example, robots, drones and satellite mediated precision techniques have been discussed in greater detail, emphasizing soil and crop management, in the following chapters of this book.

KEYWORDS

- **Agricultural Mechanization**
- **Combine Harvesters**
- **Electric Pumps**
- **Electrification**
- **Irrigation**
- **Planters**
- **Push Button Agriculture**
- **Tractors**

REFERENCES

1. Anderson, J. L. (2008). Industrializing the Corn Belt: Agriculture, Technology, and Environment, 1945–1972 Northern Illinois University Press, DeKalb, Illinois, USA, pp. 248.
2. Asouti, E. (2006). The Origins of Agriculture in South India. Liver Hume Trust. England, pp. 1–3.
3. Atwater, R. and Atwater, N. (2014). Drive through Silos. Farm Show Magazine. http://www.farmshow/printArticle.php?a-id=614, pp. 1–2 (April 22nd 2014).
4. Automatic Solutions (2014). Gates and Fencing-Solar Powered Farm automatic gate kits. http://www.automaticsolutions.com au/page/solar_kits_farm.php, pp. 1–2 (April 28th, 2014).
5. Bartlett, A. (1951). Push-Button Farming is here. Collier's Weekly http://www.unz. or/Pub/Colliers-1951sep01–00032?View=pdf, pp. 32–33 (April 30th, 2014).
6. British Pathe (2014). Push Button Farm 1914–1918. http://www.youtube.com/ watch?v=NxtCjftelns, pp. 1–8 (May 1st 2014).

7. Brace, T. A. (2006). Precision Agriculture. Thomson Delmar Learning, New York, USA. 224 pp.

8. Feldman, (1970). Wheats, *Triticum* species. In: Evolution of crop plants. Simmonds, N. W. (Ed.). Longman Scientific Company, Edinburgh, Scotland, pp. 120–128.

9. Freye, (2014). Class: Agricultural machinery. http://en.wikipedia.org/w/index.php. php?title=Claas&oldid=603180875, pp. 1–8 (May 10th, 2014).

10. Fuller, D. Q. (2005). Archaeabotany of Early Historical sites of Southern India. Institute of Archaeology, University of London, England, http://www.ucl.ac.uk/archaeology/staff/prifiles/fuller/tamil.htm, pp. 1–3 (April 9th, 2009).

11. Godwin, H. (1965). The Beginnings of Agriculture in Northwest Europe. In: Essays on Crop Plant Evolution. Hutchinson, J. (Ed.) Cambridge University Press, England, pp. 1–22.

12. Hunt, L. (2014). Push Button Irrigation Flow. Rural Weekly of Central Queensland Publishing Company. Australia, p 1.

13. Hurt, D. R. (1991). American Farm tools. From Hand Power to Steam power. Kansas State University, Manhattan, Kansan, USA, http://digital.library.okstate.edu/encyclopedia/entries/a/ag005.html, pp. 287 (May 4th, 2014).

14. Indian Mirror (2013). Indian Tractor industry. http://www.indianmirror.com/indian-industries/tractor.html, pp. 1–5 (May 12th, 2013).

15. Kipple, K. F. and Ornelias, K. C. (2000). Cambridge World History of Food. Cambridge University Press Vol 1. 234–238.

16. Krishna, K. R. (2002). Historical aspects of Soil Fertility and Crop Production Research. In: Soil Fertility and Crop Production. Science Publishers Inc., Enfield, New Hampshire, USA, pp. 1–32.

17. Krishna, K. R. and Morrison, K. D. (2010). History of South India Agriculture and Agroecosystems. In: Agroecosystems of South. Krishna, K. R. (Ed.). Brown Walker Press Inc. Boca Raton, FL, USA, pp. 1–52.

18. Krishna, K. R. (2012). Precision Farming: Soil Fertility and Productivity Aspects. Apple Academic Press Inc., Waretown, New Jersey, USA, pp. 189.

19. Krishna, K. R. (2015). Agricultural Prairies: Natural Resources and Productivity. Apple Academic Press Inc., Waretown, New Jersey, USA, pp. 522.

20. National Academy of Engineering of USA (2000). The Impact of Mechanization on Agriculture. Agriculture and Information Technology. The Bridge 41: 1–3.

21. National Academy of Engineering of USA (2014). Agricultural Mechanization Time line. National Academy of Engineering of USA, Washington D. C., pp. 1–12 http://www.greatachievements.org/?id=3725 (May 10th, 2014).

22. Navarre, P. (1968). Modernized, Push-Button Agriculture revealed on highly automated (1700). acres Dundee Farms. Toledo Blade June, 9th, (1968). Section A: pp. 22–23 http://news.google.com/newspapers?nid=1350&dat=19680609&id= RWc xAAAAIBAJ&sjid=mwEEAAAAIBAJ&pg=7119,934600 (April 29th, 2014).

23. New Holland Agriculture (2012). A long history. http://www.agriculture.newholland. com/roi/i.e.,/WNH/whoweare/Pages/alonghistroy.aspx.htm, pp. 1–8 (May 10th, 2014).

24. New Holland FIAT (2012). New Holland SKY WATCH -the First GPS/GPRS Tractor delivered at Punjab Agricultural University, Ludhiana, India. http://www.businesswireindia.com/news/news-details/new-holland-sky-watch-the-first-gpsgprs-enabled-tractor-delivered-at-p/30609, pp. 1–4 (May 12th, 2014).

25. Novak, M. (1958). Push Button Farm of the Year – Paleo future. Hutchinson News, Kansan State University, Manhattan, KS, USA, pp. 1–6, http://www.paleofuture.com/blog/2011/3/20/push-button-farm-of-the-year-2000-1958.html (April 27th, 2014).

26. O'Dell, L. (2014). Agricultural Mechanization. Oklahoma Historical Society, Oklahoma State University, Electronic Publishing Centre, http://digital.library.okstate.edu/encyclopedia/entries/a/ag005.html, pp. 1–3 (May, 4th, 2014).

27. Ohio History Central (2014). Farm Mechanization in Ohio. http://www.ohiohistory-central.org/w/Farm_Mechanization_in_Ohio, pp. 1–4 (May 12th, 2014).

28. Peterson, C. L. (2012). Technology and Power in Agriculture. In: Agricultural Mechanization and Automation. Encyclopedia of Life Support systems (EOLSSS). http://www.eolss.net/sample-chapters/c10/E5–11–01.pdf, pp. 1–14 (May 4th, 2014).

29. Pomfret, R. (2002). State-Directed diffusion of Technology: The Mechanization of Cotton-Farming in Soviet Central Asia. The Journal of Economic History 62:170–188.

30. Reid, J. F. (2011). The impact of Mechanization on Agriculture. The Bridge 41: 3.

31. Simmonds, N. W. (1970). Evolution of Crop Plants. Longman Scientific Company, Edinburgh, Scotland, pp. 476.

32. Srinivasan, A. (2006). Hand Book of Precision Agriculture: Principles and Applications. Food Product Press, an imprint of The Haworth Press, Inc. New York., pp. 684.

33. Stafford, J. V. (2005). Precision Agriculture '05. PRECISION AGRICULTURE '05. Wageningen Academic Publishers, Wageningen, The Netherlands., pp. 1005.

34. Sutharoj, P. (2014). Push Button Pepper Farm. The Nation-Business. http://nationmultimedia.com/2008/02/19/business/business_30065778.php, pp. 1–2 (May 1st, 2014).

35. The Combine Forum (2014). The Push Button Combine. http://www.thecombineforum.com/forums/9-gleaner/5970-push-button-combine-tractor.html, pp. 1–12 (May 1st, 2014).

36. The Great Idea Finder (2014). Reaper. http://www.ideafinder.com/history/inventions/reaper.htm., pp. 1–3 (May 3rd, 2014).

37. The Western Producer (2014). Now: Electricity changed the nature of farm work forever; then: Push Button farming. The Western Producer, Lethbridge, Canada. http://www.producer.com/2014/01/now-electricity-changed-the-nature-of-farm-work-forever-then-push-button-farming/, pp. 1–7 (May 1st, 2014).

38. Time Magazine (1959). The Push Button Cornucopia. Time Magazine 73 (10) 7–9.

39. Wikipedia, (2014). Push Button. http://en.wikipedia.org/w/index.php?title=Push-button&oldid=592686354, pp. 1–3 (May 7th, 2014).

40. Worley, J. W. (2012). Standby Electric Power Systems for Agriculture. College of Agriculture and Environmental Sciences The University of Georgia, CAES Publications http://www.caes.uga.edu/publications/pubDetail.cfm?pk_id=6275, pp. 1–12 (May 1st, 2014).

41. Waterloo Cedar Falls Courier (2014). Roger Clay–Clay Equipments. http://wcfcourier.com/lifestyles/announcements/obituaries/roger-l-clay/article_3aab1640-a5ea-11e1-beda-0019bb2963f4.html, pp. 1–2 (May 2nd, 2014).

CHAPTER 2

ROBOTICS IN AGRICULTURE: SOIL FERTILITY AND CROP MANAGEMENT

CONTENTS

2.1   BACKGROUND

Robots, in general, have fascinated human beings since ancient period. Robots were initially construed as thinking machines and those capable of certain activity helpful to mankind. Robotic clocks based on movement of water were among earliest automatic machines. Robots, also called 'Automatae' were mentioned in ancient literature such as Homer's Iliad. Archytas of Tarentum in Greece, it seems, designed a robotic pigeon in 420 B.C. that would flutter and work-up using a stream of air. Aristotle (322B.C.), it seems, dreamt of a situation when each of the tools that humans worked with could perform on their own and slave for them (Editor Time line, 2008). This is a clear suggestion that human slavery could be reduced using tools and machines that worked on their own. During later years, between 500 B.C. and 200 B.C., Greek intelligentsia is known to have devised several 'Automatae' driven by streaming water. Ticking clocks and machines with complex designs and connections were developed during medieval period in Arabia and China. The first humanoid 'Robot' was designed and built by Leonardo Da Vinci in 1495. Artificial flying eagles were demonstrated in Germany during 1533. These gadgets are a kind of forerunners of the present day Unmanned Aerial Vehicles (UAVS). John Dee created an automatic moving beetle in England. During 18th century, Japanese prepared several types of puppets and automatic moving toys known as 'Karakuri-ningyo.' The development of humanoid robot known as 'Android' was first attempted in 1727 in Germany. The word 'Android' is attributed to the German Alchemist Albertus Magnus. French inventor Jacques Vaucanson created an android in 1738 that played flute. During 1760, an android capable of holding pen and writing up to 100 words was designed by German inventor Frederich von Knauss (see Editor Timeline, 2008).

The development and improvements in robots, especially their ability to accomplish different tasks has been marked during past 5 decades. First numerically controlled machine was developed in 1952. The term 'artificial intelligence' was designed. Several projects that aimed at robots were funded by Rockefeller Foundation in USA. The advantages of computer-assisted robots were first highlighted by engineers at Massachusetts Institute of Technology in USA. Heinrich Ernst developed a mechanical arm that worked incessantly based on computer programs. One of the earliest robots was installed in the General Motors Inc., at Trenton, New Jersey (Editor Timeline, 2008). This robot carried out several tasks that otherwise is dangerous and difficult for human labor. The first Department of Robotics was established at Carnegie Mellon University in USA during 1970s. 'Shakey' is one of the earliest mobile robots developed at Stanford Research Institute in California, USA. Later, several types of robotic arms capable of variety of functions based on computer controls were devised. During 1990s, several types of radio controlled and/or computer programmable gadgets and combat robots were designed and tested in the USA and Europe (Editor Timeline, 2008). At MIT in Massachusetts, USA, robots capable of indulging in war, those that could conduct sophisticated and delicate surgery, those able to wend food and money were designed during 1980s. Robots capable of logging wood, small and large loads of different commodities were developed to help industries, by Honda Inc., (Editor Timeline, 2008). Robots capable of reacting to sound, language and electronic instructions were developed during 2000–2005. Estimates indicate that about 750,000 industrial robots were in incessant use during 2005. Japanese were among most prolific users of robots. Robots are regularly used in space station to accomplish different tasks. World over, robots were given different names. A few examples are as follows: Allen (robot), Elektro, Freddy-II, Jaquet-Droz automata, Johns Hopkins Beast, Karakuri puppet, Leonardo's Robot, Shakey the robot, Silver swan automata, Tipu's tiger, The Turk, UWA Telerobot, Xianxingzhe, etc. (Wikipedia, 2011). During late 1990s, Researchers at Carnegie Mellon's University tried to classify Robots into different generation based on sophistication and activities they performed. It is said that first generation robot has intelligence capability equivalent to lizard, the second-generation equivalent of mouse, third generation comparable to monkeys. The 4th, generation robots capable of several functions are comparable with humans. Such 4th generation Robots are expected swarm the Agricultural fields and different locations, and perform a series of complicated functions thus helping human race in several aspects. (RedOrbit, 2014). Incidentally, regarding derivation of word 'Robot,' it is said that the term is picked up from a play written by a Czech play write called Karel Capek in 1921 (Editor Timeline, 2008; Robotics Research Group, 2014). It was used in play called R.U.R. (Rossum's Universal Robots) (RedOrbit, 2014). Robotics is a discipline of technology that deals with design, construction, operation, and application of robots and other autonomous systems. It also includes those automatic machines that are guided by computers and pre-set programs for different activities. During past few years, farmers were enthusiastic to practice site-specific farming. It then led to precision techniques, such as soil/crop mapping and variable rate supply of inputs such as fertilizer and water. However, soon it opened up to a wide range of possibilities now considered under the head 'Autonomous Farming' using robotics, UAVs and satellite guidance (Brown, 2013a).

2.1.1 HISTORY OF ROBOTICS IN AGRICULTURE

Agricultural expanses that generate grains/forage in large scale are worldwide in distribution. So far, large-scale production has been accomplished using mechanization, and cheap human labor leading to perceptible economic gains to farmers. Effort to replace human labor in the crop belts with robots has been delayed. Robots that mimic excellent combination of human eye and dexterous hands did not meet with great success during initial stages. Robots were incapable of performing complex tasks in the farm. Robots were not comparable to human labor in terms of farm tasks that could be accomplished. Research projects and investments were not forth coming easily (Cockburn-Price, 2012). McIntosh (2012) states that delay in deploying robots in agricultural cropping situations were not due to lack of motivation. The delay is attributable to difficulty in designing and constructing safe, reliable and easy to operate suitable robots. In some locations, availability of human labor in plenty did delay even thinking about robotics in crop production. It is true that, robots are yet preferred in many endemically subsistent farming zones or those with sizeable density of agricultural farm labor oriented families. Some of the questions about robots were difficult to answer. For example, farm laborer would ask— can the robots identify ripe and unripe fruits; can they smell and judge the taste/aroma; can they distinguish between fruits afflicted with cankers, black spots, discoloration and those healthy? In the field, can robots judge between weeds, crops and volunteers from previous crops; can robots judge water needs of a crop at different growth stages, etc.

We can also identify a few stages in the history of development of robots, especially based on their ability for different kinds of farm tasks in the cereal farms of Iowa State in USA. The first generation robots were useful in scouting within the crop fields. They were autonomous or at times guided through remote control systems. They were provided with cameras to detect crops/weeds. Sophisticated sensors that detect occurrence of drought, nutritional deficiencies or pest occurrence are being actively developed. The second-generation agri-robots were those with ability to scout, identify and locate the crop, fruit or weed using cameras and appropriate sensors. They are also capable of accomplishing tasks such as weeding using plant cutting arms or small sprays of herbicides. They carry out insecticide sprays and fertilizer supply. In some cases, the robots are connected to GPS guidance system that adds accuracy to robot's movements in crop field and the tasks they perform. Further, the third generation agri-robots are those capable of accomplishing series of tasks autonomously in a crop production enterprise. Such farms are filled with fully autonomous robots. Such an idea about entirely automated, robot filled farm was touted by Monsanto Company's Farm research wing long ago in 1957. Robot's navigation is controlled by GPS guidance system, computer programs set the tasks for plowing, seeding, fertilizer supply, irrigation, weeding and harvesting. The variable-rate applicators use the digital maps provided by sensors/cameras (Mitchell, 2014). A few examples of such entirely independent robots capable of a series of farm operations are 'AgBo' made in Japan by Yoshi Nagasaki; 'AgTrakker' developed by Mathias Kasten in Germany; 'AgAnt' a small robot capable of identifying and spraying herbicides at weeds (Plate 2.20). Such third generation agri-robots form essential work horses of 'Push Button Farms' and effectively replace human labor needs and drudgery (Department of Agricultural and Biological Engineering-UIUC, 2014; Grift, 2007; 2012; Klobarandz, 2014). Forecasts by engineers from MIT, Massachusetts, suggest that in next five years, several types of robots and those with multiple functions could be swarming all through the crop fields of Great Plains and other agricultural zones of USA. A recent trend among farmers and robotics experts has been to use previously known useful farm vehicles and convert them to become effective robots that can mow the grass, plant seeds/seedlings, spray pesticides, monitor crops, harvest them, process them and move them to store houses (Sofge, 2014).

In countries like Japan, 4 different stages of agricultural robot development are easily identified. Initially, robotics was confined to industries, for example in automobile industry. So, during 1980s, agricultural engineers were confined to just exploration of various farm activities that could one day become amenable to robots. During late 1990s, autonomous and semi-autonomous robots feasible for use in horticultural plantations were devised. Robotics was popular with canning industries for sorting fruits and grading them, then to process and package them for market. Semi-autonomous robots that operate with the guidance by human labor were preferred in fields with cereal grain crops. Since 2002, a series of agri-robots are being designed to accomplish various activities that come under 'Precision Farming.' Satellite guided movement of robots in farms became conspicuous. Robots that take instructions based on digital maps of soil fertility or crop stand and GPS controls are becoming increasingly popular.

For example, combine harvesters used currently are mandatorily provided with GPS connections and ability to map grain yield (Kondo, 2014).

In general, agricultural robots developed during 1980s are termed first generation robots capable of just harvesting the fruits. Next, those developed during 1990s, for use in farms with greater versatility and rapidity was categorized as second-generation agri-robots. The 3rd generation robots include autonomous movement, vision controlled identification, location using GPS connection, grasp and detachment of fruits or rapid grain harvest, then grading and separating of products. Advanced combine harvesters operated in cereal fields are highly sophisticated robotic harvesters (semi-autonomous). They are capable of efficient harvest of grains from cereal fields, processing, grain separation, grading, storage and transfer to vehicles that transport the grains to market yard. All done in one stretch *in situ* in the crop fields. These combines are invariably GPS guided machines. There are procedures to simulate and verify the performance of robotic combine harvesters. We can simulate the robot prior to actual use. It allows us to trace robot's movements, obstacles it may face and rapidity with which it accomplishes tasks, on the computer screen (Zhang et al., 2012).

Based on reports about the status of Robots in farming in different continents, Pullinger (2013) has expressed that robots may swarm the farms in Norfolk in England. They may be functioning incessantly and performing tasks such as seeding, irrigation, fertilizer supply, pesticide spray harvest, etc. Harvey (2014) opines that 'farmbots' would in future take many of the backbreaking tasks in the meadows of England. They will be most important ingredients of large scale agriculture in United Kingdom. In Germany, driverless tractors and robots that pick and grade fruits could be popular in very few years. Wheat crop in German Plains would be tended by robots able to supply fertilizers, organic manures and irrigation. In Australia, large farms could be handled through robots that are capable of incessant drudgery. Australian wheat belt may benefit largely since they generate food crops both for home consumption and to answer export demand (Packham, 2013; Roodt, 2013). Robots are perhaps best suited to accomplish timely operation in a large scale that otherwise is cumbersome for human labor. In Japan, intensive crop production concepts are best attended by robots. Robots are capable of greater accuracy than human labor. However, computer programs that guide the

robots regarding dispensation of fertilizers, irrigation and pesticides have to
be tailored accurately. In Nordic countries such as Denmark and Sweden,
in addition to crop production, maintenance of dairy cattle is being accomplished using autonomous robots (Hagele, 2014; Szabo, 2013). Generally,
it is believed that popularization of agricultural robots depends on extent of
advantages derived by farms, in terms of removal of human drudgery and
profits. Billingsley et al. (2008), in their treatise, opine that farmers, especially in developed countries realize very well the need for further automation, use of automatic GPS guided vehicles and gadgets that accomplish in
the field. Automatic sensing, harvesting, handling and processing of mature
grains are almost a common place in Americas and Europe. Robots have also
made their mark in forestry. Harvesting by deploying relatively large sized
robots, logging and processing of wood is prominent in Scandinavian countries. In case of field crops, machine vision and GPS based guidance is being
installed on tractors and other robots. Crop scouting using satellite maps and
UAVs is getting more common. An interesting observation, about robots and
their activity in near future has been made by Sani (2012). Robots of several different types are spreading into agricultural market yards and farms in
agrarian regions. Perhaps, it is time for farmers to daily take note of robots
attending the farm duty, note the time spent and work turn out daily much like
we do with farm labor hired. Dairying too has witnessed automation starting
form milking, to pasteurization. Packing and wending of dairy products are
also done by a series of robots (Hagele, 2014; Szabo, 2013).

2.1.2 DEFINITIONS AND EXPLANATIONS ABOUT AGRICULTURAL ROBOTS

By definition, Agricultural robot or Agri-robot could be autonomous or
semi-autonomous equipment used for various procedures during agricultural crop production. Agri-robots are used mainly in land preparation,
ridging, making channels, spraying liquid fertilizers, spraying pesticides,
sprinkling irrigation water and most importantly in harvesting grains or
picking fruits. Robots are used when repetitive, hard labor and drudgery
is essential to achieve results. In the horticultural gardens, tasks such as
repetitive pruning of branches (e.g., grape vines), weeding the inter-row
spaces among trees, spraying tree canopies with pesticides, irrigating and

fruit picking are tasks accomplished by driverless robots. Agri-robots are equipment that could replace human labor requirement during crop reproduction or cattle ranching. There are indeed a wide range of agricultural robots available in the market and several are still under development. They may all swarm the farm belts of different continents in due course. They could be improving crop production efficiency and reduce need for human labor perceptibly. A large section of human labor tied up in farms could then be released to serve several other human functions. Most importantly, robots can fairly solve the problems created by farm labor shortage during peak crop season. Vangioukas (2013) defines agricultural robots as a fusion of mechanical devices with electrical and computational systems. Further, an agricultural robot encompasses three basic components. They are: (a) a sensing system to measure physical and biological properties of agricultural fields; (b) a set of decision making instruments such as computer that processes data from sensors based on programs; and (c) a manipulative arm or gadgets that actually perform the tasks in the crop field by receiving electronic signals (Harrel et al., 1988).

Agricultural robot is an excellent component of the larger concept called 'Push Button Farming.' Robots, in general, allow farmers to replace human labor and reduce drudgery. Automation of robots is primarily achieved via fuel engines or captured electric power. Sensors and computer guided implements accomplish specific activity. Agricultural Robots are amenable for use in crop fields and diversified farms with dairy cattle, poultry and other related enterprises. In some cases, entire gamut of agronomic procedures up to harvesting, sorting, grading and package of farm produce could be achieved using robots. A list of advantages of using agricultural robots includes:

a) Robots substitute human labor and farm workers to great extent;
b) Agricultural robots can accomplish monotonous and dangerous drudgery that avoids risks to humans. For example, spraying pesticides through aerial vehicles and robots avoids exposure of farmers to its fumes and aerosols. Robots trained to select uniformly good sized and quality products graded automatically using vision control and IR cameras, enhance the commercial value grains/fruits;
c) Agricultural robot harvested products carry low amounts of microbial contamination. They are hygienic compared to human labor harvested products;

d) Robots can effectively operate in any season of the year and accomplish tasks rapidly;

e) Robots can repeat almost all activities with greater accuracy;

f) Agricultural robots are generally considered less detrimental to environment and preferred;

g) Agricultural robots that allow development of Push Button Agriculture are feasible even to aged farmers, since many aspects are accomplished by machines supported by computer-based decision makers and electrical energy;

h) Agri-robots add to accuracy of various operations such as fertilizer supply, irrigation, and detection of ripened fruits/grains. Hence, they are supposed to offer consumers with healthier food items. Excessive accumulation of nutrient ions or microbes is avoided, if accurate estimate of soil properties, accurate dispensation of nutrients and low microbial contamination is achieved through agri-robots;

i) Robots can work up day and night. For example, robots that are guided with 3D laser scanner can work even through the night (Simonite, 2009);

j) Robots that keep track of fields and trees in an orchard periodically can help in deciding on timely response by farmers. Robots could be used to spray pesticides at the earliest warning of insect build up, supply water at the onset of soil moisture deficit, and harvest crop at appropriate time (Simonite, 2009);

k) Regarding deployment of autonomous robots during agricultural experimentation, it said that robots will allow extensive sampling of fields, since they can travel and pick samples in greater number and as many times. Further, robots collect data and automatically record in the attached computers. Parameters such as crop stand, growth rates, nutrient status (Chlorophyll and leaf-NO_3, leaf area index, and canopy temperature are some examples that could be expertly noted and transmitted to computers capable of analysis (Bayer Crop Science, 2014); and

l) An important observation made in the Webster's Dictionary states that a few of the semi-autonomous robots deployed in agricultural situations have already been carrying out tasks quietly in handling many of dangerous chemicals and other items (McIntosh, 2012).

Some of the difficulties in rapid acceptance of robots are:

a) Slow movement in the field/greenhouse caused by slower vision control;

b) Identification and grasp of the ripened fruits/pods needs several algorithms;

c) Agricultural robots are costly items right now. Of course, mass production and subsidies will help its popularization;

d) Agricultural robots, may at times need slight or drastic changes in the way crop is planted, its density, genotype preferred, and procedures adopted;

e) Sensors and vision control systems may be affected by sunlight and lighting pattern;

f) Crop plants with greater variability say in rooting, height, foliage/canopy formation, ripening pattern of panicles/fruits or other products are not easily amenable for grasp by robots. Fields may have to be repeatedly treated with robots, if panicles ripe at different times in a season;

g) Currently, robots are used on few food grain and fruit crops. Robots suitable for wide range of crops are yet to be developed.

Agricultural robotics as a concept did not make perceptible progress during the period 1932 till 2000 (Szabo, 2013). However, during recent years the situation has changed. Agricultural Robotics is a rapidly developing aspect, both in terms of its production by various enterprises and adoption in farms world-wide (Crow, 2012; Janmaat, 2014). A few of the robots are small, handy in the field to operate and exceedingly useful in accomplishing the tasks. Several others are complicated. They need careful preparation of the equipment, accessories and computer programs. A report by Wehrspann (2014) is indicative of the rapidity with which robotics is being sought in the realm of agriculture. Report states about introduction of five different types of robots into fields in North America by different companies such as John Deere Inc., New Holland and others. Yet, there are reports that deal in detail the reasons for delay in adoption of robotics in global agriculture, compared to other aspects such as Industry (see Kassler, 2001; Belforte, 2006). At present, only a small fraction of crop production is exposed to treatment with robots. Many of the efforts are still in drawing board stage or in prototypes. They are yet to be commercialized. Farmers have to seek robots and accrue profits out of their use in open field. In order to analyze various aspects of robot design, deployment and commercialization, Comba et al. (2010) have classified agricultural robots into four groups. They are:

a) robots could be grouped based on operative environment. Robots operative in open field and glass house are important currently. The emphasis is on accomplishing physical weeding of inter-row spaces without using herbicides, spraying pesticide based on site-specific prescriptions, irrigating fields using variable-rate applicators, etc. In the greenhouse, robots that pick fruits are being sought (see Belforte et al., 2006; Belforte et al., 2007);

b) The type of robot machine envisaged decides the rapidity with which we can design, develop and use them in the crop fields. Under this classification, most of the robots are constructed as stand-alone robotic platforms and tractors with wheels. Wheels are preferred against traction because of ease with which sharp turnings could be negotiated in the fields, although soil compaction is a problem with heavy autonomous tractors (Bulanon and Katoaka, 2010; Chatzimichali et al., 2009; Tillet et al., 2008). Robotic harvesters meant for field crops and those functioning in fruit plantations are again different in design and development;

c) The kind of operation envisaged on different crops also affects robot's acceptance and use in fields. In the field, most common functions currently considered are weeding (e.g., in Organic farming), pesticide and herbicide spray, irrigation, grain and fruit harvesting and sorting by quality; and

d) The type of navigation, guidance systems, swiftness and control systems are important in terms of popularity and economic advantages derived from robots. Robots currently endowed with GPS guidance, infra-red sensor cameras, ultrasonic sensors and facility to respond to radio signals are also possible. Versatility in movement, regulation of speed, ability to take sharp turns and respond to GPS signals are among most important aspects that decide adoption of robots (Soegaard and Lund, 2007; Comba et al., 2010).

A list of examples of Robots used in Agricultural farming (not exhaustive):

Ag Ant (Department of Crop Science, University of Illinois-Urbana Champaign, USA): AgAnt is a small, very light weight robot that moves in inter-row space of cereal crops. It is fitted with sensors to detect weeds and it can communicate with other AgAnts or robotic weeders existing in the vicinity. So, an AgAnt that detects weeds can intimate and call other AgAnts to move to area with weed infestation. Hence, these AgAnts literally work together in groups like insect ants do. They can swiftly remove weeds in the field. Researchers opine, that at any time farmers may introduce or maintain a swarm of AgAnts and remove weeds in the field (Grift et al., 2004; Plate 2.20).

AgBot (John Deere Inc., USA): Agbot is an autonomous small buggy like vehicle originally designed and produced by John Deere Inc., AgBot is lightweight vehicle that navigates using GPS signals. The robot has been fitted with extra motors and electronic control of steering, brakes and throttle. These additions allow the buggy to operate autonomously without a driver. It can identify small weeds using sensors and eliminate them. Agbot's movements can also be controlled using a computer tablet or programed to accomplish set functions.

Autoprobe Robotic Soil Sampler (Agrobotics, Arkansas, USA): This robotic soil sampler picks as many as 20 soil cores per hr. and hastens soil map preparation during precision farming and devising soil fertility treatments.

Rosphere (Technical University of Madrid, Madrid, Spain): This robot is spherical in morphology. It moves in the inter row space by shifting center of gravity, much like a football or a hamster. It seems this unique type of locomotion is advantageous over others. The Rosphere wanders all across the field. It has sensors to measure soil moisture at surface layers. The information about soil moisture is relayed through wireless to farmers. Farmers could use it to monitor soil moisture and decide on irrigation schedule (Card, 2013).

Bonirob Crop Scout (BoniRob, developed by Amazonen-Werke in association with Robert Bosch GmbH, Osnabrück University of Applied Sciences): Bonirob is a scouting robot capable of transit with in maize crop rows. It is a small robot, but it is versatile in terms of stages of the crop that it monitors/scouts. It takes note of crop growth, leaves and canopy characteristics.

Hortibot (Danish Institute for Agricultural Sciences, Bygholm, Denmark): It is a robotic platform that can be fitted with high-tech robotic arms, weeding devises and micro-sprayers. It is an excellent autonomous vehicle operative in horticultural gardens (Norremark, 2007).

Kinze Autonomous Tractor (John Deere-Kinze Company, Illinois): Kinze autonomous tractor is used for soil tillage, as planter to plant corn seeds and to act as farm cart to load and unload corn grain. This totally autonomous and driverless vehicle is guided by GPS and computer programs (Kinze Manufacturers Inc., 2013; 2014).

LettuceBot (Vision Robotics, San Diego, California): It is a robot used to thin and harvest lettuce foliage. It is also used to weed lettuce fields (Plate 2.26).

R-Gator (John Deere's MachineSync and Claas, Urbana Champaign, Illinois, USA): R Gator is a utility robot in a farm. It can be commissioned to perform different sets of farm functions. It is an auto-fill, driverless equipment with windrow guidance and GPS connection (Wehrspann, 2014). R Gator negotiates terrain using 3D lasers and radio signals. It is also used to apply fertilizer at variable rates.

RowBot (RowBot Systems LLC. Minnesota, USA): It is a small sized robot that transits driverless within the crop rows. It is a GPS guided robot with variable-rate fertilizer supply ability (RowBot Systems LLC, 2014).

Harvey Robot (Harvest Automation Inc., Billerica, Massachusetts, USA): Harvey robot is a specialized machine that picks pots with plants, moves it to locations predetermined or as guided by wireless. Harvey has been in use in nursery farms, green houses and in fields when pots are used. During past couple of years, Harvey robots might have already handled over 3 million pots in nurseries across USA (Plate 2.15). Of course, variations of this Harvey Robot could be used as porters in farms.

Spirit Tractor (The Autonomous Tractor Corporation, Fargo, North Dakota, USA): This is an autonomous tractor capable of soil tillage and land preparation based on the accessories used. It is often used with a driverless tractor, as a leader that controls the movements of spirit and clears obstacles if any.

Vitirover (Vitirover, St Emilia, France): This is small autonomous robot that works in the inter-row space in grape vine yards and other orchards. It has infrared sensors and cameras and removes weeds in the orchards.

[*Source*: Several, including AgArmour, 2014; Moorehead and Bergerman, 2012; Wehrspann, 2014; Jones, 2013; Note: Above list is indicative, not exhaustive.]

Latest reports about robotics in agriculture suggest that 'service robots' are gaining in popularity. The Service robot is a versatile autonomous farm vehicle. They have a common platform upon which different accessories can be fitted. They can sow seeds, weed the field, fertilize open field or inter-row space and they can also irrigate or spray pesticides. Such field robots, it seems, accounted for 33% of all agricultural robots sold in European nations during past year (Automatica, 2014). The driverless service robot, it seems, is relieving farm workers of their jobs in the farm. Another observation regarding agricultural robots is that advent of small robots may allow us to accomplish a wide range of activities using them. Small smart robots could literally function in variety of places at the right time (Blackmore et al., 2005). Interestingly, Blackmore et al. (2004; 2005) state that a large share of energy expended during traditional farming is garnered by procedures that correct the damage done using large tractors (e.g., soil compaction). Small robots, again, can reduce energy needed to correct the damage or mistakes that occur with large farm vehicles. Reports by Robotics specialists and industries producing this equipment suggest that agriculture must become more efficient in terms of costs on labor and several inputs required during crop production. Robots and autonomous super tractors are among most important options to improve crop production efficiency. Agriculture indeed offers great opportunity to introduce as many types of robots. Reports suggest that in China, Japan and Australia the thrust is to develop autonomous tractors with facility to attach implements capable of different farm tasks (Bayer CropScience, 2014; Makim, 2014).

2.2   ROBOTS IN SOIL MANAGEMENT

The new generation agricultural robotics envisages use of small, compact and lightweight farm vehicles capable of tillage, inter-culture and other operations related to soil management. The light weight robots actually avoid problems such as soil compaction. Since they use lasers for traction and swift movement, regular repairs and adjustments to tractors is reduced. Cockburm-Price (2012) states that lightweight robots move on wide, low pressure tires therefore cause least compaction to soil. During seeding, a robot's movements are well controlled through GPS and laser aided traction. Robots move only along the rows with optimum moisture. Therefore undue disturbance to soil is avoided.

2.2.1   PLOUGHING, ZERO (NO)—TILLAGE SYSTEMS, AND INTER-CULTURE

Tractor fitted with petrol engine is among the earliest and important semi-autonomous agricultural robots designed through human ingenuity. It has served the farmers ably in accomplishing soil tillage of different intensities plus several other farm activities. Farmers still rely on this automotive for several aspects of farming and transport. Currently, there are several models of tractors that have been converted from being a driver steered semi-autonomous robot to a GPS guided steer-less tractor-, for example, an autonomous tractor. Conversion of several types of soil tillage and transport vehicles from semi-autonomous to steer-less systems has been brisk during past 5 years.

Historically, during past 6–8 decades, several types of tillage implements have been utilized to plow land and prepare seed beds for sowing a range of crops. Tillage essentially digs, turns and freshens soil, in addition to getting it aerated and exposed to environmental factors. Tillage also helps in mixing up soil and providing a semblance of uniformity regarding texture, pH and fertility in general. Tillage equipment has been generally drawn by animal traction or automotive tractors. During recent past, automotive tractor driven tillage has held sway all across the developed world, parts of developing world and large farms anywhere. These are essentially semi-autonomous driver guided tractors, fitted with electric or petrol engines. They rely on human labor to secure good soil tilth, sow seeds as uniformly as possible and accomplish inter-row soil disturbance (earthing-up). They are not provided with GPS connectivity and may or may not record area worked out. Let us consider a few salient features of several classical tillage equipment used during land preparation and later as crop develops. For several decades now, heavy disc plows have been used to achieve deep tillage and drastic loosening of earth. It is termed primary tillage. Disc plow became a conspicuous part of conventional tillage systems in North America and Europe (Plates 2.1–2.5). The disc plow digs soils to a depth of 20–40 cm, creates large clods and loosens the soil surface for other implements to later take over. Matching the tractor that pulls, mainly its engine characteristics, disc size and depth of tillage envisaged is important (Razali et al., 2013)

PLATE 2.1 The Landoll 2510 Series In-Row Ripper (Source: Paula-Landoll-Smith, Landoll Corporation, Marysville, Kansas, USA; Note: In- Row rippers are excellent implements for use under Conservation tillage systems. They are good on hard soils that need to be disturbed and turned. Several variations of such rippers are in use for the past 6–8 decades).

PLATE 2.2 Chisel Plow (Source: Mrs. Paula Landoll-Smith, Landoll Corporation, Marysville, Kansas, USA; Note: Chisel plows are part of primary tillage systems. These are pulled by semi-autonomous driver operated tractor. They are useful when land is rugged, with excessive residue and stones. Chisel plows are used when practicing 'Conservation tillage.' It is still a very popular plow in the Great Plains of North America).

Farmers have also used a wide range of soil rippers to achieve deep tillage as part of conservation tillage systems. In-row rippers are usually rugged and durable tillage tools (Landoll Corporation, 2014a; Plate 2.1). They are used when soil is hard. These rippers till soils to a depth of 15–30 cm. Rippers induce better infiltration, disturb soil and help in getting the top layers of soil aerated. It hastens organic matter degradation and several soil transformations. Root penetration and crop establishment is better, if soil is ripped to a certain depth prior to planting.

Chisel plows are another set of primary tillage implements used frequently all over the agricultural regions of the world. Chisel plowing, again helps in digging, turning and freshening the soil layers to greater depth of 20–30 cm. There are several types of chisel plows operated based on soil types; crops to be produced and extent of soil disturbance intended by farmers (see Plate 2.2). Chisel plows are also utilized when handling

PLATE 2.3 A semi-autonomous tractor (not driverless) plowing (disking) wheat field in Kansas, USA (Source: Mrs. Paula Landoll-Smith, Landoll Corporation, Marysville, Kansas, USA).

PLATE 2.4 Discs used for Deep tillage during Conventional and Conservation tillage practices (Source: A Farm show room near Bangalore, South India; Note: Discs are used to rip open hard soils or those not cultivated for a couple of season or even virgin fields that are to be brought under crop production. Discs usually dig deep up to 30 cm and turn the soil).

PLATE 2.5   A Disc plow attached to Semi-autonomous Tractor (Source: A Farm near Bangalore, India).

fields with larger quantity of dry residue, rocks and stones strewn on it. Chisel plows are fairly efficient compared to discs regarding the speed with which soil is tilled and inverted. There are chisel coulters and larger sized plows that are amenable for adjustment using hydraulic controls in the tractors. This is semi-autonomous system that allows the driver to set coulters to achieve different depths of soil tillage (see Plates 2.1 and 2.2). Chisel plows are preferred by farmers who intend to adopt conservation tillage. Farmers with large fields also use small disc harrows and pulverizers to achieve surface loosening, softer soil tilth, level fields and prepare seed beds. These harrows are used to accomplish several aspects of land preparation. Again, hydraulic systems provided on semi-autonomous tractors allow modifications regarding depth of tillage, surface disturbance, etc. For example, in North America, Farm equipment producers opine that tandem Harrows and Pulverisers have served farmers in North American Great Plains for over 100 years now (Landoll Corporation, 2014b; 2014c).

There are other versatile implements attached to semi-autonomous tractors that perform tasks in the field. For example, a tiller cum seeder that rapidly accomplishes both tasks at one stretch is often opted while developing pastures. Pastures built using light surface tillage and rapid seeding are common in the Plains of North America. Hydraulics on the tractor allows adjustments on both tillage depth and seed rates. Micrometer could be altered to suit the seed type and planting rate envisaged (Landoll Corporation 2014d). Tractors and several types of lighter farm vehicles are consistently used to inter-culture a crop stand. Crops at seedling stage need in-season loosening of soil in the inter row space. Tractor driven Row Crop cultivators are popular in many regions of the world. These cultivators are also used prior to banding fertilizer along the crop row (Landoll Corporation, 2014e). The row crop cultivators are faster and are primarily used to achieve better soil aeration, infiltration and weed control in the inter-row space (Landoll Corporation, 2014e; Plate 2.6)

Now let us consider GPS guided autonomous steer-less tractor systems. We should realize that a good guidance is essential while using semi-autonomous/autonomous tractors, during precision farming. A clear

PLATE 2.6 The Landoll 2000 Series Row Crop cultivator for earthing up, aeration and weeding (Source: Mrs. Paula Landoll-Smith, Landoll Corporation, Marysville, Kansas, USA; Note: Row crop cultivators are excellent in disturbing the inter-row soil mildly. They rip and stir the surface layer of soil. Cultivators are used when the intension is to induce oxygenation of soil, turn the soil and induce rooting of crop. They are also used just prior to placing fertilizers in a band).

PLATE 2.7 Corn Seed Planter-A Semi-autonomous Robot used widely all over the Agricultural zones of the world (Source: Mr. David Nelson, Nelson Farms, Iowa, USA; Note: Tractor speed and settings on seed bins decide the rate of seeding. Seeding could be made accurate if GPS guided and variable-rate seed planters are used).

PLATE 2.8 Left: Touch Screen GPS Guidance console-example; Right: A Robotic Steering Control System-example (Source: Mr. David Nelson, Nelson Farms, Iowa, USA; Note: The touch screen guidance allows farmer to verify the areas already planted with seed. Planter could be selectively moved to areas that need planting. It can be used to adjust planting speed, seed density, replanting spots that were devoid of seeds, etc. The GPS guidance equipment could also be used to regulate variable-rate applicators regarding fertilizer/irrigation and water supply to entire field).

display of the field, a map that depicts area treated already and that yet to be covered is important. For example, guidance systems produced by Leica Geo-systems provides an entire view of the field, the rows, ridges and furrows plus the over-lay field map. This allows the farmer in the driver cabin or one using a remote control system to control the speed, variable-rate flow and then carefully cover the entire field (Precision Farming Dealer, 2014b; Plates 2.7 and 2.8). Such display systems are useful in recording boundaries of work, plus multiple options such as color coding of area covered, pivot guidance during irrigation, etc. The display system mounted on the tractor or any other robot has to withstand the roughness of the field and movement in the farm.

Robots to till field is not a new idea. The driverless tractor guided by GPS signals or mobile radio-based 'Real Time Kinematic' systems currently available offer an accuracy of ±2 cm regarding field operations, especially soil tillage. It is believed that RTK guided systems will become more common during field plowing. Soil management, particularly deep disking, shallow tillage and ridge formation involves careful planning and co-ordination among several farm vehicles involved simultaneously in a field. Tractor operations need to be coordinated explicitly, if we desire high accuracy and efficiency. It is interesting to note that, as swarms of driver-less tractors move across large fields during a rapid tillage exercise, inter-tractor communication may become important. Currently, GPS-RTK guided master tractor and slaves are being experimented in many locations in USA. The master tractor has instrumentation and controls, to judge the tillage progress across the fields and it controls the 'slaves' to accordingly change course, regulate tillage speed and depth. If necessary a driver-less slave tractor first removes obstacles in field such as large clods, stones, small bushes and then clears up the field for the next slave driver-less tractor to operate swiftly and till the field. Under such situations, inter equipment signal transfer and rapid responses to GPS-RTK signaling system is crucial. It is actually swarm of robots talking to each other and performing the task, for example, clearing and plowing the field rapidly (Spackman, 2014). It is believed that GPS-RTK guided coordination of driverless tractors will reduce soil compaction, improve soil structure and immensely reduce tillage costs. Let us consider an example from North Dakota in USA. The driverless tractor known as 'Spirit' is used, in groups,

PLATE 2.9   "Spirit" – The Driverless Autonomous Tractor (Source: Mrs. Ann Anderson, Autonomous Tractor Inc., Fargo, North Dakota, USA; Note: The 'Spirit' could be attached with variety of implements such as discs, coulter, tines for plowing, harrows for flattening, crushing clods and seed bed preparation. Spirit could also be attached with variable-rate equipment meant for fertilizer, irrigation or pesticide application).

to till large fields (Plate 2.9). Such fields are eventually planted to wheat. Initially, a farmer trains the whole swarm of tractors by driving to edges of the field and marking the limits for their movement and soil tillage. The tractors then move within the prescribed digitized map and keep away from edges within a failsafe distance. Repeated tillage and collisions could be easily avoided using radio signals. A remote station guides the master tractor and sets 'slaves' to perform the plowing and ridging rapidly and as prescribed through a radio signal. Radio signals could be sent out from a small hand-held gadget, say a mobile phone, to control whole set of tractors (Spackman, 2014).

A step further, a project operative under aegis of European Commission is directed at simulating the activities such as plowing, seeding, earthing-up inter rows, fertilizer supply, irrigation and harvest using computer programs. For example, simulating plowing and seeding aspects prior to actual activity will be helpful. Computer programs, which judge and arrive at most efficient systems for using a fleet of autonomous tractors and planters to follow is an exemplary idea for future. Actually, fleet of robotic machines capable of weeding have been simulated and evaluated (Emmi et al., 2013). It may not be long before farmers can simulate and evaluate the activity of several types of robots and see the effects on a computer screen. They may also be provided with best options from time to time.

Recently, in North Dakota, Agricultural Engineers released a massive sized, 300 hp tractor that operates autonomously. It has no cabin and facility for driver—no doubt it is a driverless robot. This tractor is programed using radio and laser signals and directed to conduct different activities such as disking, land preparation, formation of ridges and furrows, etc. The boundaries and location of ridges and their number are marked on a computer screen that translates to tractor activity in field via GPS signals. Usually, a pick-up van leads the tractor to correct position along the length and breadth of the field. Over all, the emphasis is on inter-vehicle communication, the concept of leader and slave tractor operates to get the job done accurately and with greater efficiency (Mattern, 2012). Noguchi et al. (2004) opine that primary idea for adopting master-slave robot systems, say a master autonomous tractor and several other slaves is to decentralize signals/commands and accomplish farm tasks efficiently without overlaps. The master-slave tractor concept utilizes algorithms known as GOTO for the master tractor and FOLLOW for the slaves that receive commands from other robots.

Auto-steer tractor fitted with GPS guidance facility and variable-rate applicator is among the popular acquisition of many a farmers in North Carolina State of USA. Many of the hitherto ordinary tractors are also being fitted with Trimble navigation auto-steer system, prior to deploying them for plowing or inter-culture operations, such as earthing-up or organic mulch application. The auto-steer technology in coordination with GPS, aligns the tractor perfectly with furrows/ridges in the field. The auto-steer system allows tractors to navigate accurately. Therefore, many other aspects such as seeding, fertilizer input and water supply too get that much accurate. It is interesting to note that auto-steer tractors with GPS and display computer screen are easy to operate. It just needs a few flips of switch and finger touch on few icons on the screen. This is mainly to align the vehicle and instruct the operation required such as plowing, disking, fertilizer supply, etc. Further, the auto-steer, GPS guided tractors are also endowed with computers that store and keep track of all the activities and data. The data management, drawing conclusions about inputs, their timing and consequences on grain harvest are easier (NC State, 2014; Reid et al., 2000). Farmers tend to conduct seeding and fertilizer application more efficiently.

Robovator is a farm robot that functions as a driverless hoe in the fields supporting row crops. It is produced by a Danish company named F. Poulsen Engineering ApS. The digital cameras recognize weeds based on height and leaf characteristics. Hydraulic tines on robot reach the weed, based on electronic pulses created by digital cameras. The Robovator moves using electrical energy between crop rows, at a speed of 4 km h^{-1}. The Robovator works even in the night using infra-red cameras. Poulsen (2014) explains that Robovator is being preferred because it does not use chemical herbicides; it quickly hoes the field using tines and loosens earth. It is economically efficient.

2.2.2   ROBOTS IN SOIL ANALYSIS AND SOIL FERTILITY MAPPING

The knowledge about soil characters, mainly its physico-chemical properties and fertility status in relation to crop production trends is essential to farmers in any agrarian region. Appropriate upkeep of soil conditions is necessary, if not; reduction in crop yield could be drastic. Soil deterioration is yet another problem that builds up, if soil fertility management is neglected. World-wide, agricultural agencies and farmers alike expend time and funds to judge soil fertility status and its variations in crop fields. This step is essential to decide on the cropping pattern, investments on fertilizers, irrigation, pests and to decide the yield goals. Soil fertility analysis, anywhere, irrespective of soil type, is a tedious procedure, laborious and could be costly, if not subsidized by governmental agencies. Soil sampling, if attempted using human labor, it is a hard physical task to dig several soil samples per field/ha. Soils could become hard in certain spots or regions. Sometimes, for a deep-rooted crop, surface sampling of soils may not provide accurate data. Sub-soil fertility has to be estimated. This requires extra human labor to dig deeper into soil horizon. The costs on labor to sample the soils and analyze soil characteristics plus nutrient availability, depends directly on accuracy with which farmers' desire to understand the soil fertility variations. Soil fertility variations are often mapped based on soil nutrient status, crop productivity trends, or textural groups, etc. Grid sampling stipulates soil samples at several points and this automatically increases costs on soil fertility analysis.

Now, there are indeed several aspects of soil and its fertility analysis that are handled efficiently using robots. Robots perform efficiently both during out-door field sampling and during laboratory analysis. In the present context, discussions are focused more towards field. Soil sampling is among the first steps that require human labor and it is also tedious. Originally, soil sampling was done using human labor and it continues to be so in many agrarian regions. Currently, there are several designs of semi-autonomous and autonomous robots that swiftly dig several cores per plot/field, pick-up soil samples, pack them into sample vials and transport it to soil analysis labs. 'Autoprobe' is an entirely autonomous robot that transits in the field using GPS signals and collects sample at locations marked by computer decision-support systems. 'Autoprobe' is produced by AgRobotics, a company in Arkansas, USA. 'Autoprobe' offers cost effective sampling of field soil. The autonomous robot has 6 ft. deep probe that collects sample in short space of time compared with semi-automatic or manual digging/sample collection systems (AZoRobotic Staff Writers, 2013; see Chapter 4, Plate 4.7). A single soil sample collected by this robot represents 20 cores drawn in the stipulated area. It is relatively a rapid procedure. Some of the advantages listed for soil sampling robots are:

a) It is capable of sampling 150 acres within an hour;
b) It generates a quality sample every 45s and sends it to labs;
c) Autoprobe accomplishes soil sampling at costs much lower than human labor; and hence profits are higher (AZoRobotics Staff Writers, 2013).

'RapidProbe' is a recently developed soil sampling robot. It doubles the efficiency of soil sampling. However, each sample is derived from only 10 different cores that the robot accomplishes. RapidProbe is supposedly cost effective compared to 'Autoprobe.' It is useful in large farms. Agricultural consultancy companies that deal with soil fertility may possess it to hasten sampling. The RapidProbe has GPS connectivity. Hence, each soil sample could be marked on a grid and identified in the field. It helps in developing a soil map for texture, physico-chemical characters estimated on the soil samples and nutrients (Lyseng, 2011).

Small robot capable of pulling several different types of soil analyzers is being envisaged in many soil research centers. Small robots are being increasingly touted for use in soil nutrient and fertilizer management. Right now, there are GPS guided, semi-autonomous or steer less (autonomous) tractors that are capable of cm level accuracy in soil sampling. Sensors actually pick up data regarding an array of radiometric, optical and electrical conductivity aspects. Soil pH, cation exchange capacity (CEC), soil moisture and organic matter content could also be measured on-the-go. For example, Veris pH manager, Optic mapper, Electrical Conductivity (EC) mapper allows farmers to obtain a soil map depicting several aspects that affects crop production. Soil NO_3 levels could also be measured on-the-go in future as probes for it are being developed. Blanket application of fertilizer-N using robotic tractor is already in vogue. Split applications of fertilizer-N could be dispensed using computer decision support (e.g., Adapt-N). Precision farming method that includes, tractors fitted with variable-rate applicators for fertilizer-N, GPS guidance and computer-based decision support have found their way in large numbers in the cereal farming belts (Adamchuck, 2010; Krishna, 2012; Lee, et al., 2010; Pocock, 2012). However, small robots that traverse the inter-row space and deal with the above aspects such as studying soil properties, including soil-N fertility status are yet to be developed. Small robots capable of precision techniques such as variable-rate application of fertilizer-N are yet to be developed. However, an autonomous robot that transits in the inter row space and judges soil properties and soil N status is a clear possibility. Several models of on-the-go soil analyzers are available to be hitched to robots. Pocock (2012) opines that it is matter of time before small robots such as 'AgAnt' are fitted with on-the-go electrodes to assess soil pH, electrical conductivity, soil moisture, organic matter content, soil NO_3 content, plant nutrient status, etc. The electrodes meant for robots need to be adapted to hard soil surface, since the original instruments are designed for fluids or slurry and may not withstand rigidity of soil surface.

2.3   ROBOTS IN CROP PRODUCTION

Crop production, in general, involves a series of agronomic procedures that are aimed at mending soils, mainly plowing, freshening, clod crushing, harrowing, ridging, etc. Next, a set of procedures are aimed at maintaining crops in the fields. They are firstly sowing accurately based on prescribed plant density and spacing. This is to ensure that rooting, photosynthetic efficiency and biomass formation are maximum. Inter-culture using hoes with tines, loosening soil for aeration, stimulating root growth, applying fertilizer (top dressing) in between rows/ridges are done at periodic intervals. All of these procedures are amenable to be accomplished using robots of different kinds. In fact, one of the earliest and most successful robots in agriculture- 'the tractor' begins with deep/light plowing of soil. Of course, currently there are areas that practice zero-tillage systems; where in, first exercise with tractor is to plant the seeds. Robots, both autonomous and steer-less, controlled using GPS connectivity are available to accomplish several tasks such as ridging, fertilizer placement, pesticide spray, foliar fertilizer sprays, harvesting and processing grains. A series of robots meant for different tasks, that are small and used individually or those medium in size (e.g., tractor) and used in groups or swarms are discussed in greater detail here. A couple of robots pertinent to maintenance of greenhouse facilities and crops grown in containers/pots are also discussed.

2.3.1 ROBOTIC SEED PLANTERS AND TRANSPLANTERS

A standard and earliest of the procedures that helped humans invent agricultural production depended predominantly on his ability to dibble seeds into earth at most appropriate depth, so that optimum aeration, soil temperature and moisture is perceived by the seed. This allows proper germination and leads to acceptable crop stand that are pre-requisites for better grain yield. For a long stretch of time, human laborers toiled like robots or zombies during dibbling seeds and thinning seedlings. Seeding techniques have evolved remarkably from primitive hand dibbling. Some of the improvements until the discovery of regular tractor driven planters are line sowing in furrows by hand, sowing in hills if seed viability and germination are in doubt, placing pre-germinated seeds, etc. Seed spacing and depth of sowing in the soil is a matter controlled by human accuracy. Planters drawn by animal traction (coulter planters) were common and continue to be so in many regions, where subsistence farming procedures are in vogue. Here, depth of sowing depended on soil tilth, compaction

and texture at each spot and coulter adjustment. Then, automotive tractors became common in most agricultural zones. Introduction of tractor
driven planters is an example of semi-autonomous robot with ability to
dibble/place seeds into furrows. The seed rate, release of seeds by hand by
farm workers decided number of seeds planted at a point and density of
seedlings. Farmers generally thinned seedlings to achieve optimum plant
population. Planting accuracy and crop stands derived out of it became
important. During past 5 decades, semi-autonomous tractor driven planters have been effectively utilized to raise crops, all across different agricultural zones.

Let us consider a typical corn seeder, a tractor driven planter used in
North American plains, rather more frequently than any other farm con-
traption. Farmers aim at maximum accuracy in planting seeds. They wish
to place single seeds, so that seed rate is reduced. It reduces labor on thinning the seedlings. No doubt, best planting accuracies depend directly on
the planter adjustments and calibrations. For bulk planting systems, delivery of seed from hopper to seed planter is an important step. Factors like
planting time, planter lubricants, ground speed of vehicle, seed treatment
and seed size (s) affect accuracy the semi-automatic planters. Based on
planter, liberal use of graphite lubricants, talc powder and adjustments are
crucial aspects (DuPont-Pioneer, 2014). Farmers are generally guided by
the Planter manuals and consultants. Currently, several types of tractor
driven planters are in vogue. Bedord (2011) recently released information
about an 'Autonomous Planting system' devised by Kinze Manufacturers
Inc., situated at Williamsburg, Iowa, USA. This concept clearly replaces
the farmer from the driver's seat of a planter. Instead, a driverless robotic
tractor does tasks of seed planting. The planter is dependent on GPS signals, automation and sensing. The farmer has to feed the tractor/planter
with GPS information about the field, mark the four corners and limits of
seeding operation. A computer-based decision system then offers a few
options that allow the farmer to seek most efficient method of planting
program. Then, the farmer has to place the tractor (robot) at the starting
point shown on the computer screen and start the planting device. This
autonomous tractor system, it seems, is versatile and could be adapted to
accomplish fertilizer application, irrigation water supply and inter-culture
(e.g., earthing-up or placing organic matter). We may note that GPS

PLATE 2.10 Robotic Seed Planters in use in a Farm in North America (Source: Kinze Manufacturers Inc., Iowa, USA; Note: An autonomous tractor with GPS guidance and variable seed rate facility operating on Mollisols (Chernozems) of USA. Steer less tractors are becoming more common in the agrarian zones of North America and Europe).

guidance allows this planter to perceive the obstacles for its movement in the inter-row space and takes suitable measures (Bedord, 2011). A few examples of such semi-autonomous robots are John Deere and Kinze Finger type planters, John Deere vacuum planters, Kinze Edgevac vacuum planter, Kinze Air delivery seeder system, etc. There are now several variations of GPS guided 'Precision planters' operative in different parts of the world. In case of precision planters, the seed rate, planter settings and spacing are all guided by computer-guided decision support systems. Farmers only select the appropriate computer programs, crops, genotypes, seed rates, planting patterns, etc. (see Plate 2.10).

A few examples of GPS guided seed planters are as follows:

Autonomous Planting System (Kinze Manufactures Inc., Williamsburg, IA, USA): Planting corn into fields using robotic tractors with GPS guidance and computer decisions (Plate 2.10)

Multi-Hybrid Robotic Planter (Kinze Manufacturers Inc., Williamsburg, IA, USA): This is an electric powered autonomous robotic planter. It is capable of changing to different genotypes/hybrids of corn seeds on-the-go and plant as per the computer instructions. This is an excellent robot for experimental stations where planting different genotypes of a crop, or different species of crops for evaluation is routine.

Vibro Crop Robotti (Kongsklide Industries, Denmark): This is an autonomous robotic platform capable of movement and ability to plant seeds in the fields. It has facility to add other robotic functions such as weeding, harvesting, etc. (Automatica, 2014).

Prospero (see Trossen Robotics Community): Prospero is an autonomous robot that navigates within crop rows using Schmart board and propeller chip. It is provided with sensors at the belly. Seeds are placed at right spot after digging a hole in the soil. The soil is moved back to cover the seed. Then, the robot makes white mark to help farmers in identifying spots that are planted. This robot can also be used to dispense fertilizers and pesticides (See Trossen Robotics Community, 2014).

Seed planters with ability to sow multi-hybrids in a field are available. The seed planter changes seeds of different hybrids based on prescriptions. The planter can be guided on-the-go to switch hybrids of different genetic background and seeds could be planted based on variable-rate prescriptions. The multi-hybrid planter in use in Iowa's (USA) maize belt is a semi-automatic robot (tractor/planter) that is powered by electrical energy and it is produced by Kinze Manufacturers (Precision Farming Dealer, 2014a). Such planters are highly pertinent in research farms that evaluate different genotypes of a crop for growth and yield traits. They could be utilized in rapidly sowing inter-crops and multi crop mixtures in different ratios and patterns in a field. Whatever the range of facilities regarding seed planting, seed mapping based on germination and seedlings establishment is essential. A digital map of seedling establishment that depicts spots not planted or not germinated will help during re-seeding. Re-seeding using GPS guided semi-automatic planter or autonomous robot is essential to fill gaps and derive an optimum plant stand (Blackmore et al., 2005). A small robotic planter with ability to reach the exact spot where re-seeding is necessary is highly desirable. It reduces on energy cost of a large tractor and reduces soil compaction and seedling damage, if any.

Farmers in Australia are used to cultivating large fields. A robotic tractor is quite useful in covering such large areas. Reports by Gomez (2013) suggest that robotic tractors are useful in accomplishing precision planting of seeds. Smaller robotic tractors are preferred over large semi-autonomous planters because they may lead to soil compaction and are expensive. In a given time of say a day, a robotic tractor now used in New South Wales, it seems covers 20% more area and plants seeds accurately. Such robotic tractors are being tested for commercial use in Australia.

'Prospero' is a small agricultural robot capable of autonomous movement through four legs. It is actually an 'Autonomous Micro Planter' of seeds in the crop fields. Its navigation is controlled by Parallax propeller chip mounted on a Schmart board. The sensors placed under the body of the robot have infra-red cameras and ultrasonic ping. Prospero, first digs a hole to a certain depth based on prescription for the type of seed (crop species). Then, places the seed and closes the soil on the seed. It marks the area where seed was planted for ready reference. Clearly, this robot is useful for regular seeding, plus may be handy in filling gaps where seeds were not dibbled properly and sprouts have not occurred. Farmers could employ a swarm of Prospero robots that communicate with each other and avoid duplicate planting. Prospero, though basically a seed planter, it could be modified to dispense fertilizers and pesticides. This prototype is eventually designed to be capable of harvesting field crops. In all, Prospero should be capable of digging, dibbling seeds, correcting plant stand by replanting in exact spots, applying fertilizers, pesticides and finally harvesting crops (Baichtal, 2011; Kiernan, 2012; Morris, 2011; Dorhout, 2011a; Trossen Robotics Community, 2014). Robots such as Prospero are highly pertinent to be used, when seeds are to be dibbled wide apart, perhaps at 1–3 feet space interval. For example, pearl millet in Sahelian zone, where farmers have been planting pearl millet and cowpea at 1 m space. Prospero has to be modified and adapted to sandy soils of Sahel. A foot pad that is a wide rubberized bowl like structure needs to be attached to each leg. Legs with flat pad allow swift movement and shifting from a location to another without getting stuck in the sand. Rubber tire with wide wheel base is a good idea, so that vehicles do not get stuck in sand. The robot AgAnt is also amenable for similar modification and use on sandy soils of West Africa and elsewhere (see Plate 2.20). Incidentally, McIntosh (2012) and Dorhout, 2011b) believe that small agri-robots once they are produced in large number and made cost effective, they will be useful in any agri-zone. Projections state that small robot like 'Prospero' may cost as low as 500 US$ a piece, if produced in mass numbers.

Rice production systems are well distributed in the tropics of Asia, parts of Africa and Southern United States. Rice production occurs mainly in submerged paddy fields. Upland paddy is grown in arable conditions. Rice needs several specialized agronomic procedures such as puddling to derive fine tilth, submergence by flooding the field with 2–5 cm of stagnating water, transplanting seedlings produced in a nursery, fertilizer application, periodic irrigation, weeding, draining fields and harvesting. Rice farmers are in a position to attempt accomplishing several of these tasks using autono-mous robots. For example, grand idea of Japanese rice experts is to develop an autonomous rice production system using robots for almost each and

PLATE 2.11 A Semi-autonomous Rice seedling Transplanter [Source: IRRI, Philippines; and Yanmar Agricultural Machinery Manufacturers, Okinawa city, Japan; Note: Robotic Transplanters with GPS connectivity are being developed by a few companies in Japan and other countries in the Tropical Rice Belt. For example, Autonomous Rice transplanter-1-Kubota SPU650 and Iseki PZ60 (see Nagasaka et al., 2010)].

every agronomic procedure required during rice production (Nagasaka et al., 2010). They envisage using robots compatible with central computer that interfaces with different robots that are used in the submerged paddy fields for different periods during a single crop season. A main computer and navigation sensors help different kinds of robots performing different tasks. They forecast that such centralized control and several individual autonomous robots each performing their tasks in the field and moving are highly cost effective. The entire autonomous rice production systems firstly has automated steer-less tractor (e.g., Yanmar EG65). Rice transplanting is commonly done using semi-automatic transplanters (Plate 2.11). An automated rice transplanter, e.g., 6-row automated rice transplanter (Kubota SPU650) shifts seedlings to main field and plants them. A second rice seedling transplanter such as Iseki PZ60 could also be used to fill gaps left by the first. Seedlings are generated on hydroponic mats that allow easy pick up by transplanters and planting in the main field (Advance Agripractice, 2014). The rice seedling transplanters have RTK-GPS connectivity. Rice seedling transplanters are autonomous robots that can plant at a speed of 1000 m^2 for every 20 min without reloading of seedlings. They also record information on location and timing of seedling transplantation and general cropping history (Gan-Mor et al., 2007; Tamaki et al., 2009; Nagasaka et al., 2009; 2010). According to McIntosh (2012), rice transplanting robots are capable of placing 24,000 seedlings into the puddled paddy fields per hour. Of course, several factors affect the rapidity, accuracy and establishment of seedlings in the puddled fields. It seems there are 1.8 million rice producing households in Japan. They have been generating rice grains using techniques that are sustainable, environmentally fairly harmonious. Robots will be an added factor that could allow them to generate rice grains consuming much less hours of human labor and accurately, depending on prescriptions made to robots (AdvanceAgriPractice, 2014).

2.3.1.1 Robots and Potted Plants in Greenhouses

Green house maintenance involves intense human labor during filling up pots, moving the pots, irrigating them individually, during harvest and transport to outside the green house and then to market yard. Robots such as 'Harvey' actually replace and relieve humans of hard labor required in a nursery with containerized plants. Often, exchequer is reduced because of costly labor needed during shifting pots/containers, spacing them properly periodically by small adjustments or moving into different locations. Typically, a 'Harvey Robot' does excellent work in shifting and spacing the pots in proper order, to achieve best photosynthetic efficiency and reducing canopy competition between plants. These considerations force farmers to opt for robots such as

2.3.1.2 Harvey

Harvey lifts soil-filled pots with plants grown to different stages, rather effortlessly; picks and moves then to pre-determined location or that as directed using mobiles or Ipads. Harvey Robots are autonomous and can be guided at the touch of a computer screen. Reports from across North America, especially from nursery managers suggest that robots such as 'Harvey' are most welcome additions, because they really reduce hard labor and drudgery by humans (Plate 2.15) (Jones, 2013).

2.3.2 *ROBOTS IN CROP SCOUTING*

Crop scouting is an essential activity done periodically in any field. After all, it tells the farmer about state of the crop, so that further agronomic measures could be conducted accordingly. Crop scouting has been generally conducted accurately using appropriate sampling techniques or randomly. Crop scouting is done using human labor, or ground vehicles and measuring devices. For example, a farmer may transit using a vehicle and sample the crop for leaf chlorophyll using chlorophyll meter or to estimate leaf characteristics and get a rough estimate of plant-N status. There are several other sensing devices used during scouting of a field. For example, instruments for measuring soil moisture, soil pH, soil electrical conductivity, etc., could be deployed during scouting. Scouting can be achieved using Unmanned Aerial Vehicles (drones) mounted with sensing and vision technologies. A wide range of parameters such as crop stand, foliage intensity, leaf area, leaf chlorophyll and plant N status, expected biomass, etc. could be estimated. Remote sensing via satellites and pictures derived from them could be periodically consulted to scout the crop. However, for small farmers and large farms alike, in future, robots may have a major role to play in seeding, then monitoring seed germination, crop establishment and crop stands. Traditionally, row crops are preferred while producing cereals, legumes, oilseeds and several other field crops. Row crops are highly amenable for both heavy and smaller farm vehicles that operate in the field. They allow easy and quick movement in the crop fields. Pedersen et al. (2008) opine that with advent of GPS-RTK controlled devices, perhaps, each and every seed and seedling that emanates from it could be accurately identified, scouted and monitored till harvest. The field and seed locations need to be constantly watched through satellite images on computer monitor. Traditional farming approaches mostly involve sowing seeds in straight or contoured ridges, line sowing and close spacing within rows, but fairly widely placed rows. Such a land preparation is compatible with the tractor driven planters. A moving tractor with its fluctuations in speed and planting devices may release seeds in rows, but accurate single placements are not possible. A thickly planted seed bed results in rows of crops that need thinning. However, robots that are small in size and guided perfectly by GPS could place seeds even within rows, more accurately on the ridges (Pedersen et al., 2008; Wilson, 2000). Small sized robots are useful in planting seeds in predetermine designs. Robots may also allow farmers to plant seeds considering, both vertical and horizontal axis of field. This may allow easier movement of robots in the field when performing agronomic procedures and in reaching seedling accurately (Weiner and Olsen, 2007). There are suggestions that using small robots that are either stationed within the field at different points or using robots that roam in the entire field, we can scout the whole field easily plus mark almost each small seedling or a bunch and notice their progress through satellite vision. Also, fields could be accurately scouted for weed eruption in different location within the field. Weeds that hinder seedling growth could be then accurately reached and removed. This aspect of accurately destroying emanating weeds and allowing the crop seedlings selectively is of utmost use, when crops are sown under No-tillage system.

As stated earlier, crop scouting is an essential aspect of farm management. Crop scouting is done manually in traditional farms. It involves human labor and accurate data collections about various aspects of crop stand. Aspects such as seedling growth, its establishment, crop health, nutrition deficiencies if any, disease/pest attack, soil erosion problems are also recorded. Crop scouting and data collection could be expensive if fields are large and it involves higher input of human labor. However, if an automated system, a robot is stationed to judge canopy at definite intervals and to note the various parameters such as growth, height, leaf color, chlorophyll content, N status, disease/pest incidence, then, cost gets minimized. Biosensors that identify occurrence of different diseases on crops are available. Alternately, small robots that move in-between crop rows could also be used (Christensen et al., 2005; Pedersen et al., 2008). Over all, we may note that autonomous robots used for crop scouting has to be taller than the crop canopy or it should be designed and used as Sub-canopy robot able to squeeze through inter-crop regions avoiding obstacle, if any. Robots taller than crop canopy will provide a good overview of the crop stand and a good close-up of leaves, canopy, crop health and disease incidence if any. A sub-canopy robot, for example, one built at Hohenheim University in Germany, cruises through the inter-row space effortlessly and takes note of several parameters about the crop (Sani, 2012). Robotics division at MIT, in Massachusetts, has offered an autonomous scouting robot exclusively for gardeners. It is called 'Smart Gardner.' It takes note of several parameters of horticultural plantations, individual home gardens, etc. (Sani, 2012).

Field scouting for weed detection is a task that could be accomplished using human labor, satellite vision, robots placed stationery at vantage points in the field or those that move between rows and across large fields. The identification of weed species depends on the sensors installed and computer-based detection systems. The foliage and other characteristics of the weed are used by computers to differentiate weed and crop plants. Sometimes, weed groups are classified and bunched. For example, there are computer programs that allow identification of over 20 weed species that occur in wheat fields. They are traced by moving robots, using leaf shape, size, canopy characteristics, etc. Weed mapping is yet another function accomplished efficiently by robots that scout the field periodically. This exercise helps in focusing weedicide application only to afflicted regions. Thus, it reduces quantity of chemical sprayed and human labor requirements (Pedersen et al., 2008; Manh et al., 2001; Sokefeld et al., 2000; Vrindts et al., 2002). According to Blackwell (2012), adoption of crop scouting robot reduces usage of pesticides immensely. Prior scouting makes farmers to apply pesticides only where required. *The reduction in usage of pesticides could be as high as 98% (Blackwell, 2012).*

In addition to scouting, robots are essential in accomplishing other tasks. Robots are also used as versatile utility vehicles within large farms. Moorehead and Bergerman (2012) believe that possessing a handy versatile vehicle capable of negotiating ruggedness of plowed/cropped fields and performing well in several other off -road situations is a boon to farmers. 'R Gator,' for example, is a robotic vehicle that transits in urban and off-road farm locations carrying a sizeable pay load. This is a robotic porter indeed. Transiting off-road rugged surface is a challenge to robots. Of course, this difficulty is equally perceived even when vehicles are managed by drivers. The robot has to detect, nudge, dodge and shift tracks to avoid obstacles such as grass tufts, large clods, rocks, trees, etc. The 'R Gator' actually is fitted with 3D lasers combined with radar to read and map the terrain and its conditions. 'R Gator' makes use of a standard controller (a computer-based decision support system). 'R Gator' could be used in fields with tall grasses/crops say tall prairie regions or fields with pearl millet, sorghum or sugarcane effectively.

'CropScout-1' is yet another robot that operates in cereal fields driverless. It can detect crop and weeds separately. It navigates between maize rows that are straight for a stretch or even meandering when seeds are planted based on contours. 'CropScout-1' has 2 mechanical rear whiskers, 6 infrared distance sensors, a gyroscope, 2 inclinometers, a pulse counter for motor speed detection and camera. Navigation is accomplished using infra-red and ultra sound sensors, a gyroscope slip detectors and wide angle camera (Achten et al., 2014). It has internal micro-computer. CropScout-1 is also provided with bumpers to withstand collision with obstacles in the field.

Over all, robots could become most common contraptions that swarm the fields, scouting for data on crop growth rates, diseases and pests, soil fertility variations, water deficits, etc. The cost of scouting robots envisaged ranges from 500 US$ to 7000 US$ depending on the range of tasks to be carried out, sensors and GPS guidance required (Pocock, 2012). Some of the scouting robots designed are small, at best 1 ft. long and they can maneuver in between crop rows and take turns with ease and swiftly. Some of the more futuristic additions to such small robots will be 'on-the go' chemical analyzers for soil properties, soil pH, electrical conductivity, plant nutrients, disease incidence, etc.

2.3.2.1 Robots and Crop Phonemics: Crop stand, Foliage, Canopy and Nutrient Status

Monitoring the field from seed germination through different plasto-chrones and growth stages is of utmost importance to any farmer. Large farms need periodic inspection of various fields that could be at different stages of growth and maturity. It is a tedious task to travel in a farm vehicle or tractors and inspect each spot that needs attention. Sometimes inaccurate data may creep in field worker's note books. It is almost difficult to get a glimpse of entire large farm, if this procedure of noting crop stand is done using human labor. Observation of crop phenotype and morphogenetic expression using high clearance tractors fitted with variety of gadgets is one way. Tractors fitted with sensors for various characters such as crop height, leaf color, crop stand, canopy characters, insect attack, etc. is a good alternative (Von Mogel, 2013). It has been suggested that robots (driverless tractors) that periodically monitor phyto-morphology could become popular with farmers in the developed nations. The cameras and sensors that accumulate quantitative/qualitative data of the crop provide useful information to the farmers. Large farms need such robots most urgently to reduce errors in judgment and dependence on seasonal farm labor. Further, Von Mogel (2013) believes that robots make rapid assessment of crop and its growth, in other words, phenotyping will be perpetually needed by farmers who operate large cropping zone. Tracking crop by employing 'Phenocopter' is another rapid and efficient method. It is becoming popular in North America. Several types of UAVs fitted with a range of sensors and cameras help the farmer to assess the crop stage, its growth characteristics, nutrient and water status, etc. This information is useful to farmers in deciding on the appropriate computer programs and yield goals. Farm inputs such as fertilizers, water and other agronomic procedures could be channeled using decision support systems and robots. Phenocopters allow farmers greater versatility in deciding on crops, timing of procedures and revisions, if any about fertilizer inputs and yield goals. The entire field on 1 ha could be aerially observed for variety of crop traits in one stretch of 6–7 passes on the field and data as digital pictures supplied to farmers in a matter of 6 minutes (Von Mogel, 2013).

Studying the phenomics of roots in the fields is indeed difficult. Perhaps not preferred because of tedious nature of the work, cost of plant root excavation could be high, 3D picture of roots are also difficult to obtain. However, to acquire knowledge about root growth and its progression, artificial conditions are created in a rhizotron. A novel pheno-typing system called GROWSCREEN-RHIZO is available. It scans for root growth in the entire rhizotron. Results suggest that root data from rhizotron could be helpful in gaining better understanding of root growth of crop species such as maize, barley or wheat (Nagel et al., 2012).

Currently, there are several models of robotic scouts being developed and few that are ready and being used in farmer's fields in North America and elsewhere. These scouts assess and collect a series of data related to crop phenology through different stages of the crop, may be at weekly or fortnightly intervals. Such data is helpful in judging nutrient and irrigation needs of the crop, as it progresses towards maturity. In addition to ground based robots that make a series of observations on the crop growth pattern and canopy traits, there are phenocopters fitted with sensors that make aerial assessments of crop canopy, its temperature, crop nutrient status (chlorophyll), crop lodging, growth, and maturity (Chapman et al., 2014). The utility of UAVs in studying crop growth pattern and several other morphological traits has been dealt in Chapter 3.

To study the rice crop growth pattern and get a detailed information of the phenomics, Sritarapipat et al. (2014) have devised a stationary robot, usually called the 'rice field servers.' The field server is a platform fitted with various gadgets that detect the plant height, leaf number, leaf area, assess growth rates, identify onset of different growth stages such as tillering, panicle initiation, seed set and maturity. Such field servers are linked to a wireless guidance system and can be directed to assess desired parameters at a particular time of the day or crop season. They can also be linked to collect data from weather stations nearby.

Crop productivity is also dependent on formation of buds, flowering and pollination that finally leads to development of seeds/fruits. Natural pollination aided by bees is an important phenomenon in case of certain species like sunflower, coffee, watermelon, etc. (Roberts, 2014; McSpadden, 2013). Rampant use of pesticides could deter bee population at crucial stages of flower. Improper bee population and reduced pollination may result in reduced seed set in case of sunflower. Reports from Department of Engineering and Applied Sciences and Harvard's Microrobitc laboratory suggest that development of 'Robobees' that are exclusively capable of inducing population by moving from flower to flower with pollen is a possibility. Such a Robobee, it seems is now under development and it is capable of flying at low altitudes and hover around selected flower or flower heads (compositae) (Roberts (2014). Perhaps, in future when Robotic scouts detect lack of bees in the air and suspect that seed set could be hampered, farmers may opt to release a swarm of Robobees to overcome the dearth of bees.

PLATE 2.12 Fertilizer supply to Maize fields Using Semi-Autonomous Tractors and Dispensers [Source: David Nelson, Nelson Farms, Fort Dodge, Iowa, USA; Note: Fertilizer supply to Maize fields (basal dosage) using semi-autonomous tractor and fertilizer-N supply set up. The tractor also has facility for steer less, GPS-aided transit in the field. Fertilizer supply could be regulated using variable-rate applicators. Farmers in the North American Corn belt utilize management zones and Precision Farming approaches to regulate fertilizer-N supply to soil].

2.3.3 ROBOTS AND FERTILIZER SUPPLY TO FIELDS

Fertilizer supply to cereal fields have generally been accomplished using semi-automatic tractors fitted with fertilizer dispensers. During recent years, large dispensers fitted with variable-rate technology have been used in many of the farms that produced maize or wheat (Plate 2.12). These are large vehicles and are prone to cause problems such as soil compaction. RowBot is a driverless farm robot useable predominantly to supply fertilizers to row crops or field crops. The 'RowBot' was developed by an environmental scientist named Kent Cavender-Bares and his brothers Charlie and John Bares (Wehrspann, 2014). It is an excellent robot suited to Corn Belt of North America. The RowBot navigates effortlessly in between rows of maize crop. 'Rowbot' travels equally swiftly both when the crop is in seedling stages and small, as well as when it is taller, mature and ready for harvest. RowBot is a versatile robot in terms of easy movement across fields with a cereal/legume crop. It avoids obstacle, if any, through forward GPS signals. The Rowbot clearly avoids trampling or any damage to cereal crop. Rowbot scouts the crop at various stages of development. The crop height is inconsequential since robot is small and fits the inter-row space. Rowbot delivers fertilizer-N in the rows throughout the season. It is an excellent vehicle to apply split dosages of fertilizer-N based on guidance from computer decision-support systems. This Rowbot is to be released for farmers in Minnesota during mid-2014 (Wehrspann, 2014; Plate 2.13). It is a small sized robot and totally avoids soil compaction.

Regarding robotics during maize production in particular, reports from Minnesota, in USA suggest that 'Rowbot' which is a small autonomous robot capable of performing farm tasks such as fertilizer supply at variable rates, irrigation or pesticides is a good proposition for use. It is an apt option during split application of fertilizer-N (Tamer, 2014). This robot navigates within crop rows even when corn crop is tall during cob maturation. We may note that Corn Belt of America is among most intensively cultivated zones. Fertilizer-N inputs are generally high at 220–280 kg N ha^{-1}. However, entire fertilizer-N is not channeled in one load. Usually 3 to 4 split dosages are applied at critical stages such as seedling, cob initiation, tasseling and grain growth. Accurate distribution of fertilizer-N as split dosages aids in improving fertilizer-N use efficiency.

PLATE 2.13 Rowbot's Robotic Fertilizer Applicator (Source: Drs. John Bares and Kent Cavender-Bares, Rowbot, Pittsburgh, Pennsylvania, USA).

It avoids excessive N inputs that could otherwise accumulate and percolate to ground water. Hence, use of 'Rowbot' to supply split-N accurately, based on decision support computer guidance is essential. Rowbots are also important because fertilizer-N placement using human labor could be costly. Use of semi-autonomous tractors with fertilizer applicators could be less accurate compared to 'Rowbot.' 'Rowbot' is small enough to avoid soil compaction. Overall, using 'Rowbot' to regulate fertilizer-N dynamics in the corn fields is a good idea. These small robots are excellent for in-season fertilizer-N supply (Tamer, 2014). General forecast is that introduction of 'Rowbots' could improve corn productivity further, preserve soil environment, and reduce harmful effects on ground water and channels. Recent reports suggest that an elaborate collaboration among Australian Research Institutes, Hitachi Zosen Corporation and Yanmar Co of Japan has led them to test an autonomous tractor that picks signals from Japan's Zenith satellite and performs tasks such as fertilizer application, inter-row cultivation and captures digitized data about plant's N status (Makim, 2014). Reports from Canadian Prairies state that yield maps are used to decipher soil fertility variation. However, robotic fertilizer applicators that utilize digitized soil-N variation maps are preferred. Maps for several seasons are over-layed and used. Robotics that helps in improving fertilizer-N efficiency is to be emphasized (Country Guide-Canada, 2015).

2.4 ROBOTS TO MANAGE SOIL MOISTURE STATUS AND IRRIGATION

Human effort to reduce drudgery and hardship during water supply to crops has a long history. Several types of contraptions and ingenious methods that effectively employ both gravity based flow and mechanical devices, that lift and spread water to different locations in the field, have been in vogue. Persian wheel, for example, helps reduce human energy needed to lift water from wells and distribute it to crops. Perhaps, this is among more successful semi-automatic robots that have been effectively used in many of the agrarian regions of the world. Most commonly, energy for such Persian wheels were derived from draft animals or human labor. Currently, electric pumps and bore-wells are used in place of a large well and a wheel with tilting buckets. There are several variations of surface irrigation techniques employed during crop production. These methods such as rivers, rivulets, lakes, tanks, irrigation channels, and pipes, use gravitational flow. They need human labor to just monitor, direct and regulate water to different locations in a crop field.

Several types of sprinkler based irrigation of crops are in use in different parts of the world. Sprinklers are used to irrigate large areas of field crops, horticultural plantations, and glass house crops. Human labor requirement is lessened to a great extent, since the sprinklers could be self-propelled utilizing water pressure. The center-pivot irrigation systems that dominate the agrarian zones of North America and Europe are among the best robots capable of uniform delivery of water (Scherer and Crawford, 2005). Crops utilize water from such center-pivot systems efficiently both through canopy/leaves and soil. The Centre-pivot system is almost the main stay for cereals and soybean cultivated in the Great Plains and Mid-west region of North America. About 90% of the crop fields in the Great Plains are irrigated using center-pivot. The Centre-Pivot robots are in use for more than 50 years since its invention by Frank Zybach in 1952. Agriculturists opine that center pivot is the most successful of the robots used in crop production. The center-pivot contraptions can irrigate from as small as one ha to 600 ha in a stretch. The length of the pipes, water flow rates and topography of fields may affect the efficiency and uniformity with which water gets distributed to crops. A corner attachment pipe is supposed to allow the center pivot to be used in any kind of field. Human labor required to maintain center pivot and irrigate crops is highly efficient when crop fields are large, say 10,000 ha or more. They are best suited to irrigate large fields that require stringent regulation of water flow, also precise and uniform water supply. Currently, center-pivot systems that are adaptable to fields with short statured vegetable crops, field crops, tall sugarcane fields, and even tree crops such as orange and apples are available. No doubt, center-pivot robots are versatile. Forecasts, by irrigation engineers suggest that center-pivots capable of precise movement in the crop field using GPS connectivity, and variable-rate water supply using computer-based decision support are on the horizon. Centre-pivots regulated by computers-based on soil moisture status, plant water content and total water requirement at different stages of the crop are also possible. Together, robotics and precision techniques may actually render crop irrigation into an easy task, removing human drudgery and excessive cost on human labor (Scherer and Crawford, 2005). Centre-Pivot irrigation system can be adapted to deliver fertilizer-based nutrients-, for example, during fertigation. However, we may note that water discharged from sprinklers is affected by wind direction and speed, leading to variation in water distribution.

There are other concepts, designs and variations in crop irrigation systems. Small size robots that are mobile and autonomous are being popularized. They suit small farms and green houses. Hernandez et al. (2013) have reported use of an autonomous spherical robot that efficiently tracks soil moisture levels in the inter-row space of a crop field. The purpose is actually to use this spherical robot to monitor soil moisture tension rapidly during precision farming. This autonomous robot has been designed to withstand ruggedness of soil, slippery conditions that occur in crop fields, endure weather conditions, etc. This spherical robot is tele-operated and cost incurred to purchase it is affordable. Recently, agricultural engineers at Technical University of Madrid in Spain have reported development of a spherical robot that rolls all across the crop field to monitor the soil moisture levels. The soil moisture data is periodically relayed to farmers via wireless. This robot is being improvised further to accomplish a few more tasks (Card, 2013). Researchers at CSIC, Spain, state that ROSPHERE's sensors are effective in collecting data on relative humidity and temperature surrounding the crop canopy. However, an additional sensor that measures the rotation speed of the sphere is required. The rate of rolling may alter based on the slippery nature and slope of the terrain (CSIC, 2013).

Blackmore et al. (2005) report that robotic irrigator has been developed for use in cereal fields. The mechatronic robotic irrigator is capable of applying irrigation water at variable-rates. It could also be used to fertigate the fields and apply fertilizers at variable rates. The variable-rate supply is achieved using stepper motors that regulate jet sprays and release of water at the nozzle, based on directions from a computer. The sprinkler jet angles could be adjusted to overcome errors that occur due to wind speed.

'Aquarius' is a small robot that is excellently suited to measure soil moisture and irrigate the crops grown in glass houses, say in pots or soil flour. The robot holds about 110 lts of water that could be dispensed into pots (Plate 2.14). Farmers have to first mark the path of the robot within the green house. The robot accurately traces the path and finds its' watering nozzle into each pot. Pots are usually inserted with a soil moisture detector. The robot's sensors pick up signals from soil moisture detector kept in pots and computers calculate the amount of water that needs to be poured into each pot (Dorhout, 2013; Harrison, 2013; Plate 2.14). Farmers can skip a pot, if need be, but the robot identifies the pot and places full dosage of water in the pots not possessing the detector. Alternatively, farmers

PLATE 2.14 Aquarius- an Autonomous Watering Robot useful in maintaining Greenhouse crops (Source: Dr. David Dorhout, DorhoutRD LLC, USA; Note: 'Aquarius' is capable of transiting within the rows of pots or plants and water them based on soil moisture measurements. It is also capable of watering all the pots uniformly, if soil moisture detectors are absent. It holds up to 110 liters of water and works up a distance of 21 km, after which electric batteries need to be charged. Aquarius is also capable of returning to watering source autonomously, get its tank filled and return to the previous spot to start applying water to pots).

can totally avoid placing detectors in the pots but tune the robots to apply uniform dosage of water to all pots. This practice is common to glasshouse grown vegetables, flowers and other ornamental species. Once, the water is exhausted, robots move on their own to the water source and refill the tank. Robots travel in all about 21 km without requiring re-charging of its electric source (Batteries) (see Plate 2.14). The robot- Aquarius can accomplish 24 different tasks at a time using the four microcontrollers. Some of the advantages in using Aquarius are:

a) it accurately measures soil moisture in each pot or a particular spot;
b) it measures soil moisture on-the-go or while passing the row of pots placed in a green house;
c) It is a durable instrument and can withstand fluctuations in temperature;

PLATE 2.15 Harvey- the Green House Robot (Model HV-100) (Source: Dr. John Kawola, Harvest Automation Inc., Billerica, Massachusetts, USA; Note: Harvey Robots are already in use in different Nurseries across USA. They are autonomous robots capable of picking, transporting and arranging the pots filled with soil and plants at different stages of growth. They replace hard work and human drudgery in green houses. They are useful while periodic shifting of pots to arrange them at proper spacing).

 d) it is inexpensive, so that individual farmers can own it;

 e) It is simple to operate and draw benefits.

The designer of Aquarius, Dorhout (2013) states that the electronics and computer programs embedded into the robots is easy to service, in case of a brake-down. It has a long life despite consistent daily usage. Further, it has been stated that for cash crops grown in green house conditions, robots like Aquarius are excellent. The usage of robots that sense soil moisture and irrigate appropriately is increasing in North America and Europe (Harrison, 2013). This robot is an excellent addition into Crops Science Departments that adopt greenhouse studies. It is equally preferred while producing flowers and vegetables. Aquarius can be used in small fields. It suits a small farmer dealing with field crops or vegetables. The water use efficiency is higher if Aquarius robot is used.

2.5   ROBOTS IN DISEASE AND PEST CONTROL

2.5.1   ROBOTS TO SPRAY PESTICIDES AND FUNGICIDES

Generalizations indicate that use of robots to scout for disease/insect attack has been efficient and economically useful. Field maps showing pest attacked regions makes robots that spray pesticides to confine to only afflicted region. In fact, reports suggest that use of pesticides gets significantly reduced because GPS guided robot sprays chemicals only at those small spots where pests are traced. Pesticide applications are routinely based on prescriptions. Blanket sprays all through the field require

PLATE 2.16   A Semi-autonomous Tractor driven Pesticide Spray System in Western Illinois, USA (Source: Ag Armour Inc., Washington, Illinois, USA).

relatively larger amounts of chemicals (Plate 2.16). Plus, farm labor involved in routine spray is comparatively higher than that when robots are used. In some cases, pesticide requirement if robotic sprayers are used gets reduced by 98% of that needed, if routine sprays are adopted. Obviously, residual effects of pesticides, undue accumulation in soil and side effects on non-target organisms will be minimized, if robotic sprayers are used (Cockburn-Price, 2012). There are actually three different kinds of robotic spraying systems practiced in different farms. The Spot spraying is the most common that takes care of pests in a region in a farm. Cell spraying controls pests/disease in a previously marked region and Micro-spraying involves application of pesticides at a particular plant afflicted with disease/pest (Christensen, 2008). In fact, there are robots that consistently track insect pests that affect different crops. The population of pest and damage could be monitored by using a mobile/ipad connected to robots moving within the farm or those placed at vantage spots. In China, robotic pesticide sprayers have been tested on vegetables such as cucumber. The robot utilizes sensors to trace locations that need chemical sprays. The quantity of pesticide needed is calculated using computer decision support systems. The intensity of disease and spread of symptoms dictate the amount of chemicals sprayed. Changxing et al. (2012) state that compared to traditional blanket spray of chemicals all across the plot/field at a definite rate that needs more of chemicals, robots use much less. The robotic spraying that concentrated only on locations afflicted with diseases required only 60% chemicals, if compared with that required under traditional pest control system. Further, unlike spray guns and pulled trolleys, robotic sprayers are supposed to deliver a fine jet of chemicals (Deveau, and Ferguson, 2010). Evaluation with greenhouse grown tomatoes suggests that robotic sprays are more uniform in the target zone. Chemical spray rates can be controlled based on computer decision support.

Citrus production is among major agriculture related enterprises in Florida. The citrus belt in Florida is attacked by several pests and diseases. Periodic monitoring of citrus groves for disease/pest attack is essential. Robotics projects at University of Central Florida focuses on integrating observations made by UAVs on canopy, foliage and other parts of trees with the data derived from ground-based robots that travel in the groves autonomously. Actually, pictures scanned by UAVs flying at low altitude are integrated with those from sensors located on the ground robots. Farmers can inspect the orchards, detect occurrence of diseases, its spread and intensity rather accurately. This facility avoids drudgery and inaccuracies that may occur by general observations made human labor (Martin, 2013; UCF, 2013). Researchers at UCF, Orlando, opine that monitoring crops such as citrus continuously with ground robots, for nutrient deficiencies and diseases could lead to early warning of crop deterioration and reduce loss. This procedure also reduces use of chemicals. The quantity and number of sprays required will be reduced immensely, if sprays are done immediately after warning, when disease intensity is still low (UCF, 2013).

Currently, sensing techniques for crop biomass, canopy characters, weed infestation and several soil properties related to crop productivity are available (Lee et al., 2010; Krishna, 2012). They are being used in site-specific management. However, detecting disease occurrence and mapping its incidence in entire field is not easy. It involves complex interaction between crop plant and sensor. A recent idea that is partially already successful is to develop micro gas chromatographs that detect the emissions caused by disease afflicted leaves and plant tissue. There are many signature chemicals (VOCs-Volatile Organic Compounds) emitted by plant pathogens that could be easily detected by the robots carrying such micro-gaschromotographs. Incidentally, such micro-gas chromatographs are really small in size. Usually about the size of small coin (e.g., a 25-cent coin in USA). Mounting a few of such gas chromatographs on small robots is not a difficult task. At the bottom line, we ought to realize that for robots to be deployed and chemicals to be sprayed at different locations in a farm, a map that depicts disease occurrence is required. Such a map with disease afflicted regions could be easily prepared, digitized and relayed to robots that take GPS signals and maps, then decide to transit to each of such disease stricken spots and spray the chemicals. This system is not new. It is easily achievable in near future (Colar, 2014). Inter-phasing robots' activity with digitized satellite imagery is essential.

The vine yards, citrus groves and apple orchards world over are treated with a variety of pesticides, herbicides and other chemicals. Farm laborers have to practice precaution and exposure to these chemicals for a longer stretch of time is not recommended. Despite using sprayers designed carefully, depending on location, weather parameters, wind speeds and direction, farmers do get exposed to harmful chemicals. Engineers at Cornell University's Robotics section suggest that it is useful to develop autonomous tractors with attachable robotic sprayers that completely keep drivers and farm workers out of reach of chemical vapors. In addition, since robots are guided by field maps showing pest incidence, pesticide usage is reduced to areas that need to be sprayed. Several different types of mobile platforms that are versatile and move across crop fields have been tested to carry robotic sprayers (Ogradnick, 2009). Some of the robots are also endowed with rapid machine vision that allows the vehicle to identify the tree, in case of forest or fruit tree in a plantation and then selectively spray pesticide to those specific trees. For example, John Deere Company is testing tractors with robotic spray devises to spray citrus trees in Florida. Such robotic pesticide sprayers are to be tested across different locations supporting vineyards.

Vieri et al. (2013) and Gonzalez de Santos et al. (2013) have discussed an interesting concept for robotic control of diseases, pests and weeds that occur in tree plantations. It involves a swarm of robots of different capabilities placed in the orchard and commissioned to spray pesticides/chemicals at appropriate stages, in exact quantities and at correct spots based on sensor vision. This procedure is called integrated pest management using swarms of robots. There are at least 6 models of robots used in the system proposed by them (Centro de Automatica y Robotica, CSIC-UPM 2012; Gonzalez de Santos, 2011; Gonzalez de Santos et al., 2013; Vieri et al., 2013).

2.6 ROBOTS TO CONTROL WEEDS IN THE CROP FIELDS

Aspects such as field scouting using robots, then detecting weeds and preparing weed maps depicting their density in a crop field are the initial steps. Automatic weed detection using machine vision and computers is possible. Weed mapping is not an important exercise in conventional cereal farms. Here, farmers tend to spray weedicides using blanket prescription. The sprayers cover the entire field uniformly. During organic farming, fields are repeatedly plowed to remove weeds and no herbicide is used. Autonomous Plant inspections using robots are necessary, if herbicide application is to be focused to areas with weed infestation. Robots locate weeds accurately using digital maps. Sometimes, single weed plant is sprayed appropriately small quantity of herbicide. Field evaluations suggest that if manual scouting and weed mapping requires 0.7 man hr y^{-1} ha^{-1}, then an autonomous robot supported weed scouting and preparation of digital maps needs 0.2 man hr y^{-1} ha^{-1}. Reports suggest that herbicide sprays made based on weed maps and computer-aided decision support systems allow a 30–75% reduction in the use of herbicides, depending on location and weed density traced in a crop field (Heisel, et al., 1999; Soegaard, 2005; Pedersen, 2003). During recent years, robots capable of micro-sprays of herbicides on individual weed plants have been tested. Herbicides are targeted on single weed plants at a right time, when still young and eradicated without affecting the crop plant. In some locations, herbicide required for manual application could be 720 g active ingredient ha^{-1}. However, if micro-spray technique is adopted, it decreases to <5 g active ingredient ha^{-1} (Pedersen et al., 2008; Soegaard and Lund, 2005).

Robotic Weeder is an important ingredient of 'Push Button Farms' that are expected to dominate the agricultural expanses in near future. Robots that weed the fields usually operate on their own or semi-autonomously with almost negligible involvement of human labor. They

PLATE 2.17   A Small-sized Autonomous Weeder (Source: Dr. Garford, P., Garford Farm machinery, Peterborough, England; www.techdibitis.com).

PLATE 2.18 Hortibot- an autonomous robot to weed fields (Source: Ms. Theresa Eveson and Steven Vale, Farmer's Guardian, Preston, United Kingdom; Note: Hortibot is a robotic Inter-row weeder for use during field and plantation crop production).

impart several advantages such as reduced or nil usage of herbicides, reduced dependence on human drudgery, reduced labor costs, accurate and rapid eradication of weeds, adoption of multiple weed control methods simultaneously based on a computerized decision support system, reduction in herbicide accumulation in soil and ground water contamination (Plates 2.17, 2.18 and 2.19).

According to De Baerdaemaker et al. (2005), weed management using robots involves two concepts of site-specific management. Firstly, weed monitoring is to be carried out using robotic scouts or satellite surveillance pictures. The digitized maps could be utilized during computer based decision making about spray location and quantity of weedicide that is to be released through nozzles. Sometimes, a robotic sprayer is equipped to react instantaneously to weed population traced. The incidence of weeds as detected by sensors triggers an automatic spray of weedicide. This system is called 'weed activated spraying.'

PLATE 2.19 Robotic Weeding Machine (Source: Dr. Philip Garford, Garford Farm Machinery, Peterborough, England; Note: These Semi-automatic robotic machines are programed to apply herbicides at specific spots).

The basic components of any robotic weeder involves electronics, sensors, computer decision support system and mechatronics that allow the vehicle a movement in the fields and ability to perform, say herbicide application or physical removal of weeds accurately and efficiently. For example, Klose et al. (2012) have described an autonomous robot called 'Weedy." It is useful during herbicide application within maize fields. This, or any other autonomous robotic weeder, according to Klose et al. (2012) should contain the following components:

a) Robotic control system: includes coordination and control of the other systems; navigation control system; and communication with user interfaces;

b) Navigation control system: includes sensor system and guidance for autonomous movement in the inter-row space; sensor system for autonomous turn in fields

PLATE 2.20 'AgAnt' – An Autonomous robot that inspects the field using Sensors and removes Weeds. AgAnt is small ant-like robot that can work in groups in a field with inter-robot communication (Source: Dr. Tony Grift, Department of Crops Sciences, University of Illinois at Urbana-Champaign, Illinois, USA).

c) Weed control system: includes weed detection system, ability to uproot, cut or spray herbicide to individual weed plants;

d) Safety control system: includes detection of obstacles in the inter-row space of crops; detection of any malfunctions in identification of weeds; automatic shut-down of robotic arm that cuts the weed or sprays herbicide; manual controls if automatic/electronic systems fail;

e) Speed and steering control system: includes control of robots' speed in the inter-row space; electronic control of robots' steering; continuous power to the motors that allow transit of robots in the fields; GPS connection to account for the area weeded.

Source: Klose et al. (2012).

A review by Engineers from University of California, Davis suggests that robots could become common vehicles in agricultural fields in near future. They operate on their own without drivers and totally avoid drudgery that is otherwise needed in un-mechanized farms and those utilizing a greater share of human labor to accomplish agronomic procedures. Slaughter et al. (2008) opine that among various aspects of robots such as guidance in the field especially autonomous movement, mapping of weeds, identification and detection of weed location, elimination of weeds without disturbance to crop plants; the first few important steps that need greater standardization are detection, discrimination and locating weeds. Weed species have to be identified accurately.

It is believed that automated robot that utilizes captured electrical energy (batteries) and cameras that identify and discriminate weeds could be useful in removing weeds, without recourse to excessive herbicides. Robots that are semi-automated and that require a certain level of guidance are being tested in many European nations (Graham-Rowe, 2007a,b). Some of these robots also called 'Hortibots' are endowed with dual purpose. They spray herbicides and/or mechanically remove weeds in the crop fields. Currently, in its place, farmers tend to spray the fields and remove a part of weed population using manual labor. Hortibots can reduce use of herbicides and delay deterioration of soil environment (see Plate 2.17, 2.18 and 2.19). It seems, currently, European Agricultural Researchers are aiming at reducing the weight of such robots, so that, they could elegantly and accurately move in between the crop rows and take turns at the end of rows. Report by engineers at MIT, Massachusetts, USA suggests that light weight robots are being developed using a light mower known as 'Spider.' A set of cameras with ability to distinguish between leaves of crop plants and weeds are fitted to the Spider. The robot is also guided using global positioning technology, so that, it covers the prescribed zone accurately.

Farmers in Iowa's maize belt state that they spend a large fraction of time trying to mend fields and reduce weeds. Automated robot that works incessantly in the field and removes weeds is indeed a boon. According to Researchers at Agricultural and Biosystems Engineering Department of Iowa State University, Ames, Iowa, weeding maize and vegetable fields using robots decreases costs on human labor and man hours required to attend to weeding. At the same, it helps in reducing soil compaction problems. Also, it primarily avoids use of toxic chemicals and herbicides. It avoids ill effects of residual chemicals, if any, in field soils and in the vegetable product. Robotics to weed crop field is totally compatible with organic farming concepts (Wooley, 2014). During past decades, robots used for weeding have been improved for various components. Firstly, they are being streamlined, made accurate and efficient. For example, Robotic weeder standardized at Ames, Iowa is being equipped with sensors and computer data and programs that allow it to select and distinguish crops such as carrots, beans, lettuce, and sweet corn. Some of these robots are fitted with 3D vision that helps the weeding arms to reach the weed accurately in real time (Wooley, 2014). A few of robots tested are light weight vehicles that move within the rows without affecting the crop. Karger and Shirdazifer (2013) states that weed detection and classification are important. Therefore, machine vision algorithms for as many local weeds should be incorporated into computers on the robotic weeders. 'LabVIEW' software is an example that is appropriate for use in maize fields. It discriminates almost all weed species common to maize fields in the Corn Belt of USA and applies herbicides selectively at small dosages on each weed plant.

There are reports from University of California at Davis, that, robots with infrared cameras and appropriate computer programs have been standardized to detect tomato seedlings at a very early stage of growth. It selectively picks weed species that has to be physically uprooted or cut. Weed targeting in tomato fields has been accurate thus showing its possible use in vegetable farms (Lee, 1998). Further, it has been stated that any plant species that is off-type, not wanted in the field is a weed. There are wide array of weed species and volunteers generally encountered in the fields. Hence, we could classify and group the weeds, crop species into sets that could be detected by infra-red sensors and then selected appropriately by the computer programs. Algorithms for different crops need to be included into computers fitted to robots (Lee et al., 1998).

Researchers at Michigan State University, East Lansing, have devised and standardized a robotic weeder that removes weeds from both intra-row and inter-row spaces. It utilizes machine vision and computer decision to remove the intra-row weed, using physical eradication methods (cutting arms). It rapidly applies herbicides to eradicate inter-row weeds that occur in greater intensity (Che et al., 2013). It appears to be a safe bet in large farms.

Reports suggest that a single vehicle based robot, that accomplishes a series of activities such as visual distinction of weeds and crops, provides a 3D view and GPS location of weed, and is equipped with computer programs that direct arms to remove the weed is a clear possibility. A robotic weeder called '*Vibro Crop Robotti*' is being released for commercial use on vegetable farms in Denmark. These robots run on captured electrical energy and work without break for 2–4 hr. They move at 2–5 km h^{-1} in the field and are efficient in avoiding human drudgery in crop fields (University of Southern Denmark, 2014). Norremark and Soegaard (2006) have reported development of an automatic weeding robot suitable for vegetable fields in Denmark. It is basically an inter-row hoeing system that is mounted on a robotic, driverless tractor with GPS guidance facility (RTK-GPS). Computer based decision systems allow it to weed the inter-rows and take turns in the field. In Sweden, crop production zones are currently getting exposed to an automatic robotic weeder known as 'Lukas.' It is equipped with infra-red cameras and computer based information to detect even minute weed plants within and in-between rows. It easily identifies crops such as lettuce, cauliflower, carrots and sugar beets as crop plants and avoids any damage to them, if left to weed the fields. Farmers who wish to produce organic crops have preferred robotic weeder- 'Lukas,' because, it avoids use of herbicides. Field trials at Halmsted University in Sweden have proved that 'Lukas' performs efficiently both in terms of weed eradication and economic benefits to farmers (Piquepaille, 2005). In due course, computer programs that allow the robot to identify and discriminate more number of crops and weed species could be added. At Osnabruck in Germany, a robot named 'Weedy' has been developed to remove weeds from maize fields. It has an integrated system that allows it to transit in the inter-row space at required speed, operate sensors (infra-red cameras), detect weed plants and apply herbicides in low quantities exactly on to weed plants. It is also equipped to move and make appropriate turns in the fields using GPS guidance systems (Klose et al., 2012). In Spain, weeding maize fields using autonomous robots has been attempted. The method described by Abadia et al. (2012) includes an autonomous tractor that is able to transit through the inter-row spaces of maize fields. On its platform, sensors, with necessary vision (infra-red cameras), GPS connection, and computers with programs to detect maize crop and weed are placed. The weeding operation is accomplished by rotary brushes and blades. Regarding, crop identifiers

PLATE 2.21   Vitirover – An Autonomous Robot used in Grape vines of Southern France (Source: Dr. Arnaud Frauchdiere, Vitirover, La gare de Saint Emilion, France; http://www.vitirover.com).

PLATE 2.22   Vitirover – An autonomous robot for weeding fruit plantations of Southern France (Source: Dr. Arnaud Frauchdiere, Vitirovoer Inc., St de Emilion, France).

and classifiers, Abadia et al. (2012) state that several classifiers such as 'Multilayer perception,' 'Ripper,' 'Random Forest,' 'Bayes' and 'Nefclass' were tested. Currently, a combination of classifier systems is in operation. This robot emphasizes on rapid removal of weeds in the inter-row spaces. It is a farm vehicle well suited for use in fields adopting precision farming systems. It removes fair amount of human drudgery and farm labor needs.

'Vitirover' is a French designed weeding robot that is being used in groups in the grape vineyards, fruit plantations and several other locations where grass cutting/weeding seems essential (Vitirover, 2014; Plate 2.21; 2.22). There are indeed several advantages that accrue to farmers adopting swarms of 'Vitirover' in vineyards of Southern France. 'Vitirover' has no installation costs. This robot does not disturb the grape plants. Since, it acts in groups/swarms and not a single robot, failures are minimal if any. One or the other robot will take over the tasks, if an instrument fails. It means, at no time, weeding is drastically stopped or affected if a machine breaks down. 'Vitirover' operates on solar energy and fuel input is nil. It does not cause emissions. Vitirover weeding robot is an excellent option if one wants to reduce usage of herbicides and still harvest similar levels of grape fruits. Farmers in the grape production region of France, now a days, frequently opt for zero tillage and do not turn the inter-row soil. This is possible because autonomous robot such as 'Vitirover' takes over the maintenance of weeds in the inter-row region. Manual labor to hoe the grape garden is almost abandoned with the advent of weeding robots. Large tractors fitted with weeding implements are not advantageous. They bring about soil compaction. The costs of purchase of tractors are high and they also consume fuel at higher rates. 'Vitirover' robot affects soil environment least since it cuts weeds and does not disturb the soil layers. Farmers using weeding robots have expressed that robots provide good control of weeds. 'Vitirover' robot allows good aeration of top layer of soil and water infiltration in the profile. Development of grape root system is unhindered in fields weeded using small autonomous robots such as 'Vitrover.' It allows optimum rates of soil organic matter degradation processes.

'Vitirover' avoids a range of negative effects that otherwise occur when mechanical weeding is adopted in grape vineyards. Soil tillage to greater depths affects earthworm population and induces erosion on sloppy terrain. Tillage equipment induces soil compaction and tractor lines. However, injury to grape vine is least if Vitirover is used. A weeding robot such as 'Vitirover' is an all season and all time equipment. It withstands rain, high temperature and sunlight. Since it has infra-red sensors, it can be used during any time of the day/night. 'Vitirover' is a weeding robot that is versatile and we can adjust the height to which the grass mat in the inter-row space is to be cut/trimmed or if weeds are to be removed entirely from base. Small robots left operative in groups in grape vineyards or groves of fruit crops help in reducing usage of herbicides to different degrees. There are reports that herbicides could be entirely reduced. In some cases, if farmers use a combination, herbicide usage could be reduced to 6–7% of original and rest could be taken care off by autonomous robots (Vitirover, 2014)

Jones (2014) has reported about a robot called 'The Organic Weed Eliminator' that transits on 4 wheels in between crop rows. It has light weight components. It removes weeds occurring both in intra and inter-row spaces of the crop fields. It is autonomous and needs no driver. This robot detects crop plants such as lettuce and beans and distinguishes weed flora. It is estimated that if major cropping tracts in the Great Plains of USA are supplied with such autonomous light weighted robots, they could reduce use of over 250 million lbs of herbicides that are used yearly in wheat/maize fields. Ground water contamination by pesticides could be effectively avoided. In addition, it reduces on labor cost. Agricultural drudgery too is reduced enormously.

There are indeed several forecasts about the way crop weeding would be handled in highly automated and electronically sophisticated farms of developed world. No doubt, it is going to be rapid, efficient in terms of identification of weed species, and their eradication. Methods adopted and extent of benefits derived by the farmers could be higher in terms of grain/forage yield. We may note that farmers utilize different methods of weeding depending on the geographic location, cropping systems, main crop species and most importantly the weed species that proliferate and intensity of their spread in the fields. Integrated weeding procedures that utilize more than one method of weed eradication are also common. In this regard, one of the forecasts suggests that tractors without drivers, meaning robots fitted with hitches that till the fields to remove weed seeds would be common. Next, a single tractor fitted with sensors to detect different weed species and computer based decision support systems to select the best possible eradication methods may become common. Such computer programs could take account of the species, growth stage, foliage and size of the weed. There are weeds that are eradicated best using small dosages of herbicides, and others that may need good drenching of leaves with herbicides. It seems, tractors that are equipped to employ different methods such as micro-spray of herbicides on individual weed plant, or removal through cutting blades and/or uprooting from ground is a clear possibility. Over all, forecast suggests development of robots (tractors) equipped with on-the-go ability to identify weed(s) and apply appropriate eradication method(s) depending on intensity and diversity. This is a situation comparable to variable-rate applicators in case of fertilizers such as N, where in fertilizer-N is applied at variable rates based soil-N fertility variation (ASA, 2014). One of the advantages of such robots able to apply only very small doses of herbicides on to a leaf or a small plant is that it avoids excessive/lavish spray of chemicals. It surely avoids ground water contamination. Robotic weeder also avoids loss of non-target plant species along the channels and bunds. It economizes on herbicide usage.

Kaur (2012) has discussed about possible use of robots in farming, in a large scale, for accomplishing several aspects of crop production, such as mowing weeds, hoeing, fertilizing, spraying, etc. Aspects of milch cattle maintenance and other functions too could be accomplished. Automatic robotic weeding of pastures using driverless tractors is one of the items.

In Australia, large scale adoption of zero-tillage systems has exposed the drawbacks of owning large sized tractors. Their use is diminished if zero-tillage is practiced. The consistent use of large-sized tractor has also led to soil compaction (Corke et al., 2013). Vehicle brake-downs result in elaborate effort to get new equipment. Single machine failures can result in delay in several other agricultural operations. Smaller robots are relatively low cost equipment's. They delay soil compaction, if any. We ought to realize that zero-tillage induces weed growth and their intensity could be high. Therefore, weeding robots are essential part of zero-tillage systems. The Queensland University of Technology at Bendee, Australia is supporting a program that allows development and use of small sized robots. They rapidly weed the fields after the harvest of previous crop. Small robots do not cause soil compaction. Aerial robots that are equipped to make surveillance of weed growth and its spread in the fields are handy. They are useful to farmers in large farm of Queensland wheat belt (Queensland County life, 2013; Fitch and Sukkarieh, 2014).

2.7 ROBOTS IN ORCHARD MAINTENANCE

Thinning and pruning orchards are tedious tasks that involve skilled farm labor. During recent years, semi-automatic mechanical thinners, trimmers and pruners have been deployed in many apple orchards. Report by Schupp et al. (2008) suggests, mechanical thinning/pruning plus follow-up hand weeding reduces cost on farm labor. The time required to prune gets reduced by 51–85% by adopting mechanical thinning in combination with hand pruning. Reports suggest that mechanical thinning reduced burden on hand pruning, plus induced more flowers and fruit set compared to plots that were entirely under hand thinning /pruning. Similarly, in peach orchards, mechanical thinning plus hand pruning reduced human labor hours by 31% compared to hand pruning. Wallheimer and Robinson (2012) have reported that intention to employ robots in grape and apple orchards is strong. Introduction of robot is said to reduce costs on labor by 20% over the present. It also avoids constraints posed by non-availability of human labor during intensive crop production. Prototype of 'Grape Vine Pruner' that is pulled in the inter-row space by a robotic tractor is gaining in acceptability. However, a robotic 'apple pruner' will have to negotiate complex tree canopy. Apple tree branches could be complex in formation. The robot could get its vision obscured by leaves. A complex algorithm is needed to decipher branches to be cut and then leave the rest. These robots are being tested on 3D models using computer simulations.

According to Moorehead and Bergerman (2012), orchard and specialty plantations are large enterprises in North America. Adopting robots and lessening costs on human labor and improving efficiency/quality of products seems essential. Orchard maintenance and harvesting are intensive operations. Robot that is versatile and re-configurable based on actual activity on hand in the orchard is a better option. Of course the robot should carry sensing and computing infrastructure. A single robot with a platform that could be added with variety of functions such as ability to prune branches, pluck fruits, spray pesticides or mow the inter-row space is envisioned. Autonomous orchard tractors are also an option during orchard maintenance.

In New Zealand, robots are being designed and developed to accomplish grapevine pruning. Grape tree pruning using robots is set to save immensely on human labor. The robots being constructed at University of Canterbury, New Zealand carry a set of 3D imaging and night vision instruments. The idea is to deploy the robots even during night. It is fitted with very sharp shears to cut the hard branches, stubs and twigs during pruning. The robotic arm is made of light-weight aluminum based alloy. We may note that periodic and timely pruning is necessary to avoid disease and pests, and to obtain proper regeneration and fruit bearing characteristics of the vine yard.

2.8   ROBOTIC HARVESTING MACHINES

Harvesting different crops then separating grains from straw/panicles, processing grains to get kernels could be a real ordeal, if the fields are large. Often, farmers find a very small window of time within which they have to accomplish harvesting and remove the food grains/fruits into safe and secure place. This necessitates large scale human labor involvement. Farmers' profits may also diminish, if harvest dates are postponed. During past few decades, machines and variety of contraptions have been devised and utilized to accomplish crop harvest. Earliest of the contraptions that could be quoted is McCormick's harvesting machine for cereal harvest in North America. Tractors fitted with harvesting devices have been in vogue for the past several decades. Currently, combine harvesters are most popular world over. They are fitted with GPS connectivity. They chop the crop, then, separate the straw and panicles/grains and process to get the kernels of say wheat or maize or rice. These are semi-autonomous robots capable of recording grain harvest per unit area, accurately. They could be made autonomous. Horticultural plantations such as citrus or apples need specialized harvesting robots. Similarly, strawberries borne on creepers need harvesters with special ability to identify, pick, detach and collect the fruits. Robots to accomplish these tasks are available currently, though a few are still at prototype or field testing stages. In case of cotton, farmers harvest the crop by spraying chemicals that cause leaf drop leaving the ripe bolls that could be easily sucked into carts. However, this is not an option when grain or fruit crops are considered. Residual chemical in the edible parts is forbidden and hence it is not attempted. Robotic machines are preferred.

2.8.1   ROBOTS TO HARVEST OF FIELD CROPS

Earliest of the efforts to mechanize and introduce semi-autonomous robots in the field to harvest cereals was achieved by McCormick in 1883. The Virginia farmer produced a mechanical reaper dragged by horse. It could cut the stalks and grains from the wheat fields rapidly. A single reaper was equivalent to 12 human reapers in terms of quantity of grains harvested. The mechanical reaper became popular to cut stalks and grains. During the

PLATE 2.23   A Kinze's Robotic Harvesting system in operation at a Corn field in Western Illinois, USA (Source: Kinze Manufacturers Inc., Williamsburg, Iowa, USA; http://farmofthefuture.net/#/slideshow/autonomous-tractors-take-field; Note: The mature corn field is harvested by a robotic combine harvester (No driver!). It unloads the corn grain harvested into Kinze robotic farm cart system. The robotic grain cart is also driverless. It is led by signals from a remote control or led automatically by inter-vehicle communication. The Robotic Grain Cart then travels across the field and unloads corn grain into a large transporting truck. It is then driven into Grain bins, storage zones or to market yard).

past century, excellent modification to reaper's design and its efficiency in harvesting forage/grains has offered us with several types of reapers. They are controlled by a driver. They derive power from internal combustion engines. Tractor driven reapers and mechanical reapers *per se* are available in plenty of models all across in different agricultural regions. Farmers adopt a range of reapers that vary in terms of grain harvesting efficiency, human labor requirements and profits. It is said, there are still a few farmers of vintage, who may like drudgery and adopt a primitive class cereal reapers akin to one devised by McCormik in 19th century (The Economist, 2009). Currently, there are Combine harvesters that house the farmer in an air conditioned cabin. The vehicle is guided by GPS signals. Operations such as cutting stalks, separating grains, threshing grains, and collecting the clean grains into a trailer are all done in one stretch (see Plate 2.23). There are also autonomous combine harvesters. They are driverless but are guided by a farmer controlling their functions through computer signals. The record for largest harvest of grains in a single day of 8 working hours is held by a Combine Harvester designed by New Holland CR9090. This equipment harvested 551 tons Red wheat grains in 8 h. These combine harvesters are

PLATE 2.24   A close up view of Kinze's Autonomous Grain Carting Vehicle (Source: Kinze Manufacturers Inc., Williamsburg, Iowa, USA; Note: This robotic vehicle is controlled by wireless or GPS guidance system. It shuttles all along the fields, from one combine harvester to another, from combine to transport vehicle, etc. The Robotic grain cart can be programed and controlled using computers located in a cabin nearby or by a technician situated on farm bund using a computer tablet. The grain cart could be controlled using touch screen system on a tablet. The Autonomous Grain Cart (Driverless!) along with Robotic Combine harvesters are excellent examples of the forecasted 'Push Button Agriculture.').

PLATE 2.25 Side view of Autonomous Corn Grain Harvesting System- The Robotic Combine harvester and Grain cart system. Source: Kinze Manufacturers Inc., Williamsburg, Iowa, USA. Note: These are driverless GPS guided vehicles. Their movement in the field is designed by a computer program. The vehicles take shortest and most fuel efficient routes to accomplish the task.

highly economical and reduce burden of human labor immensely. The economic gain is true, despite the fact that a decently sophisticated combine harvester with multiple functions costs 580,000 US$ (The Economist, 2009). No doubt, at present, Combine harvesters with GPS guidance and grain yield mapping facility swarm the large farms in North America, Argentina, Brazil, European plains, Russian steppes and Fareast. The grain quality is held intact, since many of the procedures are done sequentially without being exposed to environmental effects. We ought to realize that when a combine lifts grains into trailers, it solves problems related to logistics, economics and collection of grains in one stretch. These operations were generally cumbersome and needed several days of drudgery by farmer and his large crew of grain harvesters. The ease with which a series of operations are done rapidly, by this excellent robot is a noteworthy invention of modern times (Plates 2.23–2.25).

Autonomous robots and vehicles that complete crop harvest safely, reliably and rapidly are becoming popular in North American Plains. Soon,

PLATE 2.26 Lettuce Crop Thinner (Source: Dr. Tony Koselka, Vision Robotics, San Diego, CA, USA; Note: During lettuce production, appropriate spacing and thinning of plants is essential to maintain a good crop. The quality of foliage and value of vegetable is dependent on intra-row spacing. The above robot is endowed with sensors that allow the robot to identify less developed plant and selectively thin them).

they are expected to become indispensable. However, they have to be economically viable and profitable. They have to withstand harsh conditions encountered in the fields. The combine harvesters and accompanying grain storage vehicles that are GPS guided, haul up to 30 tons from fields planted to row crops such as corn or wheat (Jaybridge Robotics, 2014).

'Demeter' is an autonomous robot capable of harvesting cereal fields. It has a cruise control. So, it is steer-less and transits in rows and takes turns as per GPS signals and computer-based decisions. The 'Demeter' robot is equipped with video cameras and GPS for navigation in the entire field. It can sense obstacles and avoid collision. It harvests the cereal by cutting the stem. 'Demeter' literally does not require human supervision. A swarm of Demeter robots could hasten the harvesting process (Pillarski, et al., 2002).

2.8.1.1 Autonomous Robotic Systems for Harvesting Field Grown Cereals Grains

The most recent trend in farm machinery that deals with grain crops is to establish an autonomous system. The harvesting system is actually composed of a fleet of robotic harvesters/combines that cut the grain crop. They separate out forage and grain, process the grain to a certain stage, for example, winnowing and de-husking. Next, a grain cart system collects the grains into its bin. This is again a driverless robot that is controlled electronically. Its moves are coordinated swiftly and meticulously using GPS guidance, wireless systems and computer-based movement decisions. Interestingly, this driverless cart correctly positions itself by the side of the combine harvester, in such a pre-determined location that, corn grains from combines that are processed directly unload into the cart. It is a real 'Push Button Technology' that is controlled by a computer. A single technician is more than enough to handle the whole operation from a computer cabin or a pedestal in open fields. The Grain cart, once loading is done, automatically moves out of the region and transits to the main truck located on the road or farm. The robotic cart takes the shortest distance, and avoids any obstacle to its movement through machine vision and GPS guidance. The Autonomous grain cart system is a vital robot in the farm that moves from location of operation to next in highly coordinated fashion (Kinze Manufacturers Inc., 2013, 2014; McMahon, 2012; Plates 2.23-2.25). It is endowed with ability to move in the rugged conditions swiftly, avoiding obstacles. The GPS sensors are crucial components. Some of the best Autonomous harvesting systems are being developed by major agricultural engineering companies such as Kinze Manufactures Inc., at Williamsburg in Iowa, John Deere, New Holland and others. These robotic harvest systems are excellent representatives of what is in store for agricultural farming in future- after all, it is clearly a 'Push Button System.' Researchers at Kinze Manufacturers Inc., opine that selection of appropriate computer programs, setting up the autonomous Combine harvesters, seed load carting vehicle and transport systems, their inter communication systems needs skilled technician. The technician controls repeated robotic operations using a computer tablet.

In fact, currently many of the Farm Sales Depots are expressing shortage of farm workers able to control the whole system. An autonomous robotic system, like those operative in Corn Belt could reduce farm labor needed to just *one person* compared to a *few hundreds* in classical farms found in several locations of Asia. A robotic harvest system works through the night and for longer stretch of time without fatigues. The quality of grain product is guaranteed. Further, it is interesting to note that, farm engineers have developed highly improved autonomous carts, cabs and porters that can accomplish different tasks required during harvest of large farms (Eckelcamp, 2012)

Reports by Garber (2014) indicate that German Agricultural companies are trying to establish a series of robotic machines that organize and act in a highly coordinated fashion to achieve maximum harvest of corn/wheat in shortest time. The computers in the robotic combine harvester, infield transporters and trucks are all interlinked through GPS signals. The computers are programed to see that all the harvest vehicles take shortest routes and harvesting is done with great efficiency, particularly, regarding time and fuel consumption. Field trials at CLASS factory in Harseinkle in Germany indicate that the system works both day and night, it is rapid and economical. Again, it replaces dependence on farm workers. The whole harvesting system is versatile and is amenable for modification using computer tablet. Forecasts suggest that a swarm of harvesting robots could improve farm efficiency remarkably.

Selective harvesting of field crops using small robots is a possibility worth trying. This may have special advantages when farmers cultivate inter-crops, strip crops or cultivars that mature at different periods. Crop mixtures sown, their timing, maturity period and grain types are some points to note while adopting selective harvesting using guided robots. Most of large combine-headers are not amenable for selective harvesting, unless crops are sown in strips. Small robots could be versatile and take turns autonomously after sensing the right crop, if intercrops of small row ratios are grown. Alternately, it is suggested that a small robot, if allowed inside a field with inter-crop, it could be programed to select a particular crop species, harvest it entirely and collect grains into a stationary wagon (Pederson et al., 2008). Selective harvesting allows easier segregation of grains of specific crop or genotype, also products of definite quality could be easily collected. Grading of grains or pods becomes proportionately much easier, if selective harvesting is done using small robots guided by GPS (Cembali et al., 2005; Pederson et al., 2008; Abounajmi, 2004; Shinners et al., 2003)

Tomatoes are cultured in large scale in many countries, mainly to use it as fresh vegetable, as syrups and jams when canned. Tomato fruits are borne for a long stretch of time and farmers harvest them in 2–4 picking based on fruiting pattern and ripening flushes. It is time consuming to pick each fruit and store them with least damage. Human labor is needed for extended periods and at intervals. Robots are best suited since they reduce labor requirement and costs, plus fruit damage could be minimized. Earliest of the efforts to use a robot to harvest tomatoes was made in 1982 in Japan. The tomato harvesting robots are slow to identify, locate, grasp and detach the fruit. Initially, these robots could be costly, unless the vegetable farm is large and exchequer from farm products is sufficiently high. In some locations, farmers who opted to use robots had to change the plant training system so that tomato fruits are easily reached by the robots. Robots known to harvest cluster of tomato fruits, instead of single fruits has been very useful and adapts well to farms, if the plant training system is suitably changed. Robots that harvest fruit clusters are cheaper, effective and rapid in action (Kondo et al., 2009; Kondo, 2014). Robots capable of grading tomatoes based on uniformity of size, color and ripening are operated in sequence with harvesting robots. This procedure helps in rapid sophisticated and accurate harvesting, grading and packing of tomato fruits.

2.8.1.2   Robotic Lettuce Pickers of California

The California's vegetable growing belt also known as 'America's Salad Bowl' grows large tracts of fresh lettuce to serve internal and export demand. The leafy vegetables are generally thinned and maintained at proper plant density using an army of migrant workers. It is believed that machine thinning could alleviate the situation. Agricultural engineers in California forecast a great future for robots that harvest lettuce, after carefully judging their state of maturity. This robot is named 'Lettuce Bot' (Plate 2.26). This robot is equipped with software and machine vision that allows it to recognize the stage of the lettuce plant, the crowding of foliage and decide to thin the plant selectively. Careful thinning is essential for a good lettuce crop with large foliage. Undue crowding will reduce crop value in market. Thinning, lettuce has been generally accomplished using farm labor. It is a time consuming process and needs 20 workers per day for a field of one ha^{-1}. In addition to 'Lettuce Bot,' several other robots particularly light weight versions that are fitted with advanced sensors, powerful computing, electronics, computer vision, robotic hardware and algorithms are being developed. The computer decision support is among the crucial components of the 'Lettuce Bot.' The computer fitted to 'lettuce Bot' rapidly evaluates the lettuce plant, its foliage (shape size, thickness) and canopy characters by comparing it with over 3 million data points. There are also other robots offered that perform multiple tasks. For example, one developed in San Diego by Vision Robotics thins lettuce seedlings plus in the grape vine yards does pruning. The Robot is endowed with machine vision and computer-based decision support to select the lettuce seedlings, also the canes and branches of grape plant (CBS, 2013). Currently, there is one another trend among researchers involved in development of robots for use in crop fields. A robot platform is prepared that has ability for machine vision, driverless movement, GPS guidance and remote control via radio signals. Upon it, facility for various other robotic functions, like weeding, thinning or cutting forage, etc. could be added as per requirements (e.g., BoniRob) (Amazone, 2014).

2.8.1.3 Pumpkin Harvesters

Pumpkins are larger fruits borne on a creeper that traverses on the soil floor in a field. Pumpkins that need to be harvested in field are larger, heavier, and fewer compared to other vegetables such as tomato or brinjal or beans. It is said that robotic activity could be hastened to complete pumpkin harvesting. Zion et al. (2014) have recently reported about a pumpkin harvesting robot. It is guided by a digital picture of field or a complete data set of coordinates of each pumpkin in the field that is ripe and has to be harvested. The robotic harvester views the pumpkin field as a two dimensional ground to operate on. The frame, its movements, multiple fruit picking arms and lateral conveyors become operative based on co-ordinates supplied to the computer in the robot. Right at the beginning of harvest, robot is supplied with bunch/bank of two dimensional coordinates representing location of pumpkins in the field. Hence, it transits accurately near each pumpkin. Multiple arms grasp the pumpkin, detach it and place it on a conveyor belt that leads to a storing cart.

Asparagus is a vegetable common to many agrarian regions of the world. It is a short season crop. Asparagus fields are easily amenable for variety of farm vehicles and equipment. Autonomous robots that sense weed growth and physically remove the weeds are available. There are other autonomous robots that harvest the mature crop rapidly and with least human labor involvement. Robots are kept collision-free by accurate guidance in the fields. The row tracking device, it seems has a two-layer structure. The regulation of movement is achieved by decision support system that uses proportional-integral derivative algorithms. The row tracking and guidance accuracy is ± 0.5 cm (Dong et al., 2011). This is a robot with low cost guidance system.

2.8.1.4   Robots to Harvest Pastures and Forage Fields

'Grassbots' is a useful concept that helps farmers with a constant supply of forage grass that is to be harvested freshly time and again. The Grassbot is driverless forage grass harvester. It chops of grass from pastures with impunity. In Denmark, such grassbots are being used to harvest grass/legume from natural meadow growths. This is a highly economical venture, since it almost totally avoids human labor. The movement and pattern of harvesting conducted by Grassbots could be controlled, using a remote control or guided through GPS connectivity. The robots are followed by a pick-up vehicle that collects the chopped natural meadow vegetation (Mortensen, 2014). Interestingly, these 'Grassbots' are capable of working in swarms, with inter-vehicle communication. Overlaps and collisions are avoided and harvest is done at stages, based on height and maturity of pasture grass. Harvests from Grassbots are also used to feed furnaces, biogas production plants and electricity generating systems in Denmark. As an extension of this idea, perhaps it is useful to have Grassbots to maintain golf courses and other playgrounds. Moorhead and Bergerman (2012) have also reported about development of robotic grass mowers that could be deployed in orchards, fairway and in regular farms. In Australia, Engineers at CSIRO in Queensland have developed autonomous tractors (AT) that harvest pastures. These pasture movers are equipped with an omni-directional camera (EyeSee 360), a laser scanner and Leica RTK-GPS connectivity. The vehicle is steer-less, but it could also be used in semi-autonomous mode to cut grass. The robotic vehicle is basically designed using a John Deere Gator vehicle (CSIRO, 2013).

2.8.2 ROBOTS TO HARVEST GREENHOUSE VEGETABLES

Vegetable production in relatively larger scale occurs in the green house districts of European nations and other developed countries. Vegetables are produced intensively with great care regarding fertilizer supply, irrigation and disease/pest control. Human labor requirement could be high and only specialized and skilled assistants are useful in maintaining green houses. About 30–50% cost of vegetable production is incurred on skilled labor. Therefore, green house vegetable products are costlier. Yet, during recent years, European green house farming has expanded. Vegetable production is intense and high yields are recorded. Human labor for harvesting the vegetables, such as tomatoes, cucumbers, ladies finger, brinjal, etc. has increased. Hence, robotic harvesters that are engineered specifically to harvest green house vegetables are being sought. In Netherlands, for example, cucumber farming in green houses is popular and yields are high. Farmers, adopt 4–5 robots simultaneously in the green house bays to rapidly harvest cucumbers. Such cucumber harvesting robots are autonomous and possess great flexibility in speed and direction of movement. They possess 3D imaging, color sensors to detect ripening, and manipulators to detach cucumbers accurately and place them in carts. Computer vision, it seems, detects 95% of cucumbers accurately regarding ripening and guides harvesting arms. Such robots, it seems are capable of harvesting and collecting a cucumber every 30–45 s. Currently, in Netherlands, over 3000 ha of greenhouse produced vegetables are being harvested using robots (Van Henten, 2002).

Sweet pepper is an important vegetable produced in the green house districts of European Nations. This vegetable is grown in protected environment. Fertilizer based nutrients, irrigation and pesticide inputs are largely done, using computer controls. Robots are used for harvest of capsicum fruits. Robots with IR cameras, 3D vision control and computer-based decision supports help the robot to identify and grip the correct capsicum fruit (Wageningen University, 2014; European Commission, 2014; Rovira-Mas, 2009).

2.8.3 ROBOTS TO HARVEST FRUIT CROPS

Fruit bearing tree plantations are quite different from a uniform crop of cereals grown in open field. In case of cereals, panicles with mature grains are found at uniform height for robots to chop the straw and separate out grains. Fruits are borne all over the thickly leafy canopy. Robots are not easily amenable for use to harvest fruits on trees. Basic traits of a fruit picking robots are:

a) robot should be able to locate fruits on the tree in 3D;
b) it must be able to approach and reach the fruit;
c) it must detach fruit according to the criterion fixed by the computers;
d) robots should be able to move swiftly across the groves in an orchard (Sarig, 1994).

In addition, we ought to realize that fruit picking should be faster, efficient and cause least damage to fruit and its quality. The use of robots should be economical and justifiable. It is often remarked that identifying, locating, and reaching accurately, despite fruits occurring within a bunch of leaves or other twigs is a trait attached with human fruit pickers. In contrast, identification of fruits based on vision control, cameras and size is difficult. Computers that suggest the robot arm need a series of algorithms and it is a complex task for the robot arm. It becomes cumbersome, if the fruits hide behind leaves. Tree branches obstruct easy movement of robot arm. Next, while detaching, human laborer regulates his hand pressure appropriately to affect the peduncle and give a cut, if necessary, with perhaps least damage to fruits. Such a regulation of pressure and jerks in different directions needed to detach fruits via a robot is difficult. In some cases, this activity is done better, if humans or computer guidance is available.

Strawberries are grown across agricultural zones of different continents. The vines grow as lush green crop with fruits studded in between thick foliage. Fruits of different maturity stages could be traced at a point of time. Therefore, human fruit pickers will have to use their judgment about the ripening and quality of produce harvested. Human labor required to harvest strawberries is among the major costs incurred by the farmers. However, during recent years, robots that detect the stage of ripening of

PLATE 2.27 Strawberry Fruit harvesting Robot at work (Top: A robotic arm reaches out to a strawberry fruit using infrared vision, computer decisions and GPS guidance. It is a fairly slow process to pick the fruit, detach it and place it in the cart. Below: A strawberry harvesting robotic combine. This robot is autonomous. It has robotic arms that detach and pick strawberry fruits and place it on a conveyor belt, so that it gets collected into basket/ cart. Source: Robotic Harvesting LLC., California, USA).

strawberries, using color detected by stereoscopic cameras are available. The robots identify the ripened red strawberries and locate them using a 3D vision. Once, a berry is located, the robots' pincers (cutting knives) take only 9 seconds to reach the fruit and detach it to a container (basket). Automated machine vision based strawberry plant sorters are available (NREC, 2014; Carnegie Robotics, 2014). The Carnegie Robotics, LLC, have developed ruggedized cameras and processing hardware that help in identification and sorting of strawberry fruits. Current estimates suggest that, robots developed by Japan's Institute of Agricultural Machinery takes 40% less time to harvest the same quantity of strawberries compared with hand picking by human labor. For example, if 500 hr are required to harvest a strawberry region of 1 km^2 using human labor, then, only 300 hr suffices with robots (see Kondo et al., 2005; Kondo, 2014; Saenz, 2010; see Plate 2.27). There are added advantages of using strawberry fruit picking robots. For example, robots are efficient in identifying uniformly ripe strawberries based on electronic vision and stereoscopic cameras. Human judgment about ripening is comparatively less accurate. Of course, computer based decision support to decide on plucking of uniformly green still to ripen, yellow/orange pigmented types, red pigmented types are a clear possibility. The stereoscopic vision allows accurate location and handling of strawberry fruits without causing any damage to fruits. The development of a strawberry fruit picking robot, which operate based on multiple wavelengths in electro-magnetic spectrum is a clear possibility (Gross, 2012). In fact, at National Physical Laboratory, in United Kingdom, they are trying to develop vision systems and fruit colors in such a way that it correlates with sweetest and ripest strawberry fruit. Further, it has been possible to adapt this same type of robots to harvest fruit/vegetable crops with similar fruit morpho-genetics and bearing pattern. Further, Robotics Department at MIT (Massachusetts Institute of Technology, Cambridge, MA, USA) have developed robots that can pick fruits from a cache of fruit crops species such as strawberry, tomato, grapes, oranges, etc. Researchers at Department of Biological and Agricultural Engineering, at University of California, Davis have envisaged a program that develops light weight and relatively small robot to harvest and transport strawberries. It is supposed to replace human labor in identification of ripened strawberries and hand picking. The goal is to save on time and 20% of costs on human labor during strawberry production (Scheiner, 2013). It is actually a harvest aid robot that transits within farms from location to another autonomously. It is called 'Frail-bot' (Fragile Crop Harvest-Aiding Mobile Robot). In California, about 450,000 farm workers are engaged seasonally in fruit farms. Robots designed appropriately could replace a sizeable number of farm workers (The Economist, 2009). In Spain, the Centre for Agro-Industries and Technology has developed a robot that harvests strawberries. It supposedly reduces costs on human labor. A Spanish company in Southern California is developing a strawberry harvester with 24 arms. The movement of arms is regulated by machine vision and computer-based decision support. The harvester collects only the strawberry fruits hanging from the bed. Therefore, it requires modifications in the land that is prepared to raise the strawberry crop. The beds need to be raised and single row farming needs to be adopted (CBS, 2013).

According to Feng et al. (2008), strawberry is a delicious fruit popular across different sections of population. In Japan, Kyushu is an important strawberry production zone. Adoption of robots to replace human drudgery is an important task. Several types of robots have been examined for use in strawberry and vegetable fields (Ling et al., 2004; Chi and Ling, 2004). Strawberry harvesting robot described here uses fruit color in different segments of fruit to decide on state of ripening. Next, classification of fruit is based on size and shape (Feng et al., 2008). Robots mentioned here by Feng et al. (2008) are versatile and become operative in hill tops, plains and even in greenhouse conditions. To avoid excessive damage to fruits, specific stem (fruit stalk) cutting mechanism has been designed. Actual detachment of strawberry fruit is effected using a set of scissors and fingers.

Apple production is an important enterprise in the North-eastern State of Washington of United States of America. The upkeep of orchards and procedures that enhance fruit production involves cost on fertilizers, irrigation, pest control and fruit harvest. Hand picking of fruit is among major expenditures incurred on the orchards. During recent years, researchers have focused on developing robots that pick ripened apple fruits selectively. Robots equipped with GPS, sensors to locate and select ripened fruits in the canopy and harvest them carefully are being evaluated. It seems, during the past, robots have encountered difficulty in reaching the apples located deep inside the canopy. About 50% of apple fruits could only be harvested at a time. Human intervention in the form of defoliation or cutting the obstructing branches were required (Karst, 2013). Hence, robotic fruit harvesters were not fully commercialized. However, during recent past, robots with flexible arms to reach as many fruits and partial involvement of human labor has been envisaged in the apple orchards. This way, it may reduce cost of harvesting by 80%. A semi-automatic robot with minimum guidance from farm laborer seems feasible.

Apple harvest in bulk scale by shaking the tree and catching them in baskets has not been useful. It supposedly damages fruits, reduces quality and economic value. Apples are also damaged during transit. Peterson et al. (1999) have described a robotic apple fruit harvester, which is a bulk harvester of fruits. The apple trees should be grown in inclined trellises. The robot combines several mechanical, electronic and intelligent adaptive technology based on computer decision-support systems. Field testing has proved its feasibility and efficiency, once appropriate tree pruning and training systems are adopted. Fruit removal reached 95%. The detached fruits were graded accurately by robots using color and size parameters. Bulanon and Kataoka (2010) have described the development of a fruit detection and detachment system for Fuji apples grown in Japan. The fruit detection itself is dependent on machine vision and light reflectance. Any uneven surface that affects reflectance could affect accuracy and rapidity of identification. The shape of the apple is another factor that can affect robotic identification. The removal of apples from a branch is effected by twisting and then giving stiff jerk. This system of fruit detachment avoids excessive damage to fruits. Bulonon and Kataoka (2010) have reported that 20 fruits get identified accurately and detached out 22. Detachment effort by robots could go unsuccessful at times. The average size of Fuji apples harvested by robots was 77.8 mm diameter, 78.7 mm height, peduncle length 16.5 mm and average weight 238 g fruit^{-1}. Fruits of the weight up to 325 g fruit^{-1} were detached efficiently.

Robotics for use in fruit tree orchards is being developed in several European nations under the aegis of European commission. The research groups aim at developing robot platforms that can hold series of sensors to identify and locate the tree canopy and ripened fruits. They are specifically aiming at technology that allows fruit picking robots to overcome obstacles to reaching fruits and picking them without damage. Robots suited to specific farms and sites are being developed (Ecoweb, 2013; Payne, 2013; Lleo et al., 2009). Wynn (2007) has reported that Automation Centre for Research and Education has helped Belgian farmers with a robotic apple picking machine. The vision system embedded in the robot identifies apple fruits at different stages of maturity then decides about cutting the fruit.

Hardin (2013) explains that, during recent past, farmers in Belgium have tried to overcome seasonal constraints in availability of labor, especially apple fruit pickers by employing robotic apple harvesters. Most apple orchards are 3–4 meters in height and are amenable for use of robotic harvesters. Baeten et al. (2007) opine that for an 'Apple Fruit Picking Robot' camera vision and a gripper that does not damage fruit is essential. The harvester is equipped with a unique vacuum-gripper and vision control that hold on to ripe apple fruits until it is detached without damage to surface. As an alternative, a common agricultural tractor is mounted with Panasonic industrial robot to pick the fruit. The power supply for robot is derived from tractor driven generator or a dynamo. A scanning system that identifies ripe fruits in the canopy, a computer with appropriate software and touch screen facility, to direct the robots and camera inside the vacuum gripper are basic components of the robot in use in Belgium and other European nations.

In Spain, robotics has been tested for identification of ripeness of peach fruits using vision systems operated at different wave length bands. Multispectral imagery of fruits has been used to classify them for freshness and red-soft flesh. The visual imagery provides a 3D image of fruits in question that need to be picked and collected (Lleo et al., 2009).

Researchers at The University of Sydney in Australia have been concentrating on developing robots, with ability for autonomous movement in the inter-rows of almond trees and effectively take digital pictures of fruit distribution on the tree canopy. These robots were also equipped with computers and software that independently analyzed the nature of almond fruits, particularly, their size and maturity, and took decision on picking them (Sukkarieh, 2013; Table 2.1.). In some cases, aerial vehicles (drones) were also used to supplement data available with robots. In other situations, robots were provided with entire picture of canopy and distribution of almond fruits, by the UAVs equipped with infrared cameras. Multipurpose

TABLE 2.1   Few examples of Robotics Used in Fruit Tree Plantations

Crop/Location	Details on Robotic functions	Reference
Almonds/Medira, Australia	Robots with ability for autonomous transit in the orchards, sensors to detect fruits, and collect data on canopy/fruits	Sukkarieh, 2013
Apple/Sydney, Australia	Robots (un-manned ground vehicles) equipped with sensors to assess tree canopy, locate foliage, fruits and state of ripening, assess soil moisture and irrigate, spray pesticides and weed	Kershaw, 2013; Packham, 2013
Apple/Washington St, USA	Robots identify and harvest apples using sensors, GPS location and picking arms. Usually, 50% fruits are harvested in one go	Karst, 2013
Apple/Santiago, Chile	Robots with sensors are used to locate ripened fruits and harvest them using mechanical harvesters	LaBar, 2013
Apple/ Belgium	Robotic harvester with vision system to identify green or red ripened apple fruits	Wynn, 2007
Oranges	Robotic harvester has a few arms with facility to detach oranges and place them in a bin. It identifies fruits among the leaves. (Plate 2.28)	English, 2013; Zkotala, 2013
Grapes	Wall-Ye is a grape harvesting robot currently in use in French grape vine production zone. It has sensors to identify grape bunches. It harvests using sharp blades fitted at the tip of arms (Plates 2.29 and 2.30)	Murray, 2012

robots with ability to tilling soil, hoeing the inter-row spaces, deciding and irrigating fields, and spraying pesticides are also being developed. A few types of UAV with robotic spray nozzles attached have also been used in Almond farms to accomplish a few tasks without a pilot.

Apple orchards in Australia are being mended using robots with ability to perform various functions that are otherwise in the farm laborer's domain (Table 2.1). Kershaw (2013) reports about robots that transit independently in the tree rows of apple orchards, judge the fruit location on canopy, assess the color ripening stage and take decisions for harvest, based on sensor data and computer programs. Such robots are being developed. Robots with ability to operate all through the day (24 h) and all seven days, depending on the fuel/electric power stored in them are being preferred. Green (2013) states that robot keeps information about each fruit harvested such as its location, color and blemishes on it and can also grade them. Kershaw (2013) opines that using robots to weed the inter-row spaces of apple orchards in Australia can contribute immensely, by avoiding excessive use of human labor and herbicides. It seems, 70% of investment in some apple orchards is consumed by labor charges for hand weeding or semi-automatic weeding tractors. Field robotics especially those used to mend soils, and to achieve weeding and irrigation could have big market in future. Robots with high fuel efficiency are being preferred. In case of fruit pickers, use of robots seems to reduce harvesting costs perceptibly. They say, robots able to map fields, dig and till soil, sow seeds, spray weedicides and mainly harvest ripened fruits will carve out a niche for themselves, in the 46 billion Au$ agricultural sector of Australia, in future. Packham (2013) states that two robots named 'Mantis' and Shrimp' are spreading rapidly in the apple orchards of Australia. They have a series of sensors and GPS and guidance system that helps them in locating apple fruits and their state of ripening, within the tree canopy. Introduction of these robots is expected to reduce cost on human fruit pickers, that otherwise costs at least 16 AU $ h^{-1}.

According to Packham (2013), Australian farming community aims at becoming a major supplier of food grains and fruits to regions in Asia. Mathews (2013) opines that adoption of latest techniques involving robotics and precision farming is essential, if Australia has to keep its competitiveness in agricultural exports. Developing sophisticated autonomous robots that improve crop productivity, enhance economic efficiency and reduce dependence on human labor is being increasingly preferred, among various options. For example, currently robots such as 'Mantis' and Shrimp' are touted as possible replacement for human labor that costs 16 Au$ h^{-1}. Further, it is said that even the agricultural labor associations have welcomed the inclusion of robots during food grain production in fields and apples in the orchards. However, robots that are versatile and pick few types of berries and fruits are yet to be developed. Sometimes, apples picked by robots could be bruised and less appealing for consumers. There are several types of sensors, vision, laser and radars and computer programs being added to robots, so that the operations could be effectively accomplished by the robots.

Chile is among important fruit exporting countries located in the southern hemisphere. Chile exports fruits such as apples, grapes, plums and avocados. During recent years, Chilean farmers are testing a few types of robots that are endowed with ability to trim and harvest ripened apples or grapes. Robots with sensors to identify the ripened apple fruits are used, along with mechanical harvesters to hasten the process of fruit picking (LaBar, 2013; Table 2.1).

Oranges are grown in agrarian belts of different continents. Orange orchards of different sizes could be encountered. Let us consider an example. Florida in South-eastern USA supports intensively cultivated orange groves. There are on an average 46 million orange trees that bear fruits of different types such as tangerines, grape fruits, mandarins, etc. Most of the oranges were hand-picked using costly human labor. Hand picking is costlier in USA than in developing countries such as Brazil or India. Farmers in Florida have stated, right in 1980s, that they needed an alternative such as a robotic fruit picker that reduces dependence on human pickers, makes fruit picking efficient and economically more advantageous. Some of the forecasts in 1980s meant that development of fruit picking robots was in brisk progress and by 1990s, they would be found in great number in the orange groves of Florida and perhaps in many other places (Martin Marietta Corporation, 1983; Martin, 2013). However, even today, we have sizeable number of orchards in Florida that employ human labor to pick fruits. Mass removal techniques using chemicals that make abscission on fruit stalks have been difficult to adopt. The residual effects of abscising chemicals could be detrimental and USDA Food Quality Control Agencies do not permit use of such chemicals on orange orchards. Currently, there are specialized projects aimed at developing robots that pick orange fruits, after carefully assessing fruit's size, color of the rind, maturity stage, disease and pest attack (English, 2013; Zkotala, 2013; Payne, 2013). Researchers at the Engineering Department of the University of Central Florida, Orlando, USA, have developed robots with sensors and infra-red cameras that easily detect specs, discoloration, drought stress, canopy characters, diseases,

PLATE 2.28 Orange Fruit Picking Robot that operates in the groves (Source: Dr. Tony Koselka, Vision Robotics, San Diego, California, USA).

and pest attack on fruits and discriminate between mature and immature fruits before picking. In some orchards, both ground robots and UAV are being used to assess fruit bearing and disease. Robots that could pick and detach fruits at a rate of one per 5 seconds from outer canopy are envisaged. Fruits borne deep inside the canopy and with leaves distracting the sensors, required more time to trace and detach a mature fruit (Whitney and Harrel, 1989). Oranges grown in hedgerows are amenable to robotic harvest much better than other arrangements. It allows swift movement of fruit picking arm. Autonomous Robots with 3D laser ranging scanner moves swiftly and accurately in the groves. Plus, it keeps a detailed record of each and every tree as it passes the location. Such a robot records, data pertaining to canopy size, foliage, fruit count and location of fruits in the tree. This data could be utilized to drive the picking arm in the robot (Simonite, 2009; Plate 2.28). As an alternative, rapid harvest of all oranges and using electronic sensors to sort the poor quality fruits afflicted with specs, disease or insects could be a good idea. This procedure costs less time. Robotic picking of fruits

PLATE 2.29   Wall-ye, the grape vine robot searching for fruit bunches (Source: Dr. Christophe Millot- inventor and Dr. Guy Julian- Manufacturer. http://www.kare11.com/video/1860715174001/1/wall-Ye-Wine-Robot-takes-bow-in-Burgandy (June, 23rd, 2014)).

from orchards and manual and/or semi-automatic sorting of fruits for quality is also a feasible procedure.

Grapevine yards across different agricultural regions have also been mended using robots. For example, in the French vine yard region of Burgundy, robots that conduct pruning, de-suckering and clipping of fruitless shoots and twigs are gaining in popularity. 'Wall-ye' is a robot designed by Dr. Christophe Millot in Burgundy. Forecasts suggest that, it might replace a large human labor force and literally take over important agronomic procedures in the vine yard. Farmers may literally allow Wall-ye to look after their grape yards. French grape growers stand to gain in terms of economics and accuracy of farm activities. The robot 'Wall-ye' is 50 cm tall, 60 cm wide and weighs 20 kg. This robot traverses driverless in the inter-row space and conducts operations such as weeding, cutting unwanted branches, does the routine pruning and cuts grape bunches. The robots need to be programed appropriately. It carries a series of infra-red cameras to capture location of branches, leaves and grape bunches that have

PLATE 2.30   'Wall Ye' A grapevine harvesting robot in operation in a French Grape Farm (Source: Dr. Christophe Millot- Inventor and Dr. Guy Julian- Manufacturer; http://www. kare11.com/video/1860715174001/1/wall-Ye-Wine-Robot-takes-bow-in-Burgandy (June, 23rd, 2014)).

to be detached (Murray, 2012; see Plates 2.29 and 2.30; Table 2.1). It is a GPS guided robot and provides 3D picture of grape bunches. Computers decide about movement of clippers and harvest. These robots work day and night, since they are endowed with infra-red guidance and night vision facilities. They prune or harvest about 600 vines (shrubs) per day of 8 hr. Reports suggest that Wall-Ye, the robot is being sought by farmers because it avoids excessive dependence on human labor.

An automated grape pruner has been designed by Purdue University's Department of Electrical and Computer Engineering section. It has stereo vision technology. Two cameras focus on to stems/branches and take a close look at entire canopy. A computer translates the 3D vision into a set of actions, leading to pruning of grape tree branches. The cutting systems are guided by laser guns and robotic arm actually cuts the branch. (Brown, 2013b).

Coconuts thrive in the coasts and inland tropical regions of different continents. It seems, annually, over 6 billion coconut trees are harvested for nuts. India and Philippines are the major coconut producing nations. Among the various crop management methods, harvesting coconuts is tedious and time consuming. It is also risky since a picker has to climb the tall tree and precariously perch on it, to detach fruits. Each tree may bear 150–300 fruits per season. A human fruit picking laborer picks about 50–60 trees per day (Thattari, 2011). The Coconut Robots are more efficient in terms of fruits picked day^{-1}, energy and cost. There are now several types of coconut tree climbing robots equipped with fruit (nut) detaching facility (Megalingam et al., 2013). Several prototypes called 'Cocobots' have been examined in the plantations. A few of the Cocobots are amenable for wireless control and / or via remote control panel. A farmer stationed on the ground guides the robot climber and its cutting arms, using wireless and picture on LCD screen. A video camera shows the actual location of fruit and cutter arm helps in detaching. Robotic harvesters are safe. Human laborers need not climb each and every tree. It is efficient and allows very little damage to fruits if any.

2.9 AGRICULTURAL ROBOTS AND THEIR INFLUENCE ON HUMAN FARM LABOR

A group of several researchers from Agricultural Institutions and Industries state that production of specialty crops, especially horticultural fruit crops has spread gradually and grown into a 45 billion US$ enterprise in USA alone. Human labor requirement has increased because many of these fruit orchards are intensively cultivated. Farm labor with specialized skills is not easy to obtain. Hence, to overcome paucity of work force, robots with special abilities to replant the saplings, prune, spray pesticides, apply fertilizers and water are being sought. We may expect to see reduction of high labor costs. It may lead us to lessened human labor involvement even in specialized fruit orchards, in due course (Singh et al., 2009).

Let us consider agriculture in Australia during past 3–4 decades. It is said Australian farm productivity is moderate and it serves both internal demand for food grains plus a sizeable fraction that is exported. Currently, the trend is to export 45% of wheat grains produced in the country. Agricultural expansion and/or intensification, both, need proportionate increase in human farm labor. However, demographic trends suggest that since 1981 youngsters opting for farming has decreased by 40%. Further, about 25% of farms are operated by aged farmers. In addition, Fitch and Sukkarieh (2013) state that introduction of robots and satellite guided intensification of farm production, especially cereal crops are inevitable. It then allows higher crop productivity compared to present levels. Large scale adoption of robots to accomplish variety of farm tasks is therefore essential. Future farms will have to be 'Push Button' operated and full of gadgets that are autonomous and electronically highly sophisticated. Robots replace human labor and allow even aged farmers to operate their enterprises profitably. Clearly, in future, intensification of crop production could be aimed via greater use of robots. It avoids excessive dependence on human labor and reduces problems such as non-availability of farm workers during peak seasons, reduces seasonality, and reduces maintenance cost on farm worker populations. Robots could affect migrant labor population and migrations by farm workers *per se*. It may also induce erstwhile farm workers to shift professions since work opportunities in farms get reduced. Robots, after all replace human labor in fields. According to Kutnick (2013), right now, Australian farming enterprises in general, including apple orchards are profitable. Introduction of robots and UAVS may reduce jobs for farm workers and many may find their way out of farming enterprises into urban areas. Farm owners' dependence on seasonal farm labor will lessen. Farm robots are versatile and could be commissioned to work day and night and all through the year, during different seasons. Such advantages could be lacking, if Australian farms continue with human labor.

Pedersen et al. (2007, 2008) report that adoption of robotic tractors during initial plowing and ridging reduces human labor need by 5 times the original, if un-manned tractors were used. Obviously, robotics eases out several farm workers. Further, robotic tractors are capable of function for long stretches of time, up to 18 h a day, almost 2.5 times more compared to manned tractors. Farmers adopting robotics accomplish tasks rapidly and in time. Therefore, robotics gets preferred over human labor.

We may note that this situation created by robotic tractors, weeders, then grain harvesters and fruit pickers is no different to what ensued due to industrial revolution after introduction of machines. Robots will make farm operations efficient with regard to labor needs and evict quite a number of farm workers out of job. Farm labor will have to move to other remunerative activities. It would be wiser to forecast labor shifts from farms accurately and create new avenues for their recruitment elsewhere or in other aspects of farming.

The above consequences of robots on human labor have far fetching effect on migratory and settling trends of human beings. Previously known patterns of farm labor movement may get affected. Transitory or permanent settlements of farm workers will also be affected as robots take their place and induce migration. Inferences from previous studies about farm labor migration trends, may not fit the pattern seen in areas with massive introduction of robots. For example, farm labor seeking employment during planting, transplanting, weeding or harvesting seasons may just get replaced by robots. Transit of farm workers to crop production zones may be uncalled for. Instead, farm labor may shift to areas with better remuneration, for example, industries, etc. Most robots are not affected by natural weather pattern. Farm labor replacement, actually, occurs based on the activity that robots take over and displace the farm worker. Hence, robots will replace farm labor consistently throughout the cropping season.

Agricultural robots seem to be on the threshold of spreading rapidly in different agrarian regions. Hence, it is time for farm workers thriving in large farms of the developed regions and even in regions with small farms to get alert to realities. Farm workers skilled in planting, fertilizer application, pesticide application, harvesting and grading may all find fewer opportunities. Migration and shifts to other skills may be required. Mims (2001) says, 'You'd think that the most challenging, lowest-paid labor (i.e., farm worker) in the U.S. was safe from automation, but as robots become increasingly sophisticated, that could change.' There is no doubt that, as agricultural robots become more common and get equipped with technology, accrue sophistication and attain greater flexibility and accuracy to replace as many farm activities, the human labor eventually forfeits its importance during crop production *per se*. Farm worker population dwelling around farms or those migrating into agrarian areas will diminish, perhaps. The erstwhile well known migratory trends of farm laborers, their flux at the beginning of crop/rainy season and movement out of farming zones after harvest and several other historical cultural practices still continued may eventually make way. Farms filled with robots and sparsely manned may become common. It is clear that each robot introduced replaces equivalent hours of human labor. Robots are no doubt a detriment to lively hoods of farm workers dependent on wages. Yet, they are being sought vehemently, because, ultimately they reduce burden on need for physical labor of human beings. A well organized, rapid and versatile shift to other wage earning locations is the need of hour as the 'agricultural robot' as a phenomenon gathers momentum.

2.9.1 AGRICULTURAL ROBOTS AND ECONOMIC ASPECTS OF FARMING

Tractors are important farm vehicles capable of several different functions. They are helpful in accomplishing several procedures relevant immediately to crop production in the field. They also perform other farm activities depending on the hitches and attachments used, like threshing, winnowing, grinding, lift irrigation, etc. Making such versatile farm vehicles into autonomous robots has its effects on labor requirements, working hours, efficiency and economic advantages. Pedersen et al. (2008) have analyzed the economic effects of shifting from manned tractors to autonomous robotic tractors. They find that as robotics creeps in, farm vehicles become smaller, more dexterous and more of them could be commissioned in the farms. Human labor needs per farm vehicle and total required to raise a crop reduces significantly. A few items such as GPS guidance, use of computer decision supports and machine tracking may need skilled personal and costs to maintain them. Generally, shift from manned tractors to robotic ones increased cost on vehicle by 1.2 times, reduced human labor need by 5 times to 20% of the original levels. Further, farm working hours could be stretched enormously. It could be almost doubled from 8 hr. with manned tractors to 16 hr, if robotic. Robotic tractors are also operated during night (Goense, 2003; Have et al., 2004; Pedersen et al., 2007). We may note that, it is not just farm labor that gets replaced when farm owners adopt robots. This phenomenon is not exclusive to farms. In other aspects of human endeavor such as industrial production too, particularly in assembly lines of major heavy industries, robots effectively replace human labor. Several computer and related hardware industries involved in production of telephones, I-tablets, computers, monitors, etc., all employ robots and replace human labor. There are other walks of life such as truck driving, lab work, handling mail and parcels, where robots have replaced human labor in large numbers during few decades (Depillis, 2013).

Reports from North Carolina farming sector suggest that an auto-steer tractor is advantageous to farmer. It allows him both to reduce labor cost and attain greater accuracy, while performing various precision farming techniques. The auto-steer technology costs the farmers about 20,000 US$. Tractor with GPS guidance facility may cost 60,000 US$ to 100,000 US$. Despite such costs incurred, it seems, farmers of large farms that generate maize, cotton or soybean tend to break-even quickly and make profits. The reduction in farm labor cost is immense. Fertilizer and water supply is highly economized using tractors that are under auto-steer, GPS guided, and attached with variable-rate applicators (NC State, 2014).

High value crops are among the earliest choices if robots are to be economically highly successful. According to Eustis (2014), robots are already becoming popular during strawberry production because they are economically advantageous. Automated picking, sorting and grading systems are less costly than human labor. Robots used during strawberry production are capable of site-specific sprays of pesticides and herbicides. Hence they need less expenditure on chemical spray compared to usual farming techniques that envisage spraying entire field. There are actually three different types of spraying methods possible using robotic sprayers. They are:

a) Spot spraying that involves spraying on weeds traced in a location. This procedure is said to save between 60–80% of herbicide requirements;

b) Cell spraying involves using digital images and sensors that accurately guide the sprayer to apply individual marked cells or areas in a field to eradicate weeds; and

c) Micro spraying targets individual weed seedlings identified using sensors. The quantity of herbicide is reduced enormously.

All three methods of spraying herbicides reduce herbicide quantity required and hence are economically advantageous to farmers (Christensen, 2008). Robots could meet stringent regulations of hygiene and sanitary aspects. Most important aspect according to Eustis (2014) is to keep the cost of purchase of robots to lowest, so that they spread rapidly into agricultural zones.

According to European Robotics Research Program (2012), robots used in agricultural farming, particularly, those relevant to soil and crop management have to be simple to use and economically advantageous to farmers. Agricultural enterprises create only marginal profits, if compared

	USA	UK	Germany	Denmark
Robot operation time – hr day^{-1}	16	16	12	16
hr yr^{-1}	883	889	417	667
Days of operation during a season	52	54	35	42
Wages for skilled labor (Euros hr^{-1})	9	10	6	14
Electricity costs (Euros day^{-1})	8.2	9.0	6.7	9.3

Source: Pedersen et al., 2007;

Note: Area of fields under autonomous robotic systems.

with other industries such as defense or airways. Hence, to maintain competitiveness, robots should be economically feasible and profits derived out of their use have to be perceptibly higher.

2.9.2 ECONOMICS OF WEEDING USING ROBOTS—AN EXAMPLE

Pedersen et al. (2008) have reported about use of robots, skilled labor and financial aspects of robotic weeding of sugar beet fields in different countries. Autonomous robots capable of micro-spray are economical, mainly because they reduce the quantity of herbicide required. However, investment is required to procure micro-sprayers and specialized robots. Maintenance of robotic sprayers also needs investment. Robotic weeders work incessantly and are efficient because they are guided by weed maps and RTK-GPS systems. Yet, economic advantages accrued by sugar beet farmers differ based on geographic location, weed density, cost of robots and labor charges. The quantity of herbicide required gets reduced enormously and savings are proportionate. Following is a comparison of robotic weeding:

A few economic aspects of Robotic weeding of sugar beet fields in different countries, using RTK-GPS systems and autonomous robot weeders:

Now, let us consider the 'Lettuce Bot' that is gaining ground in Salinas Valley, the 'Salad Bowl region of California.' It seems over 5 million US$ has been invested in the design and development of 'Lettuce Bot.' Each robot, its accessories and skilled human force also costs, initially. Yet, agricultural engineers and farmers alike opine that robot thinning solves farmer worker's problems almost perpetually. In California, lettuce thinning and picking are predominantly done by expatriate or temporary immigrant workers. Robots could reduce dependence on farm workers. In fact, a single 'Lettuce Bot' at any time, replaces 12–20 human lettuce thinners in the field. It also does the thinning perfectly after selecting the plant accurately; using machine vision and computer based decision support system. 'Lettuce Bot' is known to induce higher yield. Currently, lettuce farms in California are exposed to shortages of farm worker supply from neighboring regions. There are, it seems, more jobs and less immigrant workers. Availability of farm workers is also affected by seasonality and economic advantages. Farm workers are often attracted to other locations with grape orchards or other crops or urban jobs at times (CBS, 2013). A comment in support of farm workers (laborers) states that 'it is really amazing to see the expert coordination between hand, eye and our thinking process and incredibly fast rate of lettuce picking what a human lettuce picker accomplishes.' Compared with it, The Lettuce Bot,' e.g., lettuce picking robots are slow and need massive improvement in rapid coordination (Torantola, 2013). However, the general forecast is that eventually agricultural robots will take over the fields, be it lettuce or soybean or maize. They will replace human laborers to the hilt. Robot, after all, accomplishes the task with least human toil and removes field drudgery.

Field evaluation about small autonomous robots (e.g., Vitirover) in use in Southern European grape vine yards suggests that, they are economically highly advantageous to farmers. Firstly, they reduce human labor recruitment and reduce drudgery to least, if not nil. Reports suggest that using Vitirover on one acre of vineyard for one year costs 833 Euros acre^{-1}. Whereas, conventional weeding and soil tillage to keep weeds from sprouting costs 1200 euros acre^{-1} year^{-1}. Ploughing vineyard using horse-drawn implement costs 6000 euros acre^{-1} year^{-1}. Spraying large dosages of herbicides to keep the vine yards weed-free costs 900 Euros acre^{-1} year^{-1}. During recent years, farmers have preferred to use robots and expend about 800 Euros acre^{-1} year^{-1} to weed their farms (Vitirover, 2014).

Agricultural robots are forecasted to flourish in the farms world-wide, irrespective of size of the farm, economic disposition of the farmers, crops grown in the farms and tasks required to be handled. The ingenuity and engineering acumen seems to allow us to develop a very wide range of robots. Further, robots that are capable of several tasks related to crop production are being designed. For example, Beck (2012) suggests that prototypes such as 'Prospero' and others that can perform tasks from seeding to harvest are most useful to farmers. Rice seedling transplantation is among most difficult function to be carried out by a robot. There are some prototypes of Rice transplanters developed. However, in each case, design and development of robot seems to be costly and sometimes time consuming. Blackmore et al. (2005) opine that robots are still slow to accomplish certain tasks. Their ability to identify the object/task and perform it is too slow, if a tractor driven by a driver is employed. The process becomes difficult for the driver and economically not a good proposition for the farm company to employ drivers/farm workers for longer periods of time. However, place the same robot on the autonomous vehicle and leave it to perform automatically, then entire process becomes highly remunerative.

Agricultural robots are yet to be perceived as an important economic factor within the realm of global agriculture. Large farms in developed countries are among the pioneers who are currently experimenting with these driverless vehicles and implements. They no doubt reduce on farm labor costs and have several other advantages. Robots are costly and require relatively higher capital investments. It takes a while to break-even the costs incurred. Adoption of robots in less-developed nations may require appropriate modifications to design and costs of purchasing robots has to be suitably reduced or subsidized. Again, these robots, may throw many a farm labor camps out of jobs and livelihood. This is not a new phenomenon. Agricultural zones and industries alike world over have experienced such replacements and retrenchment of farm labor/industrial workers, whenever new gadgets, procedures and high degree of automation have entered the area of work. Appropriate governmental measures and local readjustments will suffice to reallocate jobs to erstwhile farm labor community.

According to Tiwari (2014), agricultural robots are expected to stage a massive expansion in their use all over the world. The market for Agricultural robots is expected to zoom, as more crop production enterprises, dairy and animal farms adopt them. Currently, agricultural Robot market stands at 817 million US$ in 2013. It is anticipated to reach 16.3 billion US$ by 2020 (Tiwari, 2014; Eustis, 2014). It is interesting to note that, several established agricultural engineering companies are converting their vehicles into driverless robots and several more new companies are being initiated that produce a range of agricultural robots. It is believed that considering economics, robots may first engulf high value crops produced in large farms. Agricultural Robots are expected to ultimately become as common as computer usage.

Automation using robots is among the major factors that could revolutionize farming strategies and result in consistent improvement in crop productivity. Robotics have the ability to spread to several aspects of farming, ranging from land preparation, seeding, inter-culture, weeding, pesticide spray and harvesting and to even grading and sorting. These aspects will supposedly enhance crop productivity. Large scale adoption of zero-tillage systems, small robots plus use of GPS guided precision techniques could improve productivity. It decreases on inputs through enhanced accuracy and timely completion of farm operations. In near future, Robotics and GPS guided farm equipment could add to the efficiency and production. Production efficiency could be enhanced beyond the usual 2% increase noted all through past 2–3 decades (TORC, 2013).

Forecasts by Fitch and Sukkarieh (2014) suggest that robotics may influence agricultural productivity of Australian farms positively. Crop production increase may occur in steps, as and when autonomous robots spread into crop fields and farmers perceive higher grain/fruit yield. Farmers need to adopt whole-farm concept while introducing autonomous robots for accomplishing functions. There are several aspects of logistics and cost of autonomous robots that could turn beneficial to farmers, as time passes by and robots become less costly than yester years. We may also realize that there could be many other economically beneficial aspects of introduction of autonomous robots, which we have not understood still.

KEYWORDS

- **Agricultural Robots**
- **Autonomous Fruit Harvesters**
- **Autonomous Tractors**
- **Kinze Robotic Harvesting System**
- **Robotic Weeders**
- **RTK-Combine Harvesters**
- **Seed Planters**
- **Variable Rate Applicators**

REFERENCES

1. Abadia, D., Gonzalez, S., del-Hoyo, R., Paniagua, J, Seco, T., Montano, L., Cirujeda, A., Zaragoza, C. (2012). Crop plant identification for weeding operations in maize fields. http://ita.es/ita/bin.asp?Projectors&588&588crop-plant-identification-for-weeding-operations-in-maize-fields.pdf. pp. 1–6 (May 20th, 2012).
2. Abounajmi, M. (2004). Mechanization of dates fruit harvesting. American Society of Agricultural Engineering, St Joseph, Michigan, USA paper No 041028, pp. 1–5.
3. Achten, V., Wattimena, M., Hemming, J., Van Tulji, B., Balendonck, J. (2014). Robot at Wageningern UR. http://www. robots.wur.nl/, pp. 1–4 (July 2nd, 2014).
4. Adamchuck, V. I. (2010). Precision Agriculture: Does it make sense. Better Crops 94:4–6 http://www.ipni.net/publication/bettercrops.nst/ (December 20th, 2013).
5. AdvanceAgriPractice (2014). Rice in Japan. http://www.advanceariipractice.in/, pp. 1–5 (July 4th, 2014).
6. Ag Armour (2014). Would you trust a robot in fields. http://www.agarmor.com, pp. 1–3 (June 25th, 2014).
7. Amazone, (2014). Bonirob-Robots. http://www.go14.de/index.php?lang=1&news=26, pp. 1–3 (March, 19th, 2014).
8. ASA (2014). The future of weed control. Agronomy Society of America, WI, USA. http://www.agronomy.org/story2013/jan/mon/the-future-of-weed-control.htm, pp. 1–4 (May 19th, 2014).
9. Automatica, (2014). Service Robots in Agriculture. Sensors Magazine http://www.sensorsmag.com/electronics-computer/news/automatic-2014-service-robots-agriculture-13136.htm, pp. 1–4 (March 21st, 2014).
10. AZoRobotic Staff Writers (2013). Precision Agricultural Robotics. Robotics: Modeling, Planning and Control http://www. azorobotics.com/Article.aspx?ArticleID=113, pp. 1–4 (July 1st, 2014).

11. Baeten, J., Donne, K., Boedrij, S., Beckers, W., Claesen, E. (2007). Autonomous fruit Picking Machine: A Robotic Apple Harvesters. Proceedings of 6th International Conference on Field and Service Robotics, Chamonix, France, pp. 1–9.

12. Baichtal, J. (2011). Prospero, A Robotic Farmer. http://www.makezine. com/2011/02/28/prospero-a-robotic-farmer, pp. 1–3 (July 3rd, 2014).

13. Bayer CropScience (2014). Ripe for Robots: Automated Agricultural Helpers. http:// www.cropscience.bayer.com/en/magazine/Ripe-for-Robots.aspx., pp. 1–3 (July 4th, 2014).

14. Beck, E. (2012). Plant vs. Robots: Technology Fuels Agriculture of the future. Eco-magination. http://www.ecomagination.com/plants-vs-robots-technology-fuels-agri-culture-of-the-future.htm, pp. 1–6 (March 21st, 2014).

15. Bedord, L. (2011). Autonomous Planting System. http://www.agriculture.com/ machinery/farm-implements/planters/autonomous-plting-system_231-ar18253, pp. 1–3 (June 18th, 2014).

16. Belforte, G., Deboli, R., Gay, P., Piccarolo, P and Ricauda, A. D. (2006). Robot design and testing for green house applications. Biosystems Engineering 95: 309–321.

17. Belforte, G., Gay, P., Ricauda, A. E., (2007). Robotics for improving quality, safety and productivity in Intensive Agriculture: Challenges and Opportunities. In: Proceedings of conference on Industrial Robots: Programming, simulation and applications. Advanced Robotic Systems, Austria, pp. 23–29.

18. Billingsley, J, Visala, J., Arto, M., Dunn, M. (2008). Robotics in Agriculture and Forestry. Springer Verlag Inc., Heidelberg, Germany, pp. 1065–1077.

19. Blackmore, B. S., Fountas, S., Tang, L., Have, H. (2004). Design specifications for a small autonomous tractor with behavioral control. The COGR Journal of AE Scientific Research and Development. http://cigr-ejournal.tamu.edu/Volume6.html: Manuscript PM 04–001, pp. 1–7.

20. Blackmore, B. S., Stout, W., Wang, M., Runov, B. (2005). Robotic Agriculture-the future of agricultural mechanization? In: Proceedings of 5th European Conference on precision Agriculture. Stafford, J. (Ed.). The Netherlands, Wageningen Academic Publishers., pp. 621–628.

21. Blackwell, S. (2012). Crop Scout Robot. Department of Engineering, Harper-Adams College. Robot Living, Robotic Trends http://www.robotliving.com/page/5/, pp. 1–5 (June 30th, 2014).

22. Brown, J, H. (2013a). From Precision Farming to Autonomous Farming: How commodity Technologies enable Revolutionary impact. Robohub- Environment and Agriculture, pp. 1–12.

23. Brown, D. (2013b). Robotic vineyard pruning under development. Michigan State University Extension, http://westernfarmpress.com/grapes/robotic-vineyard-pruning-under development., pp. 1–3 (March 20th, 2013).

24. Bulanon, D. M., Katoaka, T. (2010). Fruit detection system and end effector for Robotic harvesting of Fuji apples. Agricultural Engineering International: The CIGR Journal Ms. no (1285). http://www.cigrjournal.org/index.php/Ejournal/article/view-file/1285/1267, pp. 1–13 (April 15th, 2014).

25. Card, (2013). Soil moisture-measuring robot wanders fields like a hamster ball. Environmental Monitor http://www. fondriest.com/news/soil-moisture-measuring-robot-wanders-fields-like-a-hamster-ball.htm, pp. 1–3 (July 1st, 2014).

26. Carnegie Robotics (2014). Custom Products. Carnegie Robotics, LLC. http://www.carnegierobotics.com/custom-products/., pp. 1–4 (July 13th, 2014).

27. CBS, (2013). Robot lettuce Pickers in Salinas Point to future of Farming. http://sanfrancisco.cbslocal.com/2013/07/14/robot-farming/.htm, pp. 1–5 (June 14th, 2014).

28. Cembali, T., Folwell, T., Ball, T., Clary, D. D. (2005). Economic comparison of selective and non-selective mechanical harvesting of Asparagus. American Society of Agricultural Engineering St Joseph, Michigan, USA Paper No 053003, pp. 1–7.

29. Centro de Automatica y Robotica, CSIC-UPM (2012). Development of a Robot fleet for pest management. Instituo de Ciencas Agricultura http://www.rhea-project.eu/doc/Posters/RHEA-Poster-3.pdf, pp. 1 (July 12th, 2014).

30. Changxing, G., Kai, Z., Erpeng, Z., Junxiong, Z and Wei, L. (2012). Assessment on spraying effect of intelligent spraying robot by experiment. Transactions of the Chinese Society of Agricultural Engineering 28: 114–118.

31. Chapman, S. C., Merz, T., Chan, A., Jackway, P., Hrabar, S., Dreccar, F., Holland, E., Zheng, B., Ling, J., Jimenez-Berni, J. (2014). Phenocopter: A low altitude, autonomous remote sensing robotic helicopter for high-throughput field base phenotyping. Agronomy (open access) 4: 279–301, doi: 10.3390/agronomy4020279 (July 4th, 2014).

32. Chatzimichali, A. P., Georgilas, I,. Tourassis, V. (2009). Design of an Advanced prototype robot for white Asparagus harvesting. In; Proceedings of international conference on Advanced intelligent Mechatronics. AIM, Singapore, pp. 145–247.

33. Che, J., Chen, J., Tang, L., Wang, Y., Jin, X., Chen, J. (2013). Development of a high-efficient weeding Robot in the crop fields. American Society of Agricultural and Biological Engineers, paper Number 131596766. http://dx.doi.org/10.13031/aim.2013131596766.htm, pp. 1–3 (March 15th, 2013).

34. Chi, Y., Ling, P. P. (2004). Fast fruit identification for Robotic tomato picker. Proceedings of ASAE/CSAE Annual International Meeting, Ottawa, Canada. Paper No. 043083, pp. 1–6.

35. Christensen, S. (2008). Robot spraying-vision or illusion. http://www.landburgsinfo.dk/plk07_tr., pp. 1–3 (July 12th, 2014).

36. Christensen, L. K., Upadhyaya, S. K., Jahn, B., Slaughter, D. C., Tan, E., Hills, D. (2005). Determining the influence of water deficiency on NPK stress discrimination maize using spectral and spatial information. Precision Agriculture Journal 6: 539–550.

37. Cockburn-Price, S. (2012). New Generation of Robots to transform Global Agricultural Production. European United Robotics. http://robotics.h2214467.stratoserver.net/cms/index.php?idcat=41idart=549, pp. 1–4 (March 18th, 2014).

38. Comba, L., Gay, P., Piccarolo, P., Ricauda, A. D. (2010). Robotics and Automation for Crop Management: Trends and Perspectives. In: Proceedings of International Conference on Work safety and risk prevention in Agro-food and Forest systems, Ragusa Ilba, Italy, pp. 471–478.

39. Country Guide- Canada, (2015). Improving Nitrogen Efficiency with Precision Farming. Precision Farming Dealer. http://www.precisionfarmingdealer.com/content/improving-nitrogen-efficeincy-precision-farming., pp. 1–5 (February 2nd, 2015).

40. Colar, A. (2014). Miniature Gas Chromatograph could help Farmers detect Crop disease earlier. http://www.phys.org/news/2014–05-miniature-gas-chromatograph-farmers-crop-html, pp. 1–9 (July 7th, 2014).

41. Corke, P., Upcroft, B., Wueth, G., Sukkarieh, S. (2013). Robotics in Zero Tillage. Queensland University of Technology, Australia, Research Report, pp. 12.

42. Crow, J. M. (2012). Farmerbots: A new Industrial Revolution. New Scientist 2888:8.

43. CSIC, (2013). Moisture measurement in Crops using Spherical Robots. Robotics and Cybernetics Research Centre, Universidad Politechnica de Madrid, Spain, http://digital.csic.es/bisteam/10261/90576/1/HernandezVega_Moisture_measurement_in-crops_Industrial%20Robot40_1_59_66–2013.pdf, pp. 1–28 (July 7th, 2014).

44. CSIRO (2013). Autonomous Sensing and Robotics in Agriculture- The CSIRO Journey. http://www.fertilizer.org.au/files/pdf/con/2013/Bonchis%20-%20Paper.pdf.pdf 1–11 (July 11th, 2014).

45. De Baerdermaeker, J. Ramon, H, Anthonis, J., Speckmann, H., Munack, A. (2005). Advanced Technologies and Automation in Agriculture. Encyclopedia of Life support Systems (EOLSS) volume XIX, pp. 1–24.

46. Department of Agricultural and Biological Engineering-UIUC (2014). Agricultural Robotics. http://abe-research.illinois.edu/Faculty/grift/Research/BiosystemsAuto-mation/AgRobots/AgRobots.html, pp. 1–8 (May 28th, 2014).

47. Depillis, L. (2013). Eight ways robots stole our jobs in 2013. Washington Post http://www.washingtonpost.com/blogs/wonkblog/wp/2013/12/23/eight-ways-robots-stole-our-jobs-in-2013/, pp. 1–8 (June 28th, 2014).

48. Deveau, J. T., Ferguson, G. (2010). Tips on Robotic spraying of Greenhouse tomatoes. The Grower. http://www.thegrower.org/readnews.php?id=2c4k1i5e0c3t, pp. 1–3 (July 14th, 2014).

49. Dong, A., Wolfgang, H., Kasper, P. (2011). Development of a Row guidance system for an Autonomous Robot for White Asparagus production. Computers and Electronics in Agriculture 79: 216–225.

50. Dorhout, D. (2011a). Future of Farming. Prospero Robot Farming. http://www.forums.trossenrobotics.com/showthread.php?4169-prospero-Robotic-Farmer, pp. 1–7 (July 3rd, 2014).

51. Dorhout, D. (2011b). Prospero: Robot Farmer. http://www.dorhoutrd.com/home/prospero_robot_farmre_pp 1–3 (December 9th, 2011).

52. Dorhout, D. 2013. Aquarius: The Watering Robot. DorhoutRD LLC. http://www.dorhoutrd.com, pp. 1–5 (July 7th, 2014).

53. DuPont-Pioneer, (2014). Seed Corn Plantability guidelines by DuPont Pioneer. http://www.pioneer.com/CMRobot/Pioneer/US/agronomy/tools/pdfs/plantability-guide.pdf, pp. 1–7.

54. Ecoweb (2013). Innovation: Intelligent Sensing and Manipulation for sustainable Production and harvesting of high value crops, clever Robots for crops. European Technology and Management, http://www.ecoweb.info/2821_3707_intelligent-sens-ing-manipulation-sustainable-production-harvesting-high-crops-clever-robot, pp. 1–2 (May 23rd, 2014).

55. Eckelcamp, M. (2012). Update on Kinze Autonomous Harvest System. AGWEB-Farm Journal http://www.agweb.com/article/update_on_kinze_autonomous_har-vest_system_/, pp. 1–6 (June 18th, 2014).

56. Editor Timeline (2008). A timeline of Robots. http://www.thocp.net/reference/robotics/robotics.html, pp. 1–34 (June 4th, 2014).
57. Emmi, L., Paradese, L., Ribeiro, A., Pajeres, G., Gonzales de Santos, P. (2013). Fleet of Robots for Precision Agriculture: A simulation environment. Industrial Robots An International Journal 40: 41–58.
58. English, J. D. (2013). Robotic mass removal of Citrus. Energid Technologies Corporation, Massachusetts, USA, Research Report, pp. 1–5.
59. European Commission (2014). Clever Robots for Crops: Intelligent sensing and manipulation for sustainable and harvesting of high value crops. http://www.crops-robots.eu/index.php?option=com_content&view=article&id=5&Itemid=1, pp. 1 (May 26th, 2014).
60. European Robotics Research Program (2012). New Generation of Robots to transform Global Agricultural production. European Robotic Technology Platform. http://www.robotics.h2214467.stratoserver.net/cms/index.php?idcat=41&idart=$549, pp. 1–6 (March 18th, 2014).
61. Eustis, S. (2014). Agricultural Robots: Market share, Strategies and strategies to work. http://finance.yahoo.com/news/agricultural-robots-market-shares-strategies115400601.html;_ylt=A0SO8xrxD61TzR8ArWVXNyoA;_ylu=X3oDMTEzM mU1ZmZvBHNlYwNzcgRwb3MDMGgRjb2xvA2dxMQR2dGlkA1ZJUDQ2MV8x, pp. 1–39 (June 27, 2014).
62. Feng, G., Qixin, C., Masateru, N. (2008). Fruit detachment and classification method for strawberry harvesting robot. International Journal of Advanced Robotic Systems 5: 41–48.
63. Fitch, R., Sukkarieh, S. (2014). Robotics and intelligent systems for large scale agriculture. Australian Centre for Field Robotics, Aerospace and Mechanical Engineering, University of Sydney, NSW, Australia, https://www.grdc.com.au/Research-and-Development/GRDC-Update-Papers/2014/03/Robotics-and-intelligent-systems-for-large-scale-agriculture, pp. 1–14 (May 28th, 2014).
64. Gan-Mor, Clark, R. L., UpChurch, B. L. (2007). Implement lateral position accuracy under RTK-GPS tractor guidance. Computers and Electronics in Agriculture. Computers and Electronics in Agriculture 59: 31–31.
65. Garber, H. (2014). Robotic Harvesting System. In: Service Robots in Agriculture-Automatica 2014. Sensors. http://www.sensorsmag.com/electronics-computer/ news/automatic-2014-service-robots-agriculture-13136.htm, pp. 1–4 (March 21st, 2014).
66. Goense, D. (2003). The economics of autonomous vehicles. In: Proceedings of the VDI-MEG Conference on Agricultural Engineering. VDI-Tatung Landentechnic, Hannover, Germany, pp. 234.
67. Gomez, K. (2013). Australian-designed Robotic Tractor enables precision in Planting. University of New South Wales, Sydney, Australia http://newsroom.unsw.edu.au/news/science-technology/robotic-tractor-deliver-precision-planting, pp. 1–3 (January 18th, 2015).
68. Gonzalez de Santos (2011). RHEA: Robot fleet for highly efficient Agriculture and Forestry Management. – NM P2-LA-2010–245986. Proceedings of the 5th Conference de VII. Program marco de 1+D de la Union Europea en Espana. San Sabastian, Spain, pp. 84.

69. Gonzalez de Santos, P., Ribeiro, A., Fernandez-Quintanilla, C. (2013). The RHEA project: using robot fleet for a highly effective protection. http://www.rhea-project. eu/Workshops/.../%20Proceedings_RHEA_2012.pdf, pp. 1–12 (July 12th, 2014).

70. Graham-Rowe 2007a Robotic Farmer. In: Computing, Robots, Environment, Agriculture. Jorgensen, R. (Ed.). MIT Technology Review v1.13.05.10.

71. Graham-Rowe, D. 2007b Robotic Farmer. http://www.technologyreview.com/ news/408225/robotic-farmer/, pp. 1–3 July 2014).

72. Green, S. (2013). Embracing Change. Apple and Pear Australia Limited (APAL). http://apal.org.au/embracing-change/#sthash.3oeq3oA.dpuf, pp. 1–6 (June 25th, 2014).

73. Grift, T. (2007). Robotics in Crop Production Encyclopedia of Agricultural, Food and Biological Engineering. Taylor and Francis Company, Boca Raton, Florida, USA, pp. 283.

74. Grift T. (2012). Robotics in Agriculture. University of Illinois Urbana Champaign, Illinois, USA http://abe-research.illinois.edu/Faculty/grift/Research/Biosystems Automation/AgRobots/AgRobots.html, pp. 1–8 (May 28th, 2014).

75. Grift, T., Nagasaka, Y., Kasten, M. (2004). Robotics in Agriculture: Asimov meets Corn. Department of Crop Sciences, University of Illinois at Urbana Champaign, Il USA. http://agronomyday.cropsci.illinois.edu/ 2004/Tour_A/ Robotics/, pp. 1–4 (June 25th, 2014).

76. Gross, R. (2012). Robot learns to pick the sweetest, ripest strawberries. https://www. google.co.in/#q=robot+ learns+ to+Pick+the+sweetest%2C+Ripest+strawberriespp 1–4 (May 24th, 2014).

77. Hagele, M. (2014). Robot milking Systems for Cattle. In: Service Robots in Agriculture-Automatica 2014. Sensors. http://www.sensorsmag.com/electronics-computer/news/ automatic-2014-service-robots-agriculture-13136.htm, pp. 1–4 (March 21st, 2014).

78. Hardin, W. (2013). Vision system simplifies robotic fruit picking. Vision Systems Design http://www.vision-systems.com/articles/print/volume-12/issue-8/features/ profile-in-industry-solutions/vision-system-simplifies-robotic-fruit-picking.html, pp. 1–8 (May 25th, 2014).

79. Harrel, R. C., Slaughter, D. C., Adsit, P. D. (1988). Robotics in Agriculture In: International Encyclopedia of Robotics Applications and Automation. Dorf (Ed.). John Wiley and Sons Inc., New York, pp. 1378–1387.

80. Harrison, D. (2013). Robotic Watering. http://www.greenhousecanada.com/content/ view/3542/9999/, pp. 1–5 (July 7th, 2014).

81. Harvey, F. (2014). Robot farmers are the future of Agriculture, says government. The Guardian. http://www.theguardian.com/environment/2014/jan/09/09/robots-far-future., pp. 1–5 (March 15th, 2014).

82. Have, H. (2004). Effects of automation on sizes and costs of tractor and machinery. European Society of Agricultural Engineering, Leuven, Belgium, paper No 285:, pp. 5.

83. Heisel, T., Christensen, S., Walter, A. M. (1999). Whole field experiments with site-specific weed management. In: Proceedings of Second European Conference on Precision Agriculture, Odense, Denmark. Stafford, J. (Ed.) Part 2, pp. 759–768.

84. Hernandez, J. D., Barrientos, J., Cerro, J., Barrientos, A., Sanz, D. (2013). Moisture measurement in Crops using Spherical Robots. Industrial Robots: An International Journal 40:59–66.

85. Janmaat, B. (2014). Agribiotics: Robotics if forever changing traditional food production. http://www.datafox.co/blog/agribotics-robotics-is-forever-changing-traditional-food-production/, pp. 1–4 (July 14th, 2014).

86. Jaybridge Robotics (2014). Software solutions and Systems Integration for Autonomous Vehicles. http://jaybridge.com/industries/agriculture.htm, pp. 1–2 (March 21st, 2014).

87. Jones, J. (2013). Harvey: A working Robot for Container Crops. http://www.harvestai.com, pp. 1–5 (July 8th, 2014).

88. Jones, S. E. (2014). Scientists invent Robot that can automatically weed crops. http://voices.yahoo.com/scientists-invent-robot-automatically-weed-11765938.html, pp. 1–3 (May 18th, 2014).

89. Karger, A. H. B., Shirzadifar, A. M. (2013). Automatic weed detection system and smart herbicide sprayer robot for corn fields. Robotics and Mechatronics (ICRoM). Proceedings of First RSI/ISM International Conference, Tehran, pp. 468–473.

90. Karst, T. (2013). Robotics could reduce apple harvest costs 80%. The Packer, Lenexa, Kansas State, USA. http://www.thepacker.com/fruit-vegetable-news/shipping-profiles/Robotics-could-reduce-apple-harvest-costs-80–222979711.html, pp. 1–2 (May 23rd, 2014).

91. Kassler, M (2001). Agricultural Automation in the new Millennium. Computers and Electronics in Agriculture 30: 237–240.

92. Kaur, K. (2012). Future of Farming with Robots. http://www.azorobotics.com/article.aspx?ArticleID=64, pp. 1–3 (March 15th, 2014).

93. Kershaw, P. (2013). Farms of the Future. The Australian Trade commission, Sydney, Australia, http://www.australiaunlimited.com/food/farms-future, pp. 1–4 (May 22, 2014).

94. Kiernan, W. (2012). Agribots may be swarming to a field near you!. Global Ag investing. http://www.globalinvesting.com/news/blogdetail?contentid=1465, pp. 1–3 (July 7th, 2014).

95. Kinze Manufacturers Inc., (2013). Kinze Manufacturing Continues progress on Kinze Autonomy. http://www.kinze.com/ article.aspx?id=152, pp. 1–5 (June 18th, 2014).

96. Kinze Manufacturers Inc., (2014). Kinze autonomous Grain Cart System. Torcrobotics http://www.torcrobotics.com/case-studies/kinze.htm, pp. 1–4 (June 18th, 2014).

97. Klobarandz, K. (2014). Farm of the Future. Time Magazine. http://www.content.time.com/time/magazine/article /0,9171832196,00.htm, pp. 1–5 (May 28th, 2014).

98. Klose, R., Marquering, J., Thiel, M., Ruckelshausen A. (2012). Weedy – A sensor fusion based autonomous field robot for selective weed control. https://my-osnabruck.de/ecs/fileadmin/groups/156/Verroefficientlichungem/2008-VDI-Weeding.pdf, pp. 1–9 (May 20th, 2012).

99. Kondo, N. Ninomiya, S., Hayashi, T., Ohta, T., Kubota, K. (2005). A new challenge of Robot for Harvesting Strawberry grown for table purpose. American Society of Agricultural Engineering St Joseph, Michigan, USA, 043083, pp. 1–4.

100. Kondo, N., Yata, K., Shilgi, T., Lida, M., Yamamoto, K., Monta, M., Kurita, M., Omori, H., Shimizu, H. (2009). Development of tomato cluster harvesting robot. In: In: Fourth IFAC International workshop on Bio-Robotics, Information Technology and Intelligent Control for Bio-Production Systems, Urbana-Champaign, Illinois, USA, pp. 242–244.

101. Kondo, N. (2014). Present situation and future prospects on Fruit harvesting and grading robots. Proceedings of International Symposium on Mechanical Harvesting and Handling systems of fruits and Nuts. http://www crec.ifas.isfl.edu /harvest/kon. htm, pp. 1–14 (May 26th, 2014).

102. Krishna, K. R. (2012). Precision Farming: Soil Fertility and Productivity Aspects. Apple Academic Press Inc., Waretown, New Jersey, pp. 189.

103. Kutnick, T. (2013). These Robots do more on the Farm than their scarecrow predecessors. Aris-Plex., pp. 1–3.

104. LaBar, M. (2013). Chilean team helps develop ground breaking harvest robot. This is Chile, Santiago, Chile http://www.thisischile.cl/people/?lang=en, pp. 1–8 (May 23rd, 2014).

105. Landoll Corporation 2014b Brillion Pulveriser. http://landoll.com/content/index.php/products/farm_equipment/brillion-farm-equipment/pulverizer/, pp. 1–3 (June 24th, 2014).

106. Landoll Corporation 2014c 6230/6250 Series Tandem Disc harrow. http://landoll.com/content/index.php/products/farm_equipment/agriculture/6230–6250-tandem-disc-harrow/, pp. 1–3 (June 24th, 2014).

107. Landoll Corporation 2014a (2510). Series In-Row Ripper. http://landoll.com/content/index.php/products/farm_equipment/agriculture/2510-row-ripper/, pp. 1–3 (June 24th, 2014).

108. Landoll Corporation 2014d Brillion Till N' Seed. http://landoll.com/content/index.php/products/farm_equipment/brillion-farm-equipment/till-n-seed/, pp. 1–3 (June 24th, 2014).

109. Landoll Corporation 2014e (2000). series Row Crop Cultivator http://landoll.com/content/index.php/products/farm_equipment/agriculture/2000-row-crop-cultivator/, pp. 1–4 (June 24th, 2014).

110. Lee, W. S. (1998). Robotic weed control system. University of California, Davis, USA, Dissertation, pp. 148.

111. Lee, W. S., Alchanatis, V., Yang, C., Hirafuji, M., Moshou, D., Li, C. (2010). Sensing technologies for precision specialty Crop production. Computers and Electronics in Agriculture 74: 2–33.

112. Lee, W. S. Slaughter, D. C., Giles, D. K. (1998). Development of machine vision system for weed control using Precision chemical application. Department of Biological and Agricultural Engineering, University of California, Davis, USA. An Internal Report, pp. 1–14.

113. Ling, P. P., Ehsani, R., Ting, K. C. (2004). Sensing and End effector for a Robotic tomato harvester. Proceedings of ASAE/CSAE Annual International meeting held at Ottawa, Canada, Paper No. 043088, pp. 1–8.

114. Lleo, L. Barreiro, P., Altisent, M., Herror, A. (2009). Multispectral images of Peach related to firmness and maturity at harvest. Journal of Food Engineering 93: 229–235.

115. Lyseng, R. (2011). 'RapidProbe' doubles soil sampling efficiency. http://www.producer.com/2011/09/rapidprobe-doubles-soil-sampling-efficiency/, pp. 1–4 (July 1st, 2014).

116. Makim, R. (2014). Tractor the future with Robotics. Precision Farming Dealer http://www.precisionfarmingdealer.com/content/tractor-future-robotics, pp. 1–3 (January 19th, 2015).

117. Manh, A. G., Rabatel, G., Assemat, L., Aldon, J. (2001). Weed leaf image segmentation by deformable templates. Journal of Agricultural Engineering Research 80: 139–146.

118. Martin, J. (2013). USDA Grants support Federal Partnership for Robotics Research. http://www.csrees.usda.gov/newsroom/news/2013news/10251_robots.html, pp. 1–2 (June 22nd, 2014).

119. Martin Marietta Corporation (1983). Scientists foresee robots picking Florida oranges within 10 years. The Christian Science Monitor. http://www.csmonitor. com/1983/1018/101829.html., pp. 1–2 (March 11th, 2014).

120. Mathews, L. (2013). In: Robots to drones, Australia eyes high-tech farm help to grow food. Packham, C., Davies, E. (Eds.) http://www.smh.com.au/technology/sci-tech/robots-to-drones-australia-eyes-hightech-faram-help-to-grow-food-20130527–2n5zu.html, pp. 1–4 (March 20th, 2014).

121. Mattern, R. (2012). North Dakota State University Demonstrations feature during 'Big iron.' North Dakota State University Agricultural Extension Service. ND, USA http://www.ag.ndsu.edu/news/newsreleases/2012/july-30–2012/ndsu-demonstrations-featured-during-big-iron/ html., pp. 1–2 (March 19th, 2014).

122. McIntosh, P. (2012). Agricultural Robotics: Here come the robots. Maximum Yield. com. http://www.mazimumyield.com/index.php/features/articles/item/13-agricultural-robotics-here-comes-the-agribots, pp. 1–4 (July 6th, 2014).

123. McMahon, K. (2012). Kinze's autonomous tractor system tested in field by farmers. Farm Industry News http://www.farmindustrynews.com/precision-guidance/kinze-s-autonomous-tractor-system-tested-field-farmers.htm, pp. 1–8.

124. McSpadden, R. (2013). Will Monsanto Use Robotic Bees to pollinate Crops. https://www.earthfirstnews.wordpress.com/2013/04/08/robotic-bees-to-pollinate-monsanto-crops/.htm, pp. 1–8.

125. Megalingam, R. K., Venumadhav, R., Ashish Pavan, K., Mahadevan, A., Kattakayam, T. C., Menon, H. (2013). Kinet-based wireless Robotic Coconut Tree climber. Proceedings of 3rd International Conference on Advertisements in Electronics and Power Engineering. Kaulalumpur Malaysia, pp. 201–206.

126. Mims, C. (2011). Down on the Farm, Will Robots Replace Immigrant Labor? MIT Technology Review. http://www.technologyreview.com/view/425278/down-on-the-farm-will-robots-replace-immigrant-labor/?p1=blogs, pp. 1–8 (July 3rd, 2014).

127. Mitchell, C. (2014). The Mitchell Farm: Farming the way nature intends. http://www.mitchellfarm.com/, pp. 1–25 (May 28th, 2014.

128. Moorehead, S. J., Bergerman, M. (2012). R Gator: an Unmanned Utility Vehicle for off-Road operations. Robotics and Automation Society: Technical committee on Agriculture and Automation http://www.researchgate.net/publication/260710461_IEEE_Robotics_and_Automation_Society_Technical_Committee_on_Agricultural_Robotics_and_Automation_TC_Spotlight, pp. 1–6 (July 3rd, 2014).

129. Morris, B. (2011). Future of Farming: Prospero Robotic Farmer. http://www.polizeros.com/2011/12/27/future-of-farming-prospro-robot-farmr/, pp. 1–3 (July 3rd, 2014).

130. Mortensen, C. (2014). From Meadow to Methane: Robots to increase biomass utilization. BE Sustainable. http://www.besustainablemagazine.com/cms2/from-meadow-to-methane-robots-for-harvesting-meadow-grass-to-increase-biomass-utilization/. Html, pp. 1–4 (March 19th, 2014).

131. Murray, P. (2012). Automation reaches French Vineyards with vine-pruning robots. http://www.singularityhub.com/2012/11/26/automation-reaches-french-vineyards-with-a-vine-pruning-robot/, pp. 1–4 (March 20th, 2014).
132. Nagasaka, Y., Tamaki, K., Nishiwaki, K and Saito, M. (2009). An Autonomous Rice Transplanter guided by GPS and IMU. Journal of Field Robotics 26: 537–548.
133. Nagasaka, Y., Tamaki, K., Nishiwaki, K and Saito, M. (2010). Autonomous Rice field operation project in NARO. Proceedings of International Conference on Mechatronics and Automation. IEEE, Beijing, China, pp. 870–874.
134. Nagel, K. A., Putz, A., Gilmer, F., Heinz, K., Fischbach, A. (2012). GROWSREEN-Rhizo is a novel phenotyping robot enabling simultaneously measurements of root and shoot growth for plants grown in soil-filled rhizotrons. Functional Plant Biology 39: 891–904.
135. NC State (2014). NC State, NC Department of Agriculture shows off latest tech upgrades. Precision Farm Dealer http://www.precisionfarmingdealer.com/content/nc-state-nc-department-agriculture-shows-latest-tech-upgrades, pp. 1–3 (June 3rd, 2014).
136. Noguchi, N., Willi, J., Reid, J., Zhang, Q. (2004). Development of a master-slave robot system for farm operations. Computers and Electronics in Agriculture 44: 1–19.
137. Norremark, M., Soegaard, H. (2006). An automatic weeding robot. In: Proceedings of 3rd Danish Plant Production Congress, Denmark., pp. 446–448.
138. Norremark, M. (2007). Hortibot-A Plant nursing Robot. Department of Agricultural Engineering, Aarhus University, Denmark. http://www.hortibot.com/, pp. 1 (June 25th, 2014).
139. NREC, (2014). Strawberry plant sorter. Carnegie Mellon University, Robotics program, http://www.nrec.ri.cmu.edu/projects/strawberry/htm, pp. 1–2 (June 10th, 2014).
140. Ogrodnick, J. (2009). Cornell helps develop robotic tractor and sprayer with shared grant. New York State Agricultural Engineering Society, Geneva, Hew York, USA, Cornell Chronicle, pp. 1–3.
141. Packham, C. (2013). Robots to drones, Australia eyes high-tech farm help to grow food. DIY Drones: The Leading Community for personal UAVS. http://www.reuters.com/article/2013/05/26/us-australia-farm-robots-idUSBRE94P0EI20130526, pp. 1–3 (May 23rd, 2014).
142. Payne, J. (2013). Transformational Robotics and its application to Agriculture. Robohub. http://robohub.org/transformational-robotics-and-its-application-agriculture/, pp. 1–7 (March 21st, 2014).
143. Pedersen, S. M. (2003). Precision Farming – Technology assessment of Site-specific input application in Cereals. Danish Technical University, PhD thesis, pp. 139.
144. Pedersen, S. M., Fountas, S., Blackmore S. (2007). Economic potential of Robots for high value crops and landscape treatment. Precision Agriculture Wageningen Academic Publications 08547–465, pp. 1–12.
145. Pedersen, S. M., Fountas, S., Blackmore S. (2008). Agricultural Robots-Applications and Economic perspectives. In: Service Robot Applications. Takahashi, Y. (Ed.) I-Tech Education and Publishing, pp. 369–382.
146. Peterson, D. L., Bennedsen, B. S., Anger, W. C., Wolford, S. D. (1999). A systems approach to Robotic bulk harvesting of Apples. Transactions of the American Society of Agricultural Engineers 42:871–876.

147. Pillarski, T., Happold, M., Pangels, H., Ollis, M., Fitzpatrick, K., Stentz, A. (2002). The Demeter system for Automatic harvesting. Journal of Autonomous Robots 13:9–20.

148. Piquepaille, R. (2005). Lukas, the robot that removes weeds. http://www.zdnet.com/blog/emergingtech/lukas-the-robot-that-removes-weeds/22, pp. 1–5 (March 15th, 2014).

149. Pocock, J. (2012). Automated Farm hands. Corn and Soybean Digest, 27: 1–4.

150. Poulsen, F. (2014). Robovator. In: Service Robots in Agriculture-Automaica 2014. Sensors. http://www.sensorsmag.com/electronics-computer/news/automatic-2014-service-robots-agriculture-13136.htm, pp. 1–4 (March 21st, 2014).

151. Precision Farming Dealer 2014a Agrigold, Kinze partner on Multi-hybrid planter. http://www.precisionfarmingdealer.com/content/agrigold-kinze-partner-multi-hybrid-planter, pp. 1–2 (June 4th, 2014).

152. Precision Farming Dealer 2014b Leica Geosystems introduces new entry level guidance system. http://www.precisionfarmingdealer.com/content/leica-geosystems-introduces-new-entry-level-guidance-system, pp. 1–2 (June 12th, 2014).

153. Pullinger S. (2013). In the Future, robots will tend the crops on Norfolk farms. EDP24 http://www.edp24.co.uk/business/farmingnews/in_the_future_robots_will_tend_the_crops_on_norfolk_s_farms_1_3155194, pp. 1–3 (June 4th, 2014).

154. Queensland County Life, (2013). Robocrop: Farm robots a reality. http://www.queenslandcountrylife.com.au/news/agriculture/agribusiness/genral-news/robocrop-onfarm-robots/2663726, pp. 1–4 (February 12th, 2014).

155. Razali, M. H., Ismail, W. I. W., Ramli, M. R. H., Wahid, M. A., Abdullah, A. A. (2013). Engine Revolution per minute estimation conversion for field experiment in Farm Mechanization training. Journal of Modeling, Simulation and Control. 1: 51–56.

156. RedOrbit (2014). The History of Robotics. http://www.redorbit.com/education/reference_library/technology_1/robotics-technology_1/1112942014/the-history-of-robotics/, pp. 1–4 (June 4th, 2014).

157. Reid, J. F., Zhang, Q., Noguchi, N., Dickson, M. (2000). Agricultural automatic guidance research in North America. Computers and Electronics in Agriculture 25: 155–167.

158. Roberts, J. (2014). The Bees are back. As Robots?. Harvard Project funds the Engineering of Robotic bees soon to be in flight. http://www.collective-evolution.com/author/jeffr/., pp. 1–4 (July 14th, 2014).

159. Robotics Research Group (2014). Robotics History Timeline. Department of Robotics, The University of Auckland, New Zealand, http://robotics.ece.auckland.ac.nz/index.php?option=com_content&task =view&id=31, pp. 1–5 (June 4th, 2014).

160. Roodt, D. (2013). Oz farmers study robot workers. Praag.org/?p=4642., pp. 1–4 (June 20th, 2014).

161. Rovira-Mos, F. 2009–3D vision solutions for robotic vehicles navigating in common agricultural scenarios. In: Fourth IFAC International workshop on Bio-Robotics, Information Technology and Intelligent Control for Bio-production Systems, Urbana-Champaign, Illinois, USA, pp. 221–223.

162. Rowbot Systems LLC 2014, Rowbot is mission critical. Association for unmanned Vehicles systems International Rowbot.com/blog/ 2014/5/14/rowbot-is-mission-critical, pp. 1–4 (June 25th, 2014).

163. Saenz, A. (2010). Japan's robot picks only the ripest strawberries. Singularity University, Singularity HUB, http://singularityhub.com/2010/12/04/japans-robot-picks-only-the-ripest-strawberries-video/, pp. 1–3 (May 24th, 2014).
164. Sani, B. (2012). The Role of Robotics at the future of Modern Farming. Proceedings of International Conference on Control, Robotics and Cybernetics (ICCRC2012) 43: 244–248.
165. Sarig, Y. (1994). Robotics of Fruit harvesting: A State of the art Review. Journal of Agricultural Engineering Research 54:265–280.
166. Scheiner, E. (2013). Government pays $1,123463 to develop strawberry harvest-aiding robots in USA. http://www.cnsnews.com/news/article/eric-scheiner/govt-pays-1123463-develop-strawberry-harvest-aiding-robots, pp. 1–3 (May 24th, 2013).
167. Scherer, T and Crawford, E. (2005). Centre Pivot is Agricultural Production Robot. North Dakota Extension Service and Experimental Service Bulletin, 92:1–28.
168. Schupp, J. R., Baugher, A., Miller, S. S., Harsh, R. M., Lesser, K. M. (2008). Mechanical thinning of Peach and Apple trees reduces labor input and increases fruit size. Horticultural Technology 18: 4660–1670.
169. Shinners, J. K., Binversie, N. M., Savoie, P. (2003). Whole plant corn harvesting for biomass comparison of single pass and multi pass harvest systems. American Society of Agricultural Engineering St Josephs, paper no: 036089, pp. 1–18.
170. Simonite, T. (2009). Robot farmhands prepare to invade the countryside. New Scientist http://www.newscientist.com/article/dn17224-robot-farmhands-prepare-to-invade-the-countryside.html#.U5AecvmSxyU, pp. 1–3 (June 5th, 2014).
171. Singh, S., Baugher, T., Bergerman, M., Grosholsky, B., Harper, J., Jones, V. (2009). Automation for Specialty crops: A comprehensive Strategy, Current results, and Future goals. American Society for Horticultural Sciences, USA, Poster 206: p 1.
172. Slaughter, D. C., Giles, K., Downey, D. (2008). Autonomous Robotic weed control system: A review. Computers and Electronics in Agriculture 61: 63–78.
173. Sofge, E. (2014). Three new Farm Robots Programed to pick, plant and drive. Massachusetts Institute of Technology, Cambridge, MA, USA, http://www.popularmechanics.com/technology/robots/a4993/4328685/, pp. 1–5 (April 10th, 2015).
174. Soegaard, H. T. (2005). Weed classification by active shape models. Biosystems Engineering 91: 271–281.
175. Soegaard, H. T., Lund, I. (2005). Investigation of the accuracy of a machine vision based robotic micro-spray system. In: Proceedings of the 5th European Conference on Precision Farming, pp. 234–239.
176. Soegaard, H. T., Lund, I. (2007). Application accuracy of a machine vision-controlled robotic micro-dosing system. Biosystems Engineering 96:315–137.
177. Sokefeld, M., Gerhrds, R., Kauhbauch, W. (2000). Site-specific weed control from weed recording to herbicide application. In: Proceedings of the 20th German Conference on Weed Biology and Weed Control, Stutgart-Hohenheim, Germany, pp. 345–249.
178. Spackman, P. (2014). Does the future lie in Robots or Driverless Tractors. http://www.fwi.co.uk/articles/23/01/2014/142566/does-the-future-lie-in-robots-or-driverless tractors.htm, pp. 1–5 (March 19th, 2014).
179. Sritarapipat, T., Rakwatin, P., Kasetkasim, T. (2014). Automatic Rice crop height measurement using Field server and Digital Image Processing. Sensors 14: 900–926.

180. Sukkarieh. S. (2013). Robotics and intelligent systems: The key to future growth in farming. Engineering and Information Technologies Department, The University of Sydney, Sydney, Australia. http://sydbey.edu.au/engineering/people/salah.sukkarieh. php, pp. 1–2 (May 22nd, 2014).

181. Szabo, J. (2013). Autonomy in Agriculture. A Nuffield (UK) Farming Scholarship Trust Report. Nufflield Trust, United Kingdom, pp. 1–32.

182. Tamaki, K., Nagasaka, Y., Kobayashi, K. (2009). A rice transplanting Robot contributing to credible food safety system. Proceedings of Advanced Robotics and its Social impacts., pp. 78–89.

183. Tamer, C. (2014). Rowbot systems to increase Nitrogen Use Efficiency. http://investeddevelopment.com/blog/2014/03/invested-development-invests-in-rowbot-systems-to-increase-nitrogen-use-efficiency/, pp. 1–3 (July 13th, 2014).

184. Thattari, P. (2011). Coconut harvesting Robot. Farm Sow Magazine 35(4): 1–5.

185. The Economist (2009). Fields of Automation. Economist Technological Quarter 4., pp. 1–2 (June 15th, 2014).

186. Tillet, N., Hague, T., Grundy, A., Dedousis A. (2008). mechanical within-row weed control for transplanted corps using computer vision. Biosystems Engineering 99: 171–178.

187. Tiwari, P. (2014). Agricultural Robots market 2020: Automated harvesting optimizes productivity.   http://www.bizjournals.com/prnewswire/press-release/2014/01/29/ MN55048.htm, pp. 1–4 (March 19th, 2014).

188. Torantola, A. (2013). Lettuce gaze upon the future of Agriculture. http://www. gizmodo.in/science/Lettuce-Gaze-Upon-the-Future-of-Agriculture/articleshow/ 21131308.cms, pp. 1–4 (July 18th, 2014).

189. TORC (2013). Agricultural productivity set to skyrocket with automation robotics. Torcrobotics http://www. trocrobotics.com /agriculture, pp. 1–7.

190. Trossen Robotic Community (2014). Prospero: Robotic Farmer. http://forums.trossen-robotics.com/showthread.php?4669-Prospero-Robotic-Farmer, pp. 1–7 (July 3rd, 2014).

191. UCF, (2013). UAVs and Ground Robots to detect disease in Agricultural Crops. http://www.ucf.edu, pp. 1–3.

192. University of Southern Denmark (2014). Weeding Robot ready for Agriculture. http:// www.sdu.dk/en/Om_SDU/Fakulteterne/Teknik/Nyt_fra_Det_Tekniske_Fakultet/ Radrenserrobottillandbruget., pp. 3 (May 19th, 2014).

193. Vangioukas, S. G. (2013). Robotics for Specialty crops: Past, Present and Prospects. Department of Biological and Agricultural Engineering, University of California, Davis, California, USA, pp. 1–5.

194. Van Henten, E. J., Hemming, J., Van Tuijil, B. A. J., Kornet, J. G., Meulman, J., Bontsema, J., Van Os, E. A. (2002). An autonomous Robot for Harvesting Cucumbers in Greenhouses. Autonomous Robots 13:241–258.

195. Vieri, M., Lisci, R., Rimediotti, M., Sarri, D. (2013). The RHEA-project for tree crops pesticide application. Journal of Agricultural Engineering 26: 359–362.

196. Vitirover, (2014). Vitirover: Replace the herbicide chemicals with Autonomous Robots. http://www.Vitirover.com/en/FAQ.php, pp. 1–12 (June 25th, 2014).

197. Von Mogel, K. H. (2013). Taking the Phenomics Revolution into the field. CSA American Society of Agronomy, Wisconsin, USA, CSA News, pp. 4–11.

198. Vrindts, E., De Baerdemaeker, and Ramon, H. (2002). Weed detection using Canopy reflection. Precision Agriculture 3–63–80.
199. Wageningen University (2014). WP5: Sweet pepper-Protected cultivation. Clever Robots for Crops. http://www.crops-robots.eu/index.php?option=com_content&view=article&id=22&Itemid=22, pp. 1–4 (May 26th, 2014).
200. Wallheimer, B., Robinson, K (2012). Purdue gets US$ 6 million to develop Robotic Pruning for Grapes. Purdue Agriculture News http://www.purdue.edu/newsroom/releases/2012/Q4/pudue-gets-6-million-to-develop-robotic-pruning-for-grapes-apples.html, pp. 1–2 (March 20th, 2012).
201. Wehrspann, J. (2014). Five robots coming to a field near you this year. Farm Industry News http://www.farmindustrynews.com/farm-equipment/5-robots-coming-field-near-you-year. Htm, pp. 1–3 (March 15th, 2014).
202. Weiner, J., Olsen, J. (2007). Konkurrencevenen kan udnyttes. Momentum nr 12: 28–30.
203. Whitney, J. D., Harrel, R. C. (1989). Status of Citrus harvesting in Florida. Journal of Agricultural Engineering Research 42: 285–299.
204. Wilson, J. N. (2000). Guidance of Agricultural vehicles—A Historical Perspective. Computers and Electronics in Agriculture 25: 1–9.
205. Wikipedia (2011). Historical Robots. http://en.wikipedia.org/w/index.php?title=Category:Historical-robots&oldid =45851576, pp. 1 (June 4th, 2014).
206. Wooley, D. (2014). Robotic weeding leads to big labor saving. College of Engineering, Iowa State University of Science and Technology. https//engineering.iastate.edu/research/2014/01/21/robotic-weeding-leads-to-big-labor-savings/, pp. 1–3 (May 22nd 2014).
207. Wynn, H. (2007). Vision system simplifies robotic fruit picking. Vision Systems Design 126 (8), pp. 43–45.
208. Zhang, Z., Noguchi, N., Ishi, K. (2012). Development of a Combine harvester Model for navigation. http://www.cigr.ageng2012.org/documentos/orales/O-SPC04.pdf, pp. 5–15 (June 20th, 2014).
209. Zion, B., Mann, D., Levin, D., Shilo, A., Rubinstein, D., Shmulevich, I. (2014). Harvest order planning for a multi-arm robotic harvester. Computers and Electronics in Agriculture 103: 75–81.
210. Zkotala, A. (2013). Using Robotics to detect Citrus disease. Department of Engineering, University of Central Florida http://www.ucf.edu/using-robotics-to-detect-citrus-disease., pp. 1–2 (June 5th, 2014).

CHAPTER 3

DRONES IN AGRICULTURE: SOIL FERTILITY AND CROP MANAGEMENT

CONTENTS

3.1 A BRIEF BACKGROUND ABOUT DRONES (OR UNMANNED AERIAL VEHICLES)

3.1.1 HISTORICAL ASPECTS OF DRONES

Historically, a variant of drone technology that utilizes a trained Golden Eagle to hunt animals for food was practiced by several tribes in Central Asia and Russian Steppes. It is called Falconry. Their basic techniques and purpose match with several aspects of present day agricultural drones. Trained eagles scout, survey and hunt animals such as wolves, rodents, ovine and fowl (see Plate 3.1). Incidentally, Golden Eagles are launched and guided over vast scrub lands just like present day fixed-wing drones that are catapulted using launchers. Drones are then allowed to glide, picturize and provide data about crops (food source). Farmers may finally, at the end of crop season, pitch in and harvest the crop. This last step is an event comparable to an eagle's master picking the hunt.

Now, let us consider in this chapter, 'Drone Technology' as we know it today and its applications in crop production. First, a generalized

PLATE 3.1 Left: An Eagle launcher on horse-back; Right: A trained Eagle surveying the terrain for animal food (Source: Kazakhstan Tourism Web site; http://backpackology. org/2012/11/21/kazakh/; Note: Food searching and procuring techniques have after all not changed a great deal! An ancient form of Drone technology for food source survey and procurement was in vogue in Central Asia and Southern Russian Steppes and it is still practiced by the tribes. Compare Kazakh and Kirghiz tribes who opt for Falconry to surveillance the terrain, identify food source (rabbits, rodents, wolves, fowl, etc.), and go for catch. Now, under a present day drone technology, a fixed-wing drone is catapulted using a launcher and allowed to survey the terrain with crops and bring information and details on crop stand and grain productivity. Later, farmers can move into fields and harvest the crop. Not much of differences in techniques and purpose).

chronology for first use or introduction of drones indicates that British Royal Air Force conducted the first launch of Unmanned Aerial Vehicle (UAV) from an aircraft carrier known as 'HMS Argus.' The first test flight conducted in 1922, it seems, lasted for 39 minutes in the air. Military use of drones began with gunnery practices conducted by the Mediterranean fleet of British Air force in 1933. Germans used drones to release cruise missiles, during 1944. First combat drone was manufactured and employed by United States Navy in 1944. They were used to drop bombs at appointed locations and take a return flight, if possible (Tetrault, 2014; Arjomandi, 2013). Regarding peaceful uses, United States Meteorological Department used drones for the first time in 1946 to prepare daily weather reports and collect data. During 1955, drones were regularly used by United States Military services to survey for military installations and conduct reconnaissance of adversaries. In 1964, UAV development was conducted under a code name 'Red Wagon.' 'Predators' are drones that are piloted by remotely placed personnel, may be even 7500 miles away from point of action. The other most popular drone in use in USA is known as 'Global Hawk,' whose movement and activity is regulated via satellite-based instructions (Singer, 2009a, b). The development of drones in USA or any other nation has been occurring in spurts followed by a lean period (Schwing, 2007; Keane and Carr, 2013). The first Drone helicopters were commissioned by United States of America in 1960. They were tested at NATC Patuxrnt River, in Maryland. Kennedy (1998) has opined that drone technology in USA attracted very low interest from user groups such as military, weather stations and agriculture right until 1980s. However, during recent years there has been a spurt. Drones are being introduced in big number in USA Military and Agrarian zones.

A major hurdle for long distance endurance was overcome with Trans-Atlantic flight of drones from Scotland to New Found Land in Canada in 1998. During 2001, first Trans-Pacific flight of UAV was possible between Edwards Air Force Base in California, USA to Royal Air Force base in New South Wales in Australia (see Arjomandi, 2013). During 1950 to 1970s, in the United States of America, drones program was mainly directed at military objectives. Drones, such as Ryan Firebee series were fitted with jet engines. Drones like Lockheed D-21 were used for surveillance (Tetrault, 2014). Israel was the first nation to use aerial drones effectively during combat in 1982 (Gale Encyclopedia of Espionage and Intelligence, 2014; Lerner, 2004). Drones were made of fixed-wing until 2000 A.D. Rotary winged Drones, in other words 'pilotless helicopters' were built to scout and release missiles/bombs. Drone helicopters are being effectively utilized to scout, pick images of soil and crop, and prepare digital maps. Notable, is the Yamaha's RMAX helicopter drone that has found consistent use in the rice fields of Japan. Currently, over 50 nations possess full-fledged drone technology (Gale Encyclopedia of Espionage and Intelligence, 2014) and are using them for military, remote sensing, meteorological assessment, survey of natural resources, commercial surveillance, domestic policing, transport, exploration of oil and natural gas, scientific research and most importantly considering the context of this book in agricultural crop production (EPIC, 2014; DARC, 2013). In the predominantly agrarian regions of North America, Europe and Australia, drone mediated farming and machinery linked to drone technology is considered as the most notable advancement in agriculture during this decade (Wilson, 2014). Use of drones is among most recent trends, and it seems to be spreading briskly. In due course, drones may become highly helpful in global farming. According to some experts in agriculture, the spread of drones in crop production zones is comparable to revolutionary technologies of recent times such as personal computers and internet (ASME, 2012; Dronelife, 2014; DeAngelis, 2014). Agricultural applications of drones began much later compared with academic test flights or those meant for military purposes (Table 3.1).

3.1.2 DRONES: DEFINITIONS AND EXPLANATIONS

According to Gogarty and Robinson (2012), most common acronyms used for Unmanned Aerial Vehicles in different aspects of human endeavor, including agriculture are:

- UVs (Unmanned Vehicles): These are vehicles operated without direct contact of humans, such as drivers or pilots. Some variants of UVs in use are UAVs (Unmanned Aerial Vehicles); UCVs (Unmanned Aerial Combat Vehicles).

TABLE 3.1 Countries producing or using Drones for different purposes, including Agriculture

Asia

Armenia (Krunk UAV); China (Aisheng series, Al Bird KC series, Hexiang series, HLKX Dragon series, HTEC series, CHSC CH series, Yintong Copters); Croatia (BL M-99 Bojnik); Iran (Ababil, Sarir, Talash); Israel (Aerostar, IAI Ranger, Innocon Mini Falcon, EMIT Butterfly); Japan (Yamaha RMAX, Yamaha R-50, Fuji RPH 2); Jordan (I-Wing, Jordan Falcon); Malaysia (Eagle ARV System); Pakistan (Hawk MK-V UAV, Vision MK 1, Shooting Star, Rover UAV); Philippines (Raptor, Knight Falcon, Pteryx UAV); Saudi Arabia (Saker 2, PSATRI UAV); Serbia (IBL-2000, Nikola Tesla, Vrabac UAV, Pegaz); Singapore (ST Aero Fantail, ST Aero Lalee); South Korea (KL KUS-TR, TR-100); Taiwan (Chung Shyang II UAV, UAVER); Thailand (Athena, Mercury).

Africa

Algeria (Alfager L-10, Amel); Egypt (ASN 209); South Africa (Denel Dynamics' Seeker, ATE Vulture); Tunisia (Tati Nasnas, TATi Super Jebel, TATI Rotary Wing SAR).

North America

Canada (Precision Hawk Lancaster for Agriculture); United States of America (Global Hawk, Predator, Trimble UX5, Venture Surveyor).

South America

Argentina (Aero Vision Arcangel); Brazil (AGPlane UAV for agricultural scouting; Triba Brazilian Mini for agricultural and civilian purposes); Chile (Stardust II for aerial imaging of farms, Sirol 221); Columbia (Navigator X2, Iris UAV), Mexico (QAE 108–100, for aerial photography); Peru (CEDEP-1, Pegaso).

Europe

Austria (Schiebel Camcopter S 100); Belarus (Grif); Belgium (Gatewing X-100 UAV for surveying and mapping); Bulgaria (RUM-2MB taret, NITI, UtRUM); Czech Republic (Sojka III, HAES Scanner); Finland (Mass Mini UAV); France (Lehman aviation's LA 100 for aerial photos, LA300 for mapping natural resources and crop); Georgia (Unmanned aerial system); Germany (AseTee aerial copter for mapping, AseTee Humming Bird for aerial mapping); Greece (HAI Pegasus); Italy (Selex Galileo, Mirach 100); Netherlands (Higheye automatic Copters); Norway (Cruiser 2, Aerobot Canard, Homet PD, Black Hornet Nano); Portugal (Antex-M, Quadcopter UX 401, I-Sky M6), Romania (IAR-T, Argus-S, Hirrus); Russia (Aero Robotics Shark, Dozor, Kamov, Tupolev, Sukoi Zond); Slovenia (C-Astral Bamor); Spain (EADS Atlantis, Paroca Robatis PRUAV-401, SCRAB-11); Switzerland (Aeroscout Scout B1–100, RUAG Ranger, eBee, Swiss UAV); Sweden (SHARC, SAAB TUAV); Turkey (Atlantis Aero Seeker, TAI Gozcu, TAI martu); United Kingdom (BAE Ampersand, Skylynx, Barnard Microsystems UAV).

Australia

Australia (AAI Corporation's Aerosande for weather data, UAV vision 18 Aeolus, V-Tol Quadrotor); New Zealand (Kahu Hawk, Angelray).

Source: Wikipedia-list of Unmanned Aerial Vehicles and websites of several others, for example, Barnard Microsystems, Volt Aerial Vehicle; Trimble Navigation Systems, 2014.
Note: Names in the parenthesis are few examples of UAVs used by the countries.

- Drones: Drone is a very common and generalized term used across different continents, including military and aerospace engineering. The Oxford English Dictionary defines 'Drones' as pilotless aircraft or a missile directed by remote signals or computerized programs. In biology, 'drone' is a term that commonly refers to 'worker' honey bees that fly incessantly across flower beds and collect honey. Humans hired to accomplish hard and repetitive tasks are also termed drones, at times.
- Aerial Robots: These are autonomous vehicles, unmanned and designed to function based on pre-designed computer programs or based on wireless or satellite-guided instructions.
- Unmanned Aerial Vehicles (UAVs): These are defined as aircrafts without pilot or crew passengers. However, UAVs are capable of controlled and sustained flight for long distance/duration, just like a common aircraft (Wikipedia, 2014). The UAVs differ from missiles. The cruise missiles are lost after their release and accomplishment of tasks. Where as a drone, say that utilized for surveillance of crop fields or transport of commodities, returns to the base after accomplishing the task.

First, let us consider a few generalized applications of drones, not exclusive to agriculture. They include:

- Surveillance: Monitoring a geographic area, surveillance and inspection of large buildings and surroundings, surveillance of personnel at large labor sites;
- Maintenance and Inspection: Structural inspection, patrolling major industries, bridges, dams, buildings, etc. Surveying networks and interconnected installations (e.g., electric lines across countries and regions);
- Survey: Survey for natural resources, like soils, water resources, vegetation, etc.
- Transport, Media and Public events: Movement of goods from one point to another, surveillance of truck movements, ships, etc. Making films from vantage angles and sky, photography, crowd monitoring, telecommunication and tourism, etc.

3.1.3 CLASSIFICATION OF DRONES RELEVANT TO AGRICULTURAL CROP PRODUCTION

The degree to which drones are autonomous is a useful criterion to classify them. First group known as *Semi-Autonomous Drones* are provided

with instructions by operators. Critical decisions such as flight path, photographing land and its topology, natural resources at a particular point,
or making a video-graphic observation of an event on land surface (e.g.,
river flow, floods, crop growth, volcanic eruption, etc.) are decided using
remote control. *Autonomous Drones* are those whose flight paths are pre-
determined. They are provided with computer programs that direct their
flight paths and activities such as photographing a crop field; large farming zones, natural resources, rivers, army movement and location, etc. (see
Gogarty and Robinson, 2012). Drones could also be used in swarms, as
part of an integrated system that rapidly makes observations and offers
digitized photographs of agricultural fields. Drone swarms could also be
guided mid-way in the air using wireless radio waves or satellite-mediated
computer-based decisions.

Agricultural drones can also be classified based on flight craft and the
type of controller attached with it for full operation. Any drone can be firstly
grouped as fixed-wing craft (see Plates 3.2–3.7) or roto-copter (see Plates
3.8–3.13). Multi-roto-copters are said be stable. They takeoff within short
space and easy to fly. Since these copters can hover and even stay still
in air for a while they can pick good still pictures of crops. Fixed-wing
drones fly at greater speeds and cover longer distances above the fields
and still offer good battery efficiency and life (DMZ Aerial Autonomous
Scouting Robotics, 2013) (Table 3.2).

Now, based on controllers, agricultural drones could be grouped into
those guided using a semi-autonomous flight controller or those by fully
autonomous flight controllers. The semi-autonomous flight controller
requires a human skilled operator to manually supply inputs. Skilled
personnel may press keys to move the drone to left, right, forward, etc.
These systems are known as semi-autonomous because they do carry
several pre-programed and self-correcting systems that are autonomous. The on-board computer aided auto-correction systems are important for safety of the drones and in making them move at right direction
and speeds. Auto-correction also considers factors such as wind speed,
its direction and GPS connectivity, although manual operator controls
flight path, etc. There are few drones with on-board computers that carry
out corrections such as emergency landing or return home functions.

PLATE 3.2   Trimble UX5 Aerial Imaging Solution along with a remote controller tablet (Source: Ms. Silvia McLachlan, Coordinator Agricultural Division, Trimble, Sunnyvale, California 94085, USA; Note: Trimble UX5 Imaging Solution is a fixed wing, fully autonomous drone (Unmanned Aerial System), capable of aerial imagery of natural resources, soils, farm topography, crops, crop growth and nutrient status using chlorophyll measurements. The flight path of Trimble-UX5 can be predetermined using the tablet with limited safety maneuvers available during in-flight operations).

While these advances in drones, their operation and functions are good examples of our progress towards Push Button Technology, we have now flight controllers for a second set known as autonomous flight controllers. The second type of drones is grouped based on use of totally autonomous flight controller. This autonomous flight controller has simple touch

PLATE 3.3 Swinglet CAM (Source: Dr. Mathew Wade, Sensefly, Cheseaux-Lausanne, Switzerland; Note: Swinglet CAM is a small Fixed-wing Drone useful for Mapping Natural resources and Agricultural crop scouting. The entire drone and accessories fit into a small suitcase and are easily portable to any location).

PLATE 3.4 Top: An eBee Agricultural Drone (Middle Top: A series of sensors/cameras that are fitted to drone to obtain digitized maps. From left to right, the cameras used are *SR110* for Near Infra Red (NIR), *S100 RE* for Red-Edge band width photography, *S110RGB* for photography at visible band widths red, green and blue; *multiSPEC 4c* for multispectral imagery and *thermoMAP* is used to record crop canopy temperature and moisture stress if any. Middle Bottom: An electronic tablet (controller) that depicts the satellite imagery of a crop field and allows farmer to program the route that drone should fly. Lower most Bottom: An aerial imagery of crop fields in an European setting, showing variations in spectral reflectance and crop productivity. Source: Dr. Mathew Wade, SenseFly- A Parrot Company, Cheseaux-Lausanne, Switzerland).

PLATE 3.5A Precision Hawk – An Agricultural Drone of immense utility to Farmers (Source: Dr. Lia Reich, Precision Hawk, Noblesville, Indianapolis, Indiana, USA; Note: Precision Hawk is a fixed wing agricultural drone. It is a light weight drone capable of several tasks related to crop production. For example, it can be used for aerial imagery of topography, soil type, water resources, natural vegetation, crop scouting, assessment of crop growth, canopy traits, nitrogen status of crop, weed infestation, pests/disease affliction, crop ripening and even yield forecasts).

PLATE 3.5B Precision Hawk Scouting a Cereal field in Indiana (Source: Lia Reich, Director of Communications, Precision Hawk, Noblesville, Indiana, USA).

PLATE 3.6 Gatewing's X-100 Drone (Source: Silvia McLachlan, Marketing Division, Agriculture Division, Trimble Inc., Sunnyvale, California, USA; Note: 'X-100' is a fixed-wing, light weight (2 kg) drone useful in crop scouting for crop stand, growth, chlorophyll and nitrogen status, disease and pestilence. It provides day-to-day imagery of crop growth. It allows farmers to decide on work schedule. It relays field maps directly to computers that process the digitized images).

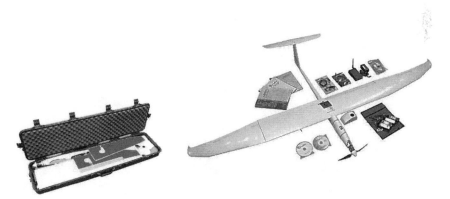

PLATE 3.7 CropCam—An Agricultural Drone (Source: Robot Shop Inc., http://www.robotshop.com/ca/en/cropcam-unmanned-aerial-vehicle-uav.html?   utm_source=google &utm_medium=base&utm_campaign=GoogleCanada; Note: CropCam is a very light and low cost fixed-wing drone that is useful in scouting crops).

PLATE 3.8A An Autocopter—An agricultural drone (Source: Mr. Donald Effren, President, Autocopter Inc., North Carolina, USA; Note: Autocopters are versatile drones that are gaining in popularity in North American Plains. They are relatively light weight and capable of monitoring, spraying pesticides and applying liquid fertilizer).

PLATE 3.8B An Autocopter (Source: Dr. Donald Effren, President, Autocopter Inc., North Carolina, USA; Note: Autocopters are used to scout and make aerial imagery of farms and also spray fields with fertilizers/pesticides).

PLATE 3.9 Yamaha RMAX Drone (Note: RMAX is a popular drone among Japanese Agriculturists. RMAX flying at low altitude over a soybean crop. It is spraying pesticide on the crop. Source: Yamaha Motor Co, Australia, http://rmax.yamaha-motor.com.au/ agricultural-use; Barnard Microsystems Inc., http://global.yamaha-motor.com/about/ business/sky/).

PLATE 3.10 Venture Aerial Robotic Drones- two close-up views in the air (Source: Dr. Rory Paul, Chief Executive Officer, Volt Aerial Robotics Inc., Chesterfield, Missouri, USA. Note: The Quadcopters shown above are highly versatile flying machines with ability to transit at low altitudes above the crop. They could be programed or remote controlled using radio controls. They are effective in obtaining imagery of natural resources, soils and crop stand).

PLATE 3.11   Left: An Agricultural Drone scouting Soybean crop for Pests and Diseases; Right: A venture Quadcopter (Venture Outrider -ProS3/Volt Aerial Robotics) hovering over crop fields. Source: Dr. Rory Paul, Volt Aerial Robotics, Chesterfield, Missouri, USA and United Soybean Board, Iowa. Note: These small, rotary, Agricultural Drones fly at very close height of 10–15 m above the soybean or rice paddy and collect pictures depicting crop health, disease and insect pest attack.).

PLATE 3.12A   A Hexa-copter drone useful in crop scouting (Source: Dr. Rory Paul, Volt Aerial Robotics, Chesterfield, Missouri, USA).

PLATE 3.12B   The Hexacopter flying at low altitude above a Maize field (Source: Dr. Rory Paul, Volt Aerial Robotics, Chesterfield, Missouri, USA; Note: These copters have sensors that pick images both at visible and near Infra-Red range. They fly very close to crop to collect images of crop, its growth, chlorophyll content of leaves, disease and pest occurrence, if any.).

PLATE 3.13   A Multi rotor (Octocopter) Drone in the Air, showing the scanning camera/ sensor

TABLE 3.2   Examples of Agricultural Drones and their predominant use in Crop Fields

Fixed-Wing Versions

Name: Wave Sight Mapper

Manufacturer: Volt Aerial Robotics Inc., Chesterfield, Missouri, USA

Description: The Wave Sight mapper is a fixed-wing aerial unmanned vehicle. The wing span is 2.3 m and length is 1.36 m. It weighs 9 kg. The entire drone fits in a small suitcase and is easily transportable. It is launched using a catapult. It is fully autonomous with regard to navigation. The drone flies at a speed of 45 km h^{-1} for over a hr. at a stretch without brake. It has two fixed cameras for extended mapping of soil/crop expanses. The Visible range and Near Infra-Red cameras have 20 Mpxl resolution. It also has facility for radiometric aerial mapping and geo-tagged location information. For close-up pictures, the cameras are provided with zoom facility. It can cover an area of 3000 acres per flight without brake. It has aerial mapping facility. Its flight path can be programed using computers and GPS connectivity. Major agricultural applications are in natural resources, especially soil and crop mapping; crop growth mapping, disease and pestilence detection.

Name: Trimble UX5 Aerial Image Rover

Manufacturer: Trimble Navigation Systems, Sunnyvale, CA 94085, USA

Description: Trimble UX5 is a fixed-wing drone capable of aerial imagery of crops. It is an all-weather and all-terrain vehicle. Trimble is fully automated, so needs no piloting or remote controlled signals. Its flight path could be easily programed (Plate 3.2) It produces quality images of crops at various stages of growth. The aerial robot has GPS connectivity and its pictures can be visualized immediately on computer screen. It has a simple data processor that is easily handled by farmers. Trimble UX5 drones, it seems, are being accepted rapidly across farming zones in North America. Trimble's fixed-wing drone is used for regular scouting of large agricultural farms, detecting pest and disease damaged spots, crop growth rates and nutrient deficiencies, if any (Caldwell, 2014; Trimble Navigation Systems, 2014).

Regarding its specifications, Trimble UX5 is a light weight drone of 2.5 kg (5.5l b). Its wing spans 100 cm and wing area is 34 dm^2. Its dimensions are $100 \times 65 \times 10$ cm^2. The drone is powered by electric batteries of 14.8 v. Propulsion is via electric pusher propeller. The cameras are crucial for this drone. It has a Sony 16 Mpxl APSc camera and Infra-red sensors. The drone is tough bodied and flight endurance is 50 minutes at the minimum without brake and flies up to 60 km or 37 miles in one stretch. Gatewing Company's X-100 is another fixed-wing drone used in farming (see Plate 3.6).

Name: Swinglet CAM

Manufacturer: SenseFly Ltd, Route Geneva, Cheseaux-Lausanne, Switzerland

Description: Swinglet is very handy and easy to launch within seconds after removing it from the case. Swinglet CAM is among the lightest of agricultural drones in operation in several regions of North America, Europe and Fareast (Plate 3.3). Swinglet CAM is used to derive NDVI. It is used to derive imagery that helps in estimating plant biomass, crop growth rate, nutrient deficiencies, water deficits, disease and pest occurrence.

TABLE 3.2   Continued

Regarding Technical specifications of Swinglet CAM, it has a wing span of 80 cm. It is a small drone and weighs about 0.5 kg (1.1 lb). It operates on power drawn from lithium polymer battery. Its flight endurance is 30 min per flight. It can travel at 36 m h^{-1} (10 m s^{-1}). It can maintain radio link up to 1 km radius. The drone is equipped with 16 Mpxl camera that is electronically integrated and can transmit aerial images directly to ground computers. It is has 3 cm resolution. It has GPS connectivity. Swinglet drone covers an area of 6 km². It is also equipped with 3D flight planning, flight simulator and visualization. The flight mission can also be controlled mid way. Swinglet has facility to avoid collision if multiple drones are in sky to surveillance crop field. This drone also comes with easy data management and aerial image storage facility.

We should note that entire drone is packaged in a small suitcase. The package has complete set of drone accessories, a still camera, 2.4 GHz radio modem and lithium polymer batteries, spare propellers, remote control tablet and of course a user manual (see SenseFly, 2013). SenseFly's recent model known as 'eBee' is a more versatile model with greater number of functions helpful to farmers (see Plate 3.4)

Name: Precision Hawk

Manufacturer: Precision Hawk Inc., Noblesville, Indiana, USA

Description: Precision Hawk is a fixed-wing drone useful for aerial imagery (Plates 3.5A and 3.5B). Precision Hawk has been deployed in agricultural farms in North America and other agrarian regions. This drone helps farm experts and farmers to perform a series of tasks such as detecting crop type, planting density, canopy cover, leaf area index, soil type, soil moisture, nitrogen deficiency, and yield monitoring. Precision Hawk has also been used to prepare maps of water resources in the area, water shed, general land cover, natural vegetation, forest biomass estimation, forest health and occurrence of disease/pest, carbon mapping, etc. Precision Hawk also has several other uses in fields closely related to agriculture such as farm asset surveillance (Plates 3.5A and 3.5B)

Regarding specifications of Precision Hawk, there are user manuals and list of accessories that allows farm technician to start off using the drone quickly. The Precision Hawk has a wing span of 4 ft., a pay load of 2.2 lb and it weighs 3 lb. The accessories include ground kit, a Laptop/Tablet that controls the flight or designs flight path of the drone. The ground kit also receives images instantaneously as the drone flies over the farm and relays the spectral data. The data processor is a 4 × 750 MHZ Linus CPU. A regular drone carries a visual camera with 4 cm resolution, when it flies at 8 m hr^{-1}. There are cameras, if fitted to drones provides farmers a resolution of 6 mm per pixel. The flight endurance is 45 min per flight. Precision hawk surveys about 500 acre per flight. However, we can alter speed by using accessories that enhance or retard drone's flight speed. Interestingly, Precision Hawk allows a live, high definition video streaming via internet. It has GPS connectivity. Power management is done by SmartReflex TM 2 technology (see Precision Hawk, 2014a,b; Plates 3.5A and 3.5B)

Name: CropCam UAV RB-Cro-01

Manufacturer: CROPCAM, Stony Mountain, Manitoba, Canada

TABLE 3.2   Continued

Description: CropCam is an unmanned aerial vehicle capable of imagery of soil, crop, water resources and pests/diseases. It has been useful for general geographical surveys of natural resources, forests, oil and gas. CropCam is a small drone that is profitable and fits into to a rather small suit case (Plate 3.7). This drone picks up digital images and transmits to ground control (computer PC). It also takes commands from a remote radio controller within 2–3 km radius. Therefore, its flight path can be altered as per immediate requirements. CropCam is utilized more commonly for aerial mapping of crops health at different intervals, so that appropriate fertilizer schedules could be decided.

Regarding specifications, CropCam is 4 ft. in length, with a wingspan of 6 ft. and a weight of 5 lb (2.5 kg) (Plate 3.7). The flight endurance is 55 min. CropCam can transit at 60 km h^{-1} and still conduct aerial imagery. The drone can fly without straying even when wind speed is 30 km hr^{-1}. CropCam cannot fly with ambient temperature of less than −20°C. The drone is generally fitted with a visual camera that makes video images and relays to ground control. The transmitters are very small and <50 g in weight. The drone consumes about 100 milliamperes for every 10 minutes. It is powered by Lithium Batteries of 11.3 volts. Over all, CropCam drone is an inexpensive and efficient alternative to piloted airplanes and even satellites, since clouds do not obstruct their operation.

Rotary Copters

Name: RMAX

Manufacturer: Yamaha Motor Co. Japan

Description: It is predominantly meant for scouting, fertilizer and chemical sprays. The maximum pay load is 28–31 kg. It holds about 21–24 kg pesticides/fungicides and flight duration is about 60 min without refueling. The copters' weight is about 94 kg. It makes excellent optical observations from 150 m above the crop/soil surface. Works on regular gasoline and has a two stroke IC engine. Aerial images from its infra-red cameras can appear immediately on PC screen. The copter RMAX has GPS connectivity. Major agricultural uses are in the area of application of chemicals (pesticides, fungicides). RMAX is used in seed distribution in land reclamation and forestry zones. Currently, RMAX is supposedly the mainstay during insect control, liquid fertilizer spray and scouting of rice paddies, wheat and soybean fields in Japan and Far East (Barnard Microsystems Ltd, 2013) (Plate 3.9).

Name: Yintong' Agriculture Crop Protection UAV YT P-5 and YT P-10

Manufacturer: Yintong Aviation Supplies, Zhuhai city, China

Yintong agricultural drone is a compact, small copter with single tall rotor blade. Its flying weight is 15 kg. Its maximum payload is 5 (five) liters of pesticide. Its dimensions are 15.2 cm x 58 cm x 71 cm and total length of the copter is 217.5 cm. It has an endurance of 35 min. The copter works on 44.4 V 8Ah/2.3 kg batteries and a single charging lasts for over 30 min. The copter flies close to the crop at just about 1.5 m above it. The copter is really very efficient with regard to pesticide spraying. It sprays about 30–45 acres h^{-1}. Therefore, it serves farmers with large farms rather excellently. Yintong's other copter model is slightly larger, heavier and covers about 45 to 60 acres h^{-1}, for example, about 300 ac day^{-1}. It holds 10 lts of pesticide and maximum payload is 10 kg.

TABLE 3.2 Continued

Such small agricultural drones are said to be useful in many of the agrarian zones of China. Yintong's drone is meant for pesticide spray as well as dusting. It is a low altitude copter and hence applies pesticides with greater accuracy. It is emission free, low on maintenance and cost efficient. Some of the advantages listed are: It suits any terrain. It is useful even in hilly Tibetan terrain. It is said Yintong copters are excellently suited during Precision Farming. It avoids human contact with pesticides and reduces risks to nil. The copter reduces pesticide usage by 50% of original, if human labor was used. Currently, the electric battery powered Yintong drones are preferred by large farms that produce rice, wheat, fruits and cotton (Yintong Aviation Supplies, 2014)

Name: Venture Outrider and Venture Surveyor – a Quad copter

Manufacturer: Volt Aerial Robotics Inc., Chesterfield, Missouri, USA

Description: Venture Outrider and Venture Surveyor are useful as agricultural drones. The Quadcopters are made of ruggedized body to withstand harsh environments of dry tropics. These copters are entirely autonomous with excellent navigation capabilities (Plate 3.10). The cruise speed of the agricultural drones is about 16–30 km hr^{-1}. They are capable of GPS guided flight for 15 to 30 min at a stretch. It is possible to extend flight time that helps in mapping and aerial imaging of over 100 acres at a time. They have Visible and Near Infra-Red cameras of 20 Mpxl resolution for aerial mapping. They are endowed with zoom cameras for picking close-up pictures of soil and crop. It has a computerized flight mission planner. The Venture Quadcopters are easy to operate and need minimal training (Volt Aerial Robotics, 2010).

Name: Hexacopter

Manufacturer: Precision Drone LLC, Indiana, USA (Plates 3.11, 3.12A and 12B)

Description: Precision Drone is a hexacopter that is versatile and flies low over the crop fields. It is an excellent instrument to surveillance large farms for crop growth, weed infestation and pest/disease attacked zones. This Hexa-copter is fitted with Visual and NIR cameras. The videos that it picks are better than resolution possible with naked eye. The high quality video provides 11 Mpxl resolution. The pictures are provided with GPS coordinates. Hexa-copter's flight path could be either pre-programed or altered using remote control. In the event communication with ground control fails, this hexa-copter is fitted with computer programs so that copter safely returns to original takeoff point and lands safely. A remote controller that monitors its battery level, flight speed, altitude, path covered and regions yet to be photographed are depicted on the screen/tablet. This drone can fly without brake for 35–45 min and cover large areas of crop field. The drone is fitted with 6 batteries for electric power and has a 16 GB memory card that records digitized pictures/video for use by the ground control. The digitized information could also be instantaneously relayed to ground robots/tractors so that appropriate weed/pest control measures could be adopted. Fertilizer application at variable-rate is possible based on digitized information collected by the drones (PrecisionDrone LLC, 2014; Plates 3.11, 12A and 12B).

TABLE 3.2   Continued

Name Octocopter

Manufacturers: Aerial Drones, Lancashire, United Kingdom

Description: Octocopter is a light weight drone with 8 blades. The copter is 1.2 m in diameter and electrically powered using 6 batteries. The drone has GPS connectivity and the cameras could be controlled by ground station. Each picture derived from drone is provided with GPS coordinates. The copter has wireless connection through which its flight path could be controlled, if necessary. The octocopter is usually fitted with Lumix GH3 or Cannon EOS 5D Mark-III camera. Remote photography could be done using hand-held controller. Computer program for predetermined flight is also fitted to octocopter. The octocopter provides pictures taken from 150 to 500 m above the crop. The drone has carries a payload of 5 kg (see Plate 3.13)

Parachutes

Name: SUSI-62 UAV

Manufacturer: Geo-Technic, Neutsr 40, Linz an Rheine, Germany

Description: SUSI is a robust and safe UAV—a parachute (Plate 3.14). It has a relatively larger payload and endurance for daily use above agricultural fields. They can float above the crop fields for a long period with slow drift. SUSI is an autonomous vehicle and can be programed to cover the crop fields accurately. The parachute movement ensures that entire field is covered. The frame of the SUSI 62 UAV is 62 m^3. It is made of steel and withstands rough conditions. It is powered by a two-stroke IC engine. The frame has space for sensors, a few DSLR cameras, Near-Infra Red cameras, and Thermal sensors. SUSI is easy to control using radio waves. It can also be linked using GPS. The SUSI parachute-UAV needs a small run way of 10–50 m or a cliff to launch. The maximum flight speed is 50 km h^{-1}. Atmospheric wind speed is an important factor that affects SUSI. It can be operated if wind speed is less than 6 m s^{-1}. The video signals, images and weather parameters above the farm could be transmitted to ground control or a computer/tablet using satellite link. Images could be obtained instantaneously and agronomic procedures could be altered, if needed (Plate 3.15). According to Thamm (2011), SUSI 62 a parachute UAV is now successfully deployed above crop fields in Europe, Africa and Australia. Further, we may note that many of the functions of this SUSI 62 UAV are controlled easily with the *push of a few buttons or touch of screen on IPad (tablets) or computers*. It is an auto-pilot vehicle; hence, it reduces need for human skilled laborers to scout the fields. Its flight path can be determined with touch of screen and pin-pointing the route. In fact, it is a good example for how a UAV can easily fit the larger concept of 'Push Button Agriculture.'

Note: The list of drones, their types and variations available for use by farmers is really exhaustive. However, here in the above table only few types are mentioned.

PLATE 3.14 SUSI 62 UAV – An autonomous (pilotless) parachute UAV (Source: Dr. Hans Peter Thamm, Geo-Technic, Germany, E-mail: Hp.thamm@gmail.com; thamm@ geo-technic.de).

PLATE 3.15 An image of a Plantation from SUSI 62 useful for Precision Farming (Source: Dr. Hans Peter Thamm, Geo-Technic, Germany; Note: Management zones could be easily marked and managed using an overall image of the field and its soil type, or crop growth pattern.).

screen controls to show flight path, pin-point locations that need detailed imagery, close-up pictures of weeds/crops, etc. The drone takes of autonomously, accomplishes tasks as per plans relayed using flight controller. A few of the advanced drones are capable of object detection and avoidance. A tall tree, windmill, electric pole or a tall building is recognized ahead and in response, copter takes deviation (DMZ Aerial Autonomous Scouting Robotics, 2013). Actually, flight controllers could be used to set pre-programs and instruct drones to automatically avoid interferences.

Based on current applications known for drones, Gogarty and Robinson (2012) classify them into micro and small drones; medium altitude drones and high altitude long endurance (HALE) drones. Medium altitude long endurance drone such as 'Reaper' is capable of reaching heights of 1.6 km and flying for 36 hr. without refueling. HALE drones fly at altitudes of over 9 km and for long durations during surveillance both at day and night. HALE drones are most commonly used for surveillance of international borders and military movements (Kaiser, 2006).

Drones could also be grouped into normal takeoff types and those with ability for rapid short or vertical takeoff drones. There are vertical takeoff surface launched rapid deployment UAVs (RDUAV) and air-launched rapid deployment drones (Cheng, 2008).

Drones could be utilized in wide range of situations that unfold during peace time, wars, natural resources and weather scouting, and now agricultural crop production that seems to be gaining in priority world over. Drones differ in their characteristics and performance. They could be classified into groups based on; a) weight of the entire drone; b) endurance and range of activity of drone; c) maximum altitude that drones can gain during survey sorties; d) wing loading; e) engine type and f) power/thrust it generates (Arjomandi, 2013) (Table 3.2). Based on the characteristics, drones could be sub-grouped as follows:

- *Weight:* Drones are grouped based on weight into super heavy weight versions that are about 2 tons (e.g., Darkstar, Predator and Hawks). Drones that weigh 200 to 2000 kg are grouped as heavy weight vehicles (e.g., Outrider, Fire Scout). Drones of medium weight ranging from 50 to 200 kg are also popular (e.g., Raven, Phoenix). There are Micro-Drones that are small in size and weigh less than 5 kg per unit (e.g., Dragon Eye, FPASS and Silent Eye). Micro drones could be really small. For example, Robobees being developed at Micro-robotics laboratory of Harvard University, Massachusetts, USA are really small and weigh few pounds. Drones that are heavy usually possess turbo jet engines or turbo fan engines. Such powerful engines help in generating required thrust and rapid transit. Small engines run by electric source are usually fitted to light weight drones that drift at low speed.

- *Endurance and Range:* Drones are capable of flying and staying in the sky for different lengths of period. The period for which they stay afloat and distance they transit to a certain extent decide the tasks for which they can be adopted. For example, surveying large farming stretches may require drones to be afloat for longer period and also travel and cover larger areas. Long endurance drones stay afloat for over 24 hr. They cover a distance of 1500 km to 22,000 km. For example, Global Hawk endures and flies for over 20,000 km. Medium endurance drones perform for 5–20 hr. in the air (e.g., Predator; Silver Fox). Drones that stay afloat for less than 5 hr. are used for short flight of reconnaissance and general upkeep of farms (e.g., Pointer; eBee).

- *Maximum Altitude:* Drones are capable of gliding at different altitudes. Their utility decides the range of altitude they are directed to reach. During crop scouting, drones capable of low altitude, short/long flights and with ability to hover and capture detailed pictures are preferred. The Low altitude drones reach heights less than 1000 m (e.g., Pointer, DragonEye). Medium altitude drones reach 1000 to 10,000 m. High altitude drones fly at altitude >10,000 m (e.g., Darkstar, X-45).

- *Wing Loading:* Wing loading is characteristic that is derived mathematically as ratio of weight of drone to its wing area. Low wing loading drones such as 'Seeker' have a wing loading value of < 50 kg m². Medium wing loading drones have values ranging from 50 kg m² to 100 kg m². High wing loading drones have wing loading values > 100 kg m² (e.g., Global hawk).

- *Engine type:* Most commonly used engines in different drones are Turbofans, Turboprops, Push and Pull, Two stroke Piston engines, Rotary, Electric and Propellers (Arjomandi, 2013). Drones are commonly fitted with electric engines. Lighter drones use electric motors run on power derived from batteries or solar panels.

We may note that considering the context of this book, drones that could be used for agricultural purposes are either totally autonomous or semi-autonomous. However, all of them, either guided by human technicians or left programed, thoroughly qualify to be part of 'Push Button Agriculture.' After all, these drones are controlled literally at the push of button or using computer key pad or a mobile. They reduce human drudgery remarkably. Here, we should note that, human labor hours required to scout, measure, collect data and prepare accurate maps of farmer's fields are more. It is a tedious effort and involves capital investment. Further, we may note that the drones prepare soil or crop maps rapidly and accurately for use by farmers.

3.1.3.1 Some important Characteristics of Drones

Drones used in agrarian zones of North America, or elsewhere in Japan have undergone a rapid development and refinement process. The improvement in design, utility, size, cost and popularity have all affected the drones found in the market for use in agricultural farms. Irrespective of variations in specifications and characteristics of the drone used by farmer, following points are to be noted (Dobberstein, 2014). They are:

a) Most, if not all, agricultural drones are relatively small in size. Many of them are small enough to fit a suit case when folded. Therefore, it is easy to transfer them in a compact and packed condition from one point to another.

b) Drones could be launched from a place. They are either pulled by a catapult, a vehicle such as pick-up van, or from a high cliff like vantage point. Several others, especially, copters attain a good vertical lift into air.

c) Drones can be controlled using radiometric connection. They usually have GPS connection and so are amenable for direction and control using remote controller, laptop or an I-tablet. Drone's flight path and destination points could be programed using computers.

d) Agricultural drones are predominantly used for high definition images of soil, crops, weeds, water resources, crop health, for example, disease and insect attack. Drones can also make a video picturization of events happening in crop fields. They can transmit a live video movie to ground controls, say a computer screen or tablet. They provide details on farm operation from time to time as it progresses.

e) Drones, depending on type of cameras fitted, may provide 2-D and 3-D pictures of fields. The 3-D pictures depict subtle changes in elevation and depth of a location. For example, soil erosion, breaches in irrigation canals, loss of top soil, and devastation to crops after a thunder storm can be perfectly depicted using a 3-D picture or video.

f) The resolution of soil maps, crop growth and disease infestation shown in pictures ranges from 12 inches to very sharp pictures of ¼ inch accuracy.

g) The endurance of agricultural drones is generally small compared to those used in military or transport. The agricultural drones stay afloat for 20–45 min per flight without re-fuelling. The rotary types (copters) have longer endurance. The fixed-wing drones have short flight time of 20–30 min. Agricultural drones fly about 8–10 km from the point of release and still stay within the limits of radio or GPS control systems.

h) Drones that are larger and heavier withstand windy situations better. Small copters are flown for scouting when wind speeds are 20–52 kmh^{-1} and not beyond. Gusty winds may blow away copters. The fixed wing drones must fly into or out of wind and not across, if good pictures are to be retrieved.

i) The market price of a low cost agricultural drone is about 500 US$, but depending on fixtures and sophistication of cameras, video facility and GPS connectivity, drones costing 100,000–120,000 US$ are available. A RMAX copter drone could be costing 60,000–65,000 US$.

Let us consider parachute-drones. According to Thamm (2011), a parachute drone SUSI 62 is a robust and safe aerial vehicle. It was developed, since classical airplane derived aerial photographs were costly and such campaigns were not possible in all agrarian zones. Aeroplanes with sensors and near-infrared cameras may not be available in many areas (Grenzdorffer et al., 2008; Thamm and Judex, 2006). A parachute drone must however pass a few tests and requirements to be adaptable as a drone. They are; parachute drone should be robust, easy to operate and maintain, should take a higher payload, and should have a long flight time (endurance). It should travel at a relatively slow speed to be able to picturize the ground features clearly. It should have built in safety features. The parachute should also carry a series of sensors, cameras and other equipment. For example, SUSI 62 has an option of 8 kg payload for sensors. It has DSLR cameras, Near Infra-Red imagers, multi-spectral and thermal sensors.

In agricultural farming, drones are also used in swarms (groups). Swarms of few drones flown simultaneously help to accomplish the tasks rapidly. Imagery can be obtained instantaneously on the computer screens and large areas of farms can be covered in short period. However, flying multiple drones that operate above the same crop field, needs good programming skills, sharp maneuvers by the drones and rapid processing of digital data (Axel and Sandro, 2014).

3.1.3.2 Multispectral Imaging Systems

The crux of the drone technology in agricultural crop production is in the usage of high resolution cameras/sensors that supply digitized information/pictures to computer decision support systems. Drones only provide the vantage points above the crop field that is not easily reachable by human scouts and rapid transit above crop fields. At present, several different multispectral imaging systems are in vogue (see Huang et al., 2010b). Some of the imaging systems are low cost, while others need high investment. Imaging systems could be manual, remote controlled or pre-programed and automatic. Cameras could pick single photos or a composite that could be latter arranged appropriately, so that a big image of the field is deciphered. Cameras are connected to GPS and computers that could immediately process the digital codes and produce a correct picture of the field. There are however, three most commonly used imaging systems adopted in agricultural drones. They are: (a) Low cost ADC Camera, (b) Geospatial Systems MS 4100 Camera and c0 Custom built TTAMRSSS System.

Following is the list of characteristics of the three imaging systems commonly used in agriculture (see Huang et al., 2010a, b) (Plates 3.16–3.19):

Tetracam ADC Camera: Sensors are equipped with 3.2–5.5 Mpxl cameras. They could be exposed manually or through automatic systems. The camera covers a band width of 520–600 (green), 630–700 (Red) and

PLATE 3.16 Cameras fitted to drones are high definition DSLR capable of picturizing land, soil and crops at visual band and NIR range. These are usually 12–18 M pxl cameras (Source: Public domain websites showing different photographic equipment and reviews).

PLATE 3.17 Cannon 550 DSLR camera (Source: Public domain websites showing different photographic equipment and reviews; Cannon 550 D is a Digital SLR camera, with 15–18 Megapixel sensor, HD video and High resolution monitor. It is used to capture images at visible range. Digital imagery is processed using matching photogrammetric software to create digital crop surface models.).

PLATE 3.18 A Multispectral Camera that can be fitted to drones (Note: There are several types and models of multispectral cameras used on drones; Source: Commercial advertisements in web sites.).

PLATE 3.19 A Camera for Thermal Imagery of crops with fixtures that fit the agricultural drone (Note: Cameras capable of imagery at infra-red band width are used to assess crop water status. Most drones carry a complement of cameras that includes thermal imaging. Source: commercial websites.).

760–900 (Near Infra-Red). Image sizes are flexible but usually of 2048 × 1536 Mpxl. Image Digital count ranges from 8–10 bit. These cameras pick one image every 3–10 seconds. All of these cameras are linked to GPS and provide co-ordinates on each and every picture exposed. These are low cost cameras (Plates 3.16 and 3.17). Over all, a Tetra cam ADC camera has slow imaging speed, images are relatively of low quality, GPS triggering is limited but good for low flying copters.

Geo-spatial Systems MS4100: These image systems comprise of three multispectral CCD (charge coupled devices). They could be operated manually or automatically. The cameras are endowed with ability to pick images at Blue wavelength of <460 nm; green wavelength of 540 nm; Red wavelength of 600 nm and Near Infra-Red wavelength of 800–900 nm (Plate 3.18). The image size is usually 1920–1080 pixel with digital count of 8–10 bit. These cameras pick one image every 2 seconds. The drones/cameras are linked to GPS and each image is marked with GPS co-ordinates. The cost of the Geospatial systems is around 20,000US$. Over all, a Geospatial system MS4100 is good in spectral coverage, image resolution is high, pictures are of good contrast and the systems have accurate GPS connectivity.

Multi-spectral Imaging systems for Remote Sensing: These systems are equipped with cameras that operate at Visual and Infra-Red ranges plus Thermal cameras (Plate 3.19). They could be mostly manual and operated using remote controller. The cameras take images at Red band width of 655–665 nm, near Infra-Red at 830–870 nm and thermal picturization occurs at 8–14 μm. The image size ranges from 1024 x 1024 pixels with an image count of 12 bit. Cameras pick one image every one second. The cameras are connected with GPS and pictures are provided with GPS coordinates. The cost of the cameras is about 80,000 US$. Over all, a Multispectral Remote Sensing system has built in flexibility to answer farmer's requirement. The image quality is high. Imaging is relatively fast and contrast is good. However, they are expensive; require larger aircraft or drone and images may need processing using computers.

Ehmke (2013) has pointed out that currently most farmers prefer to fit at least a set of three cameras given that payload of a small drone is just 3.3 lb. The three sensors for crop evaluation are visual high resolution radiometer, a thermal camera to track plant temperature and hydration and a laser scanner to measure plant height. Infra-red photography is becoming most popular because it offers insight into plant health and water status of crop.

Computer programs that decode digital imagery, analyze and offer different options to the farmer are immensely important. There are several models and computer programs used by farmers. For example, System Approach for Land-Use Sustainability (SALUS) helps in processing data acquired by drones and forecasts about crop, soil, water and nutrient conditions across different climatic scenarios and crop yield goals. 'Agpixel' is a newly developed software program by a company known as 'Roboflight.' This software allows easy conversion of pixel data in images into information about plant growth vigor (NDVI). The NDVI noted for entire crop field could be compared using 'Agpixel' and then farmers could decide on fertilizer-N and water supply to crops. It seems most drone companies; over 130 of them are utilizing 'Agpixel' or similar software to convert pixel data to imagery (Ehmke, 2013).

3.2   DRONES, NATURAL VEGETATION AND CROP GROWTH MONITORING

Drones, in future, may affect human life in variety of ways depending on profession and enterprises that employ these gadgets. LeMieux (2012) suggests involvement of drones in innumerable aspects related to human governance, patrolling, surveillance of installations, oil pipe lines, roads, traffic, meteorological observations, water resources, natural resources, geographic mapping, preserving events in images, crop production, etc. However, within the context of this book, we are concerned only with drones in farming enterprises. In the following pages, detailed discussions on agricultural applications of drones are made available. However, a comparative study of satellites, piloted aircrafts and drones seems pertinent here. Dobberstein (2014) has made comparison of satellites; fixed-wing piloted aircraft and Unmanned Aerial Vehicles with regard to their usefulness to agricultural farming situations. Firstly, satellites are advantageous when farmers or agricultural experts have to cover large areas, say while dealing with farming strategies for counties/districts regarding cropping systems, fertilizer or irrigation supply systems to farmers, etc. Satellites could be used to surveillance any agrarian area of the globe. The major disadvantages are that resolution is not great since the cameras are located far off. The best resolutions available currently are still only 60–63 cm. Satellite pictures are obscured if clouds interfere. In case of fixed-wing aircrafts with human pilots, first the plane is large and needs extra fuel. A pilot who is trained to conduct agricultural surveys has to be employed. The process could be slow and it needs an airport or tarmac to gain lift. Again, clouds could interfere with farm surveillance. Unmanned Aerial vehicles that are now becoming very common with farms in North America and Europe are highly versatile and deployable from anywhere within the farm itself. They are small and do not need a pilot. Since UAVs fly very close to the soil or crop surface, the pictures are very clear and resolution is indeed quite high at 4–10 cm, but if one wants resolution can reach < 1 cm. However, drones are not permissible yet for commercial operation in USA. They have to be used as farmers' personal gadgets. Initial costs could be more, but it seems mass production and availability of variety of copters and fixed-wing drones is reducing its costs in the market. Drones, if not managed properly during landing could result in damage of cameras and cost may be high.

One of the most important uses of agricultural drones is to conduct crop scouting using multi-spectral imagery. The drones provide Near Infra-red imagery that helps in developing Normalized Difference Vegetation Index (NDVI). The NDVI is a useful parameter while deciding on crop spread and its health. Imagery using Near Infra-Red band width can also inform farmers regarding any water stress that the crops may be suffering.

$$NDVI = NIR\text{-}Red/NIR\text{+}Red$$

where, NIR is Near Infra-Red and Red is wavelength band in Red region of electro-magnetic spectrum.

NIR is one of the more common wavelength bands used for studying crop health. Vegetation in general shows strong reflectance for NIR, since pigments that absorb visible light do not absorb NIR. The leaf tissue actually interacts with NIR and disperses it causing about one half of NIR to be reflected. Plants with healthy leaf tissue reflect more of NIR, but unhealthy and discolored tissue reflects less of it. Therefore, in a crop field, if crops are suffering from water deficiency or diseases/pest attack leading to yellowing or discoloration of plant leafs, canopy reflectance of NIR is accordingly affected. Canopy with non-green or discolored patches reflects less of NIR. Drones with NIR camera show up bright patches wherever crop is healthy because reflectance is high.

Some of the other most promising uses of drones are in crop scouting with high definition (HD) imagery, data collection on soil fertility related crop growth variation, irrigation and water way management, crop growth mapping and disease/pest detection.

3.2.1 DRONES IN MEASURING WEATHER PARAMETERS AND DEVELOPING FORECASTS

Drones are in vogue to assess weather conditions for several years now. Drones may continue to be used during weather forecasting exercises. In fact, reports by National Aeronautics and Space Agency, USA predict that UAVs might often augment weather data accrued by using satellites and piloted aircrafts. They would be adding accuracy to weather data collected and forecasts. Major capabilities of drones include observing and collecting data about geological processes, including agrarian regions. Drones could be used to observe on aerosol and gaseous emissions from agrarian regions. Periodic observation on stratospheric ozone chemistry, tropospheric pollution and air quality could be done using drones with appropriate instruments in the payload. Water vapor changes, vegetation, nutrients in coastal atmosphere, emissions from fires, O_2 and CO_2 levels could all be observed and changes noted using drones. Further, it has been predicted that drones could be very handy in collecting and maintaining weather related data pertaining to each farm (Carr, 2004).

In general, weather above agricultural farms or vast agrarian stretches could be studied and data collected periodically for use by individual farmers, scientists and policy makers. Several different contraptions and methods have been adopted, so far, to assess agricultural weather related parameters. Currently, drones are among the most important gadgets that aid weather forecasting. A few other methods involve, use of

Weather balloons, Remote controlled Aircrafts, Piloted aircrafts, Doplar Radars stationed at different vantage locations, Geo-stationery operational Environmental Satellites, Polar orbiting Environmental Satellites, Sounding Rockets, etc. (Scott, 2011).

Drones meant for weather data collection and forecasting over any of the farms or crop production zones could be small and weighing less than 50 lb. It may cost 10,000–50,000 US$ per unit with all necessary accessories. Keber (2013) states that drone could be effectively used to study storms, their movement and alert farmers about impending disaster, if any. Drones could be good tornado and storm chasers in the Central Plains and other regions of North America (Jouzapavicius, 2013). Tropical cyclones and weather patterns too have been monitored using drones (Jarvis, 2014). Currently, tropical countries with monsoon rain pattern are using drones to collect detailed data, pictures and forecast the rains, so that farmers could accordingly plan their planting and harvesting schedules. Drones could be flown over the fields with crops to assess soil moisture and other parameters. Galimberti (2014a) says drones have the capability to augment data and improve prediction accuracy about weather patterns, which were already forecasted by using satellite accrued data. Agricultural crop production depends on weather patterns to a very great extent. Any, possibility to modify weather to farmer's advantage is most sought. Galimberti (2014b) opines that in future, farmers in groups or even individual farms may venture out with drones equipped with accessories to modify the weather. For example, if expected rain gets delayed or water deficit in soil is severe, drones could be used to seed the clouds, with dust or solid CO_2 and induce precipitation in the region.

3.2.2 DRONES IN NATURAL RESOURCE MAPPING

Researchers at Mechatronics lab, University of California-Merced believe that in the near future, improvising and producing drones (UAVs), and software related to drone usage will be a major thrust in several aspects of human endeavor. They state that drones could be regularly used in remote sensing of natural resources, environment monitoring, crop production and precision agriculture in particular. Multiple drones or swarms could be used to study natural resources (UCMERCED, 2014). Wharton (2013) opines that drones could be small vehicles, but they could have large impact in rapid and periodically frequent assessment of natural resources, particularly soil, vegetation and water resources. Drone aided surveillance of water resources is an important task in many of the dry land tracts of the world. Digital maps that depict water resources should be most useful to farmers.

Horcher and Visser (2005) state that drones were initially confined to military applications, but as production of drones has become more common, it has found its greater potential in surveying natural resources and agrarian zones. Monitoring and maintaining natural resources is indeed an important aspect that could have strong impact on various other aspects such as agriculture, industry, housing, economics, etc. Drones used in scouting and surveying natural resources vary in size, aerial photographic capabilities and other accessories. Regarding weight of drones, Micro UAVs (<10 lb), Mini-UAVs (5–20 lb), small UAVs (20–80 lb) are lighter but high altitude long endurance UAVs (HALE) could be as heavy as 12 tons. All of these classes are being used to map natural resources using a range of cameras fitted to them. The aerial imagery of course varies with cameras and altitude at which drones fly while capturing images. Using appropriate software, images picked by drones could then be decoded, tailored and clear idea about distribution of natural resources such as water or fertile soil types, vegetation and crops could be obtained. Mosaicked images of forested zones and crops have been used to monitor loss of natural stands, stored timber, soil erosion, fluctuations in crop belts, etc. In some parts of Southern United States of America, soon small drones flying at close altitudes to pine stands may become the main stay for US Geological Survey to gather data on natural resources, forests, shrub land, and wasteland. Natural resource management in these locations will immensely depend on aerial imagery derived using drones (Linehan, 2012; Jensen et al., 2014). Drones with their ability for multispectral and thermal imagery are supposed to provide us with better understanding about water bodies, water movement, streams in mountainous regions and plains, influence of season and temperature on fluctuations of water resources, water springs, ground water (e.g., Ogallala), ice sheets melting, etc.

Reports by European Commission predicts that drones could find innumerable civilian uses, particularly in the areas of natural resource maintenance, estimation of water resources, influence of weather on rivers, streams and ground water, fluctuations in crop production zones, industrial surveillance, road and traffic management (Doward, 2012). Drones could be used in the surroundings of nuclear installations to monitor soil and air quality and its influence on close by crop production areas and human population. In the core agricultural sector, drones are expected to be most effective in adopting precision farming and in deciding cropping systems based on aerial imagery of natural resources that are supplied to farmers. Digital mapping of wild life sanctuaries, wild life population, air quality above sanctuaries are other functions that could be accomplished using drones.

In the present context, we are more interested in utility of drones in developing aerial imagery that depicts land cover and its changes. Mapping areas with natural vegetation and crop production zones is of utmost relevance to policy makers. Drones could provide detailed images with high resolution that shows botanical features of natural vegetation. Particularly, dominant crop species and others could be marked. Similarly, periodic drone flight could supply information on cropping systems and changes in dominant crop species preferred by farmers in a particular zone. Regarding individual farms, it is possible to allocate and manage land resources more accurately and efficiently, if drone derived imagery is available.

Drones could also be used effectively to monitor climate change parameters and their influence on natural process such as large scale gully erosion, sheet erosion of farm land, loss of top soil fertility, nutrient depreciation, nutrient imbalances and their consequences on crop growth and grain productivity. Drones are excellent for monitoring floods and their effects on water resources, river flow, erosion of river banks and consequences on crop production in large expanse or a small farm. We may also try to use drones effectively to collect data about atmospheric parameters such as relative humidity (Rh), CO_2, NO_2 and N_2O emissions, SO_2 contamination, etc. We may note that drones are best suited to collect gaseous samples from various altitudes above the crop at various intervals during crop season. Drones have also been used to obtain samples from atmosphere above the crops to assess dust particles and microbial load.

3.2.3   DRONES IN SOIL FERTILITY AND VEGETATION MAPPING

Soil degradation due to erosion of top soil caused by wind or water flow is wide spread in agrarian regions. In a large farm, human scouts will have to be deployed to accurately spot the degradation (erosion), assess the extent of damage and correction required. It is a time consuming and costly process. Aerial photography and photogrammetry assesses soil erosion more accurately and efficiently compared to skilled farm workers. Remotely sensed data from satellites and piloted aircrafts often form the basis for land and soil mapping. Locations with soil maladies and erosion effects are marked clearly using satellite imagery. However, during recent years, with advent of low-flying drones, farmers are able to accrue data about farm land, its topography, fertility variations, gully and sheet erosion of soil, etc. Digital data derived using sensors could be used to develop digital elevation models that shows the extent of soil degradation and erosion more accurately with high resolution imagery (Moritani, 2010).

More common suggestion is to obtain detailed soil maps using satellite imagery and integrate it with ground measurements. However, currently, we can also obtain detailed high resolution soil maps using drones. Integrating these sources will provide better idea about topography, soil type variations and the extent of soil erosion, if any (Sarapatka and Netpoli, 2010).

Drones have found excellent use in surveying and mapping of forests, scrub vegetation, and crop land. Let us consider few examples depicting drone use in studying forests. Drones, it seems are cost effective during aerial survey of Indonesian thick tropical forests. The aerial imagery is aimed at pin-pointing intensity of general forest cover, deforestation, river flow, soil erosion, and biodiversity of plant species. Spectral imagery from drones flying 100 ft. or 200 ft. above the forest has provided useful data on natural resources (Koh and Wich, 2012). Some of the prototype drones and popular models of drones have been flown at 15–25 km hr^{-1} speed for 25 minutes to capture data from specific locations in the forest zone. The aerial imagery had a resolution of 10 cm pxl^{-1}. The drone was programed using 'ArduPilot Mega (APM).' It includes GPS, data logger, sensors for temperature and air speed. Drone missions with programed flight paths to make grids of 50 ha plots gave good idea about terrain and contents. Drones were flown at 180 m above the forest and geo-referenced pictures taken every 10 seconds. These drones were also used in Indian Forests to scout human activity, detect soil erosion, loss of biodiversity, etc. Drones flown at 200 m above the forests gave clear pictures of forest clearing, crops grown in the cleared areas, understory flora and even fauna, localized fires, smoke and small constructions if any. The resolution of aerial imagery depended on focal length of camera lenses and flight altitude of the drone. Following is an example as depicted by Koh and Wich, (2012).

Flight Altitude	Focal length	Resolution of Image
m above ground	cm	cm
200	4.1	7.4
	5.7	5.3
	6.9	4.4
100	4.1	3.7
	5.7	2.7
	6.9	2.2

Source: Koh and Wich (2012).

Drones have been useful in monitoring soil erosion and its impact on soil productivity. To quote an example, in Morocco, UAV has helped in obtaining data that adds to details provided by satellite imagery about topography and soil erosion. D'Oleie-Oltmanns et al. (2012) state that a fixed-wing drone (*Sirius I*) equipped with Panasonic digital camera was used to survey the cropping zone. Drone was allowed to fly at low altitude above the crop/soil surface and obtain imagery of erosion, leaching loss of surface soil and crop damage. Maps derived from multispectral analysis helped farmers in site-specific correction of soil erosion. Drones actually provide accurate geo-referenced data about spots affected by erosion. The photogrammetric processing using appropriate computer software actually helps farmers in obtaining digitized terrain model of the cropping zone. They can utilize both 2D and 3D imagery to develop apt erosion control programs and earth work needed could be assessed accurately. In comparison, to scout and assess soil erosion affected zones of a large farm, requirement of skilled human labor is exorbitant and the procedure is time consuming.

Worldwide, soil erosion is an easily recognizable soil malady that results in loss of top soil and its fertility. Damage to soil fertility can be severe, if torrential rains induce large scale and intense sheet or gulley erosion. Soil loss also occurs due to wind erosion, when soil aggregation is insufficient. It causes dust bowl like situations and results in loss of soil fertility and crop productivity. Several factors related to weather may act in conjunction resulting in loss of fertile soil. Land management is an important aspect of any of the agrarian regions, irrespective of cropping intensity of productivity. No doubt, satellite-mediated imagery has helped immensely in identifying soil erosion, loss of fertility and resultant reduction in crop yield. Satellite techniques are well suited to assess large cropping belts although resolution of imagery is relatively low. However, drone-based-survey of farming zones using Visual and NIR cameras for land degradation related problems, soil fertility and crop loss, etc. seems a very good proposition. Drones could be cost effective and flown periodically to assess need for land and soil fertility management. Drones could also be used to monitor the progress in erosion control measures and their effectivity. For example, there are innumerable reports about occurrence of gully and sheet erosion in sandy West African cropping zones. Generally, satellite mediated imagery and human scouts have helped in identification of such maladies. There are indeed several agricultural agencies that have compiled and published reports about satellite imagery as a means to study land degradation and crop loss, particularly in less developed regions of Asia and Africa (see Junge, 2008). Drones could supply high resolution, close-up imagery of topography, soil surface, loss of top soil fertility, reduction in crop stand, loss of seedlings due to sheet erosion, etc. Drones could also be used to assess soil moisture and nutrient interactions in many dry land areas of the world. In parts of Sahelian West Africa, where soil is very low in organic matter content and sandy in texture, drone-based assessment of natural vegetation, soil organic-C, crop residue and mulches could be highly pertinent. In fact, effect of crop residue and mulches on soil erosion and crop productivity could be studied in detail, using drones to collect aerial imagery and other relevant data. Again, drones could effectively provide detailed information on drought and its intensity, along with its impact on nutrient recovery and crop growth. Over all, it is believed that drone technology could be utilized to provide information on land use. Agricultural agencies and farmers could then alter and adopt suitable soil management procedures.

3.3 DRONES IN SOIL MANAGEMENT AND CROP PRODUCTION PRACTICES

3.3.1 LAND CLEARING, PLOUGHING, RIDGING, CONTOURING

Aerial photographs derived using satellites and piloted aircrafts have provided excellent information on topography, distribution of different soil types and their boundaries. Similarly, drones have been used to study soil taxonomy, prepare soil maps and mark distribution of different soil types. Soil survey aimed at studying the feasibility of soil types for agricultural crop production, crops that suit the soil type and yield forecasts could be accomplished using low flying drones. Satellite imagery has problems related to resolution. For example, it is said smallest area whose soil type could be studied or deciphered using satellites are still larger for farmers. Drones could be used to survey soil at higher resolution using visual and infra-red photography. Aerial imagery from drones could be utilized to decide aspects such as land clearing, deep digging, leveling, land filling, contouring, developing drainage channels, roads and infrastructure as related crop production in a small area. Imagery derived from drones could also be used to plan plowing schedules, ridging and planting seeds. Contours could be formed based on slopes and elevation data provided by drones.

3.3.2 DRONES, SOIL FERTILITY AND FERTILIZER SCHEDULING

Soil nutrient deficiency related reduction of crop yield is common across most agrarian regions. Fields not fertilized exhibit grain/forage yield reduction. As a consequence, fertilizer supply to crops, especially N, then P and K has been among most researched topics. Farmers world-wide apply major nutrients to prop up grain productivity. Fertilizer application to crops has often depended on soil type, its inherent nutrient status particularly available forms of major and other nutrients. Soil fertility specialists have adopted a variety of different soil chemical analyzes, previous data on crop response to nutrient supply and yield goals envisaged, in order to prescribe fertilizer inputs. Soil-N is the most important fertility factor. Fertilizer-N supply has therefore received priority over other nutrients. Fertilizer-N requirements are relatively higher and its supply has to be accurate, timely and placed close to root. Fertilizer-N is fairly costly and its efficiency is of utmost importance. Fertilizer-N dearth or excessive supply has detrimental consequences on crop production. Soil chemical analysis to obtain accurate knowledge about N distribution and variations within a field is a tedious procedure. It involves series of soil sampling, including subsoil layers, then soil processing and chemical analysis using instruments. This data has to be then plotted to obtain soil-N map. During precision farming techniques, fertilize-N is channeled based on soil-N map. However, there are other indirect methods such as estimation of leaf chlorophyll using aerial photography. The soil-N fertility is related to crop growth and its chlorophyll content. There are computer programs that decode the aerial images. Remote sensing of crop growth and chlorophyll status is a technique adopted in many agrarian zones. Remote sensing techniques are relatively less costly compared to regular soil chemical analysis procedures. They offer timely results, excepting in situations of excessive cloud cover. Remote sensing offers less accurate information since it depends on spatial resolution of imagery, but still good enough to decide on fertilizer-N supply schedules. Hand-held instruments that determine plant-leaf chlorophyll, in other words leaf-N, have also served farmers in deciding on split dosages of fertilizer-N, on a standing crop. Low flying parachutes and balloons fitted with cameras have also been used to derive plant chlorophyll data (Pudelko et al., 2012). Soil fertility management using drones that supply aerial images of crop-N status is a recent technique. It has potential to become most common and useful method to assess plant-N status and therefore in deciding fertilizer-N inputs to crops (Buerkert et al., 1996; Boike et al., 2003). Zhu et al. (2009) state that during recent past, soil fertility specialists and farmers alike have obtained information on crop biomass accumulation and nitrogen status using drones. Drones used for the purpose are entirely autonomous with pre-determined flight path or those controlled using remote controller tablets/PC. Let us consider drone-based methods of plant-N status estimation as described by Zhu et al. (2009). Field trial conducted at a farm near Hangzhou city of Zhinjiang Province, using 4 different levels of fertilizer N supply to rice crop has shown that, correlation between ground-based leaf/plant-N estimated and aerial imagery derived plant-N status is positive. Aerial imagery based predictions on plant-N was 78–91% accurate compared with ground-based plant leaf analysis. The drone copter (Hercules-II) obtained imagery at 60 m altitude above the rice crop with a resolution of 0.02 m^2. Images were stored as JPG fields and decoded using computer programs. The canopy spectral reflectance was actually measured using a spectrometer. While ground-based leaf analysis was estimated using SPAD of Minolta, Japan. The hyperspectral reflectance differed with N supply levels. The reflectance decreased with increasing N supply to rice crop. The peak reflectance was measured in the green band width (500–560 nm). Zhu et al. (2009) state that considering good correlation between ground-based leaf-N and aerial imagery derived plant-N status, there is clear possibility to use aerial imagery and predict fertilizer-N requirements for the rice crop. There are several studies conducted in different agrarian regions, using several different crop species that support the idea of using satellite based or UAV derived aerial imagery to decide on fertilizer-N supply to crops (see Blackmer and Schepers, 1996; Scharf and Lory, 2002; I'nen et al., 2013).

Measurement of crop N status is an essential step while devising fertilizer-N schedules. Knowledge about crop-N status is also useful in deciding or revising grain/forage yield goals. Drones have been tested periodically to obtain imagery that is indicative of crop-N status. It is possible to estimate crop-N status using optical measurements. Leaf nitrogen content is mostly localized in chlorophyll. Therefore, measuring leaf reflectance that depicts chlorophyll content has direct relevance to leaf-N/crop-N status. Mostly, greater the visual reflectance lower is the chlorophyll content. Next, increase in Near Infra-Red Reflectance (NIR) corresponds to higher leaf area index and green biomass. Satellite imagery that provides information on NDVI and crop reflectance is a possibility. However, resolution of satellite imagery is low and clouds may interfere with accuracy of measurements. Instead, use of drones to study crop reflectance seems feasible. For example, several researchers have compared leaf chlorophyll content values derived from ground-based digital sensors with those derived from low-flying drones fitted with multispectral sensors. The idea is to use canopy reflectance data from drones and/or satellites to manage fertilizer-N supply to crop. The aim is actually to obtain a mathematical relationship between NDVI values derived from Micro-drones and those from ground-based sensors. Several researchers have compared NDVIs, leaf-N status and canopy growth. They found a linear correlation ($R^2 = 0.834$) between NDVIs derived from two different methods. Therefore, NDVI and chlorophyll reflectance derived using drones could be used to manage fertilizer-N supply to crop (see Reyniers et al., 2004; Reyniers and Vrinsts, 2006; Filella et al., 1995; Han et al., 2001; Aguera et al., 2011). In Nebraska, field evaluation of maize for its response to fertilizer-N and water supply has been conducted using drones. Drones helped the researchers in obtaining NDVI, Leaf area index and Leaf-N status. Drones (Octocopter) were used to collect data about maize hybrid's response to water and fertilizer-N and their interactions (Krienke and Ward, 2013).

Rice crop has been monitored using drones with a purpose to judge its N status and then take accurate remedial measures to avoid N deficiency or excess. Swain and Zaman (2010) suggest that assessment of leaf radiation using low-flying drone has potential to detect N deficiency. It could be used to monitor the crop for N status (see Table 3.3). Excessive-N inputs could be avoided. Excess-N usually finds its way into irrigation channels, leaches into lower horizons of soil and contaminates ground water. These maladies could be avoided by using drones, periodically, to monitor leaf N status. As stated earlier, leaf chlorophyll is actually an indirect measure of leaf-N content of rice plants. Digitized data about leaf chlorophyll could be used on-the-go or via maps obtained previously to apply fertilizer-N using variable-rate techniques. This leads to a uniform rice crop. Drones could also be used to spray liquid fertilizer at variable rates. Biermacher et al. (2006) have reported that detection of crop-N status using drones fitted with sensors and applying fertilizers using variable-rate applicators has given 22–31 US$ ha^{-1} *more* profit to rice farmers.

TABLE 3.3 Methods to Monitor Rice Crop for Nitrogen Status Using Low Altitude Remote Sensing

System	Sensor Equipment		Applications		
GPS	Laser	Camera	Large Areas	Small Areas	Digital Mapping
Air Craft	Yes	Yes	Film, Digital	Yes	No	Yes/No
Drone Helicopter	Yes	Yes	Digital	Yes	Yes	Yes
Terrestrial Perches	Yes	Yes	Digital	No	Yes	Yes

Source: Swain and Zaman, 2010

Mattern (2013) reports that UAVs have been tested on large expanse of soybean crop to detect spectral differences and their correlation to soil nutrient deficiencies. A step further; drones have also helped farmers to trace soybean zones that need nutrient application. Drone derived imagery and digital maps can pin-point areas that need extra nutrients. Drones could then fly to those exact locations at low altitude over crops and release accurate quantities of fertilizer. Drones are usually equipped to release liquid or granular formulations of fertilizer with good accuracy. For example, a micronutrient spray of Zn could be accomplished effectively using UAVs. First, drones are used to scout the fields, collect data (digital maps) and use computer-based decision support system to apply Zn accurately. Systems that allow on-the-go decoding of digital maps are useful. A computer decision support system immediately assesses fertilizer need accurately. Fertilizer release is accomplished using variable-rate applicators.

Researchers from some of the companies that produce agricultural drones suggest that these copter/fixed-wing versions produce aerial maps depicting NDVI, pestilence and disease incidence that are highly relevant. They find that excessive use of fertilizers containing N and P, and rampant use of pesticides is often uneconomical. The excess chemicals accumulate in soil to the detriment of soil environment and microbial activity. Chemicals also find their way into ground water sources. Runoff from crop fields can be a problem when fertilizer-based nutrients accumulate at a location farther away. Hence, drones are used to periodically fly over the crop, monitor its growth and nutrient status. To quote an example, Robo flight is a company that produces drones fitted with high resolution visual and NIR cameras. The drones produce imagery with plant growth pattern, NDVI and canopy variations. These traits are directly linked to N and P status of the crop. The digitized version of the imagery could be used on-the-go or later to spray nutrients through variable-rate applicators (low volume spray nozzles). Precision technique mediated by drones reduces use of fertilizers and other chemicals, rather significantly. In case of fertilizers, drones help firstly by offering soil fertility maps (plant growth) depicting subtle changes. Then, drone applies fertilizers at appropriate rates based on variations depicted in digitized maps. Fertilizer usage could be reduced by 30–70% of original, if blanket applications are adopted. In case of pesticides, sometimes, it has been stated, that drones operated using precision techniques reduce chemical usage by a whopping 90–95% of original levels, if blanket applications are adopted. Drones also hasten fertilizer supply exercise. In one of the examples, over 600 acres of farm was imaged in just 1 hr. of drone flight over the fields. Fertilizer-N applied to locations as per imagery improved crop response. It has reduced fertilizer-N requirement by 25% compared to blanket application (Ellerbroek, 2014b). At this juncture, we should note that aerial (foliar) spray of fertilizer-N (urea) is done at very low concentrations of 0.2% urea dissolved in water. Crop response to foliar spray is significant. It means fertilizer-N requirement is drastically reduced, if foliar spray is practiced using drone. Split-N is usually applied as foliar spray. Therefore, drones could be effectively used to spray split-N to cereals and other crops.

Wheat Scientists in Toowoomba, Australia have evaluated the effect of fertilizer-N application using aerial images derived from a low altitude helium balloon. The helium balloon was used as a platform to house a set of cameras capable of visual, NIR and thermal imagery. Wheat cultivars were grown at two levels of fertilizer-N input, namely nil and 120 kg N ha^{-1}. The wheat genotypes were actually grown in plots that had similar fertilizer-N treatment of zero and 120 kg N ha^{-1} for the past 10 years. Jensen et al. (2007) reported that statistical analysis of data obtained from the aerial drone (balloon) and ground measurements showed high correlation between multispectral imagery and crop's response to fertilize-N. A high correlation ($R^2 = 9.1$) was seen between wheat crop foliage, leaf-N and final grain yield. Hence, it is possible to conduct regular evaluation of large number of wheat genotypes for their response to fertilizer-N supply using drones fitted with imaging devices. Further, periodic flights by drones (copters) that collect data on crop growth and leaf-N could be used to assess variation in soil-N/crop-N and then take appropriate measures using precision farming techniques, for example, variable-rate fertilizer-N supply.

Sunflower production depends immensely on extraneous supply of nutrients through fertilizers. Nitrogen is the key element that is needed in higher quantity. At the same time, soil-N fertility variations are rampant in sunflower production zones. Hence, accurate knowledge about soil-N level, fertilizer-N that needs to be replenished and expected yield is necessary. Precision agriculture or site-specific N inputs have been adopted in many agrarian regions that support sunflower production. The fertilizer-N input at a spot is often guided by the estimation of soil/plant-N status using sensors. Drone based sensors such as visual, NIR and thermal cameras provide information on NDVI of the crop, biomass, leaf area, leaf chlorophyll content and moisture status. The NDVI values are related to crop-N status and based on yield goals and background soil-N, farmers can adopt site-specific fertilizer supply. The variable-rate applicators are used to supply fertilizer-N based on digitized data about crop-N status or sometimes on-the-go techniques are adopted. Drones are used to obtain the crop-N status maps. Aguera et al. (2011) have used ADC Lite Tetra digital cam mounted on micro-drones (MD4–200) to obtain aerial imagery and crop-N status. They have used both visual (520) and NIR and Red-Edge (845–920 nm) to obtain imagery. The data revealed that crop-N status correlates significantly with NDVI. This system was used for over 32 field plots with sunflower crop (Aguera et al., 2011).

Adamchuk (2005) and Adamchuk and Jasa (2014) state that, ideally farmers/researchers think of rapid on-the-go analysis of soil for various parameters such as electrical conductivity, electromagnetic responses, pH, soil-N status and moisture. A variable-rate applicator could be proportionately activated to apply fertilizers/irrigation accurately. However, if this facility is not available, farmers could obtain a soil map marked with fertility variations.

Now, let us consider an interesting experiment involving organic matter supply using marine algae such as *Ascophyllum nodosum, Macrosystis sp, Durvillia sp, Porphyra sp* and *Sargassum sp. Ascophyllum nodosum* is a common organic matter source that is rich in mineral nutrients. The effect of organic matter on grape vines has been examined using Compact Airborne Spectrographic Imager (Gil-Perez et al., 2010). Narrow-band hyperspectral imagery has been used as indicator of nutritional status of grape vines. Narrow-band indices correlate with seaweed fertilization. Periodic observations made using drones helped farmers in judging the effect of marine algal application on grape vines. In future, it should be possible to conduct tedious and large scale experiments using drones to collect data. Drones are rapid, less tedious and need fewer skilled farm workers.

3.4   DRONES IN AGRICULTURAL CROP PRODUCTION PRACTICES

3.4.1   DRONES AND CROP SCOUTING

Drones used for crop scouting vary enormously with regard to several specifications and their abilities to scout and derive digitized pictures of soil or crop health. Drones could be either fixed wing types or copters. Each type and specific model sold has its specific advantages in performing the tasks (Table 3.2). Generally, it is accepted that using drones allows a rapid and good view of the farm. A bird's eye view of large farm or a close-up, both are possible. In general, advantages of using an agricultural drone to oversee the farms are as listed below:

   a) Drones can detect potentially yield limiting factors and alert farmers;
   b) Drones can save time required to scout a large farm;
   c) Drones allow a better return on investment compared to hiring human scouts;
   d) Drones are easy to operate. They could be programed to fly a pre-determined path over the large farms and/or its path could be altered using remote controller;

e) Drones could also be used for security related surveillance of farm borders, like intrusions, damages, etc.;

f) Drones, some of them, provide excellent pictures of disease/pest attack at various points in the farm. Farmers can accordingly schedule chemical spray. Most importantly, chemical sprays get confined to only areas afflicted with pest. It then reduces cost on agronomic operations; and

g) Drones are safe compared to manned flight over large farms (Precision LLC, 2014; Silva, 2013).

The general forecast by manufacturers of agricultural drones, farmers with large land holding, agricultural agencies of different nations/regions and soils/crops experts is that, in the very near future, usage of drones would increase. Scouting for crop diseases, pests, weed infestation or searching for locations affected by soil erosion after a thunder storm will become a common practice. The autonomous drones used recently are capable of both exacting 2-D and well defined 3-D imagery. This will allow farmers to pinpoint areas afflicted. Remedial measures could be adopted accurately with no excessive usage of chemicals (Dobberstein, 2014). We may note that each sortie of an agriculture drone that brings latest pictures of the farm proportionately reduces human labor needs and drudgery of loitering all through the large farm, sometimes repeatedly, to get greater insight into problems that afflict the farm. Further, drones help famers in getting timely information about happenings in the crop field (Paul, 2014). At present, usage of drones to scout for diseases and pests is gaining in acceptance in agrarian zones of North America, Europe and Australia. However, use of UAVs on commercial scale is yet to be permitted in most of these nations. Agricultural UAVs are therefore regarded as one of the many farm gadgets and implements and put to use in private farms.

Drone based high resolution remote sensing and crop growth monitoring is an effective crop management suggestion in many locations (Knoth and Prinz, 2013). Let us consider a few examples related to usage of drones in scouting. Autocopter is a remote sensing drone now in operation in parts of North Carolina and neighboring states of USA (Plates 3.8A and 3.8B). This copter is well equipped with three powerful cameras, namely a full sized visual camera, multispectral camera, a DSLR and most importantly a High Definition (HD) videographer. This drone runs for 2 hr. without halt and is connected to GPS. Farmers growing maize/cotton or maize/groundnut intercrops are placed at an advantage when they scout their fields, because data from Autocopter can be instantaneously processed. The digitized data acquired is channeled into Ag Leader SMS, John Deere's Apex and SSST Farm management programs. The farmers are shown the entire picture of scouted field after ortho-stitching. In addition, information on NDVI, crop growth and health are relayed without delay to farmers. The HD video allows farmers to see the crop in detail as the drone flies past scouting the field. According to some farmers in North Carolina, cost for Autocopter and data processing programs is easily payed-up by the rapid scouting, timely treatments with fertilizers/pesticides and the resultant enhanced crop production (Plates 3.8A and 3.8B).

Potato fields in Oregon are being scouted on a daily basis using remote controlled drones. It is aimed at helping farmers in managing crops and sequencing agronomic procedures efficiently. The Boeing Aerial robots keep a watch of potato fields that are 50 acres. A few farms that are over 1000 acres are also under watchful eyes of drones. They supply the farm company/farmer with daily pictures of crop growth, its health, nutrient needs and general farm topography. Based on digitized pictures, farmers could then decide on fertilizer supply, irrigation, and soil erosion measures if any (Oregon State University, 2014).

There are several models of fixed-wing drones produced by companies in Europe that are deployed for scouting the crop. Field mapping is an important task carried out by the drones such as "X-100" produced by Gatewing Inc., of Belgium. These drones are utilized to map the crop field on a daily basis. Field maps are continuously consulted and decisions on agronomic procedures that should follow are considered. The 'X-100' drone is a small, lightweight (2 kg), fixed-wing type that flies at low altitudes over crop fields (Plate 3.6). The data capture is coupled with computer software that automatically processes imagery and delivers a geo-referenced field map to farmers. Such aerial imagery is useful to even farmers not conversant with drones and computer programming (Ball, 2012). The 'X-100' imagery is also being used efficiently to draw daily work charts and completion reports, using internet facility. Trimble UX5 drones are similar equipment that could be deployed in any commercial grain production farm or plantation (Plate 3.2).

In Oregon, drones are scouting farming zones and forest for water deficits, if any. In case of crop land, scouting using drones is becoming prominent. Mainly because, it avoids walking through a 1000 ac farm to detect spots that are suffering from water shortage. Drones are also scouting forests for tree health, presence of invasive species, the changes in under story flora, and fauna. Drones also suggest regarding density of plant species and their distribution. It depends on cameras fitted and computer programs that detect plant species based on their spectral signatures (Walker, 2014). Commensurate with recent trends in agricultural cropping that emphasizes on precision farming, there are manufacturers who specialize in producing drones that suit best for precision farming. For example, 'Observer QX1 System' is a regular drone programed to fly determined flight paths and observe the crop periodically for several traits that are necessary for precision techniques. Farmers are provided with data on water deficits if any, fertilizer needs and pest attacks if any (Farm Aerial Drones, 2014).

Parachutes or para-gliders are also used as drones once they are fitted with accessories for navigation, GPS connectivity, visual and infra-red cameras for crop imagery. Let us consider an example from Wheat production zones of Polish plains. Pudelko et al. (2012) have reported use of para-glider for aerial imagery and scouting of wheat/barley for soil fertility variation, fungal diseases, and infestation with weeds such as Couch grass. The para-glider was used to photograph crop fields from 20–700 m altitude from ground level. The glider had a cruising speed of 15–20 km, a flight endurance of 45 minutes and payload of 3 kg. The plots photographed were relatively small at 36 x 36 m for the purpose. Therefore, a high resolution camera such as Sony DSC with 8 Mpxl was necessary. The aerial images were decoded using appropriate computer programs. The drones provided higher resolution pictures of 'hot spots' and focal points of 'Eye spot disease' that afflicted barley crop. The aerial imagery also showed areas with low soil fertility. Soil variation is actually depicted by erratic crop growth pattern and yellowing or discoloration of vegetation (see Jensen et al., 2007; Leon et al., 2003; Pudelko et al., 2008; Thorp and Tian, 2004). Pudelko (2012) further suggests that aerial imagery derived via low flying para-gliders have definite advantages over satellites, since they can supply high resolution photographs of crops/soil.

At the bottom line, we may note that there are farmers who could assemble their own drone copters for a really small cost. They can then fit a camera that takes close-up shots. There are others who use hobby auto-pilot flying machines and fit them with GPS connectivity and camera. We ought to realize that, even these self-made drones pick accurate images. Each image comes along with accurate GPS coordinates and time of exposure (Ghose, 2013)

3.4.2 DRONES, SEED PLANTING, AND CROP GROWTH MONITORING

Drones could be adopted right from the first agronomic procedure during crop production. Drones could first make a surveillance of the terrain, vegetation, soil types and water resources. The digital information could then be utilized to clear the land of all vegetation such as shrubs, grasses weeds, etc. Then, land preparation activities such as bull dozing, if necessary, deep plowing and loosening top horizon of soil is done. Drones could provide aerial imagery to guide the farmers to adopt contour planting or straight ridges and furrows. Drones also provide information to farmers about soil type, fertility and irrigation facility. Such details are used by farmers to decide on crop species, genotypes and planting programs, seedling density expected, irrigation schedules, split application of fertilizers, etc. Drones provide excellent imagery of entire fields and emergence of seedlings. The aerial imagery and digitized data of gaps created by erratic germination and lack of seedling establishment can be identified accurately using GPS co-ordinates provided by drone imagery (Plate 3.5B).

In areas with field crops, drones could be regularly used to monitor seed germination. Seed emergence could be erratic; sometimes seeds could get dislodged or moved away from place of dibbling. Seedlings that emerge may be uniform or many a times get stunted. Fields sown on a large scale using tractors with automatic planters too may show up gaps in plant stand. Some of these could be very clearly observed using drones fitted with multispectral cameras. In the field, seedling establishment is affected by weeds that germinate and establish simultaneously. They interfere with water, nutrient and light interception by main crop. Drones with ability for discrimination of weeds, several species of them common to the location and main crop species are useful. Drones are used for regular survey of seed germination, seedling and crop stand establishment. Any gaps are clearly visualized using drones (see Plates 3.5A and 3.5B). Drones could be used to assess planting density in a location.

Drones have been used to broadcast seeds of different tree species to develop vegetation in wasteland, also during forest reclamation and regular plantation development. Drones with slightly larger payload to hold seeds and ability for longer endurance are often used to disperse aerial seeds. Remotely controlled aerial seeding is gaining in popularity during development of pastures programs. The drones carry small loads of seeds treated with Rhizobium, if it is leguminous tree species used in agroforestry programs. Seeds are also treated to protect them from diseases and pests. In a large forest belt, it is necessary to mark the area that has been broadcasted with seeds. Therefore, drones that apply seeds also keep accurate GPS coordinates of different tree/shrub species sown in the forest replanting zones. Forest re-vegetation zones are being regularly sown using drones in many developing countries (MSS Drone Division, 2014). Drones could also be used effectively in many of the forest planting programs to transport seedlings to spots based on GPS coordinates provided, just like a ground vehicle or human porter does. Drones can reach relatively difficult spots with greater ease compared to ground vehicles or human scouts. In wet land setting, rice seedling could be transported to the transplanter with greater ease, during transplanting. There are of course, autonomous and semi-autonomous rice seedling trans-planters in vogue in some of the Southeast Asian nations.

Crop phenotype is dependent on interaction between its genetic constitution and environmental parameters that it encounters, including those amended by farmers. Crop phenotypic expression decides several aspects of farming and farmer's decisions. Periodic observation of crop phenotype enthuses farmers to forecast forage/grain yield more authentically. This induces them to supply fertilizers, water and pesticides appropriately. They can set, revise and reach higher yield goals based on authentic information of crop phenomics. Crop growth monitoring and procuring authentic data on phenomics forms the basis for several agronomic and soil management practices, during the crop season. In season, rapid assessment of crop is necessary, if farmers have to go versatile and adopt suitable fertilizer/ irrigation schedules to reach yield goals with optimum input efficiency. Phonemics allows farmers to get a clear idea about the progress of crop towards forage/grain yield goal set by him. Phonemics involves tedious effort by several skilled farm workers collecting data on different crops/ genotypes sown by farmer in the field. Large farms with several crops/ genotypes are not easy to handle. Phenotyping of crops at different stages becomes costly and requires human drudgery and time. Rapid pheno-typing is essential for decisions on several in-season farm operations. Sometimes, ground-based robots too become inefficient and slow. Ground robots may still not give farmer an overall imagery of crop morpho-genetics. Hence, use of drones with ability for rapid phenotyping using sharp high resolution imagery becomes essential. Perry (2012a) reports construction of a copter drone that is equipped with accessories specifically for rapid scan and phenotyping of crops. Accessories are sensor-based pheno-typing, reflectance measurements, thermal measurements of crops/canopy, NIR imagery and NDVI values. Perry (2012a) reports that for a small farm of 5 ha, drones with short flight of 30 min with autopilot features and pre-programed flight path is suitable. The multispectral imaging and thermal cameras accomplish measurement of crop height, canopy, leaf area, NDVI and temperature rather rapidly. These drones also produce video output at 25 frames per second. The thermal infra-red sensor measures radiometric surface temperature in the 8–12 μm range. Following is the overall ability of drone described here. It includes acquisition of remote sensed data relevant to crop phenotype and growth pattern:

Indices: Normalized differential vegetative index (NDVI), canopy chlorophyll concentration, chlorophyll index, crop water stress index

Phenotypic Information: green vegetative cover, senescence zones, canopy leaf growth, chlorophyll content, leaf nitrogen status, crop water stress, drought affected areas, crop drought tolerance

Source: Abuzar et al., 2009; Fitzgerald, 2010; Perry et al., 2012b; Zarco-Tejada et al., 2012

Several crop species have been evaluated and studied using narrow and broadband vegetation indices. Firstly, researchers will have to fix the optimal wave band-width while trying to analyze crop growth and biomass accumulation using NDVI. Let us consider a study dealing with six different field crops namely, barley, wheat, lentil, cumin, chickpea and vetch. Thenkabail et al. (2002) estimated several phonological traits such as leaf area index, wet biomass, dry biomass, plant height, plant chlorophyll and nitrogen status, and canopy cover using remote satellites and piloted airborne cameras. Drones fitted with multispectral imaging cameras have greater flexibility and accuracy for studying many of these phenotypic traits. Field studies that evaluated several crops using hyperspectral band width range of 320 to 1010 nm suggest that, it is preferable to use simple two band width vegetation indices and/or multispectral band vegetation indices. For most crops, NDVI is best estimated using at least using 4 band widths around 520 nm (green), 675 nm, 720 nm and 905(NIR). However, there are 12 optimal band widths to acquire data on NDVI for different crops (Thenkabail et al., 2002). They are:

Blue – 490 nm (30 nm);

Green 1 – 520 nm (15 nm); Green 2 – 550 nm (25 nm); Green 3 – 575 (15 nm);

Red 1 – 660 nm (20 nm); Red 2 – 675 nm (15 nm); Red-edge 1 – 700 nm (5 nm); Red-edge 2 – 720 nm (15 nm); Near Infra-Red NIR – 845 (120 nm); NIR peak 1 – 905 nm (15 nm); NIR peak 2 – 920 nm (15 nm); NIR Moisture Sensitive – 975 nm (10 nm)

Source: Thenkabail et al. (2002).

Note: Values in the parenthesis indicate range from median.

Farmers grow different crop species. They also plant different genotypes (cultivars). In a large farm keeping track of crop growth parameters is an important task. In this regard, drones could be flown to collect information on phenotypic status of crops. The imagery could also be studied on-the-go on computer or tablet. Agricultural experimental stations dealing with plant breeding and testing genotypes can effectively utilize drones to collect data about subtle effects of genes introduced on phenotypic expression. Drones are apt to collect routine data on plant height, canopy cover, leaf area, chlorophyll and leaf N status of several genotypes (Huang et al., 2013). Berni et al. (2009a) state that one of major limitations of satellite imagery is the lack appropriate resolution. During crop production, farmers need close-up shots with higher resolution. The other problem is the unfavorable re-visit time of satellites. Hence, small light weight robots with multispectral images are highly useful. In Spain trials using drones to estimate leaf area index, leaf chlorophyll content and NDVI have been successful and they effectively replace satellite imagery, plus offer better accuracy. Such drones are capable of high resolution pictures of 20 cm resolution picked up at 400–800 nm band width.

Leaf area index (LAI) is an important parameter that depicts ability of natural vegetation or crops for photosynthesis, respiration, CO_2 emission and biomass accumulation. The LAI can be estimated using remote sensing. Hyperspectral data from satellites provides useful information on LAI. Let us consider an example pertaining to field crops such as maize, potato and sunflower. Duan et al. (2014) report that LAI was measured using UAVs (drones) and compared with *in situ* ground based measurements using leaf area meter. There was correlation between LAI measurements derived from the two different methods. Regarding, LAI measurements derived from drones, it is said, data collected from two different angles gave greater accuracy to LAI estimates.

Crop phenomics has also been studied by comparing 'Crop Surface Models (CSM)' with data accrued by small drones that are equipped with visual range cameras. Rapid comparison of CSM with digitized data provided by drones to decision-support computers, helps farmers in deciding fertilizer top-dress and irrigation scheduling (Bending et al., 2013).

Schultz (2013) states that, drones have been used to study the phenomics at periodic intervals. This has helped sugarcane and corn producing farmers in Louisiana to decide on quantity and distribution of fertilizers. Areas with inadequate soil fertility could be easily corrected by supplying extra quantities.

Horticultural crops are monitored for canopy growth which is an important phonological trait that determines photosynthetic efficiency, biomass formation pattern and fruit yield. Knowledge about plant growth rate and canopy is almost essential for deciding various agronomic procedures. In some farms, canopy cover and water usage pattern is used as indicator for forecasting crop yield. Remote sensing satellites and drones are utilized to derive data on NDVI of horticultural crops. The NDVI values are linearly related correlated to photosynthetic light interception; biomass accumulation and canopy cover (Trout et al., 2008). Hence, drones can provide crucial information needed by farmers regarding biomass accumulation Drones with multi-spectral imagery are utilized to obtain variable rates of canopy growth. Computer programs that analyze NDVI and canopy cover data and relate them to fruit yield are useful to farmers, while deciding on inputs and fixing yield goals (Trout and Gartung, 2006).

3.4.3 DRONES IN SCOUTING FOR CROP DISEASE INCIDENCE AND CONTROL

Farm companies in North America are accepting drones as a new tool to study crop health and to detect diseases. These aerial imagery based techniques are more efficient than human scouts. The drones controlled using a computer or hand-held I-tablet captures a good view of field and details of crop all over a 360 degree, and depending on altitude of flight covers an area of 20 km^2. It is rapid and decisions to spray chemicals are naturally quick and more accurate (Koontz, 2013).

Let us consider an example, where in, UAVs have been used to scout fields and to mark 'management zones' with differential biomass production. Picard (2014) has reported that in case of canola produced in Manitoba, the crop is exposed to attack by a few fungal diseases. The fungal attack varies with location and crop vigor. Fungicide treatment prescribed may often over estimate need for sprays. Therefore, site-specific techniques that utilize aerial imagery of disease zones and NDVI estimates were used. It helps to identify areas/fields that need higher fungicidal dosage and those with good crop vigor and do not require fungicidal treatment. The canola producing zone was classified into 'management zones' and fungicide application was accordingly calibrated. Management zones given high dosages of fungicide were completely devoid of fungal disease and yield difference compared to unsprayed check zones was 13 bu ac^{-1}. The fungicide requirement was markedly reduced, if management zones demarcated using drones were employed. Most importantly, periodic checks of disease incidence using drones helped in identifying zones prone to disease, at a rather early stage. Hence, fungicidal requirement and human labor cost to control disease was lessened to a great extent.

Researchers at Kentland Farm, Virginia Technological University have been deploying UAVs to detect occurrence and spread of crop diseases caused by fungi such as *Phytophthora* and *Fusarium* (Dobberstein, 2014).

The spread of disease could be monitored at very early stages. Therefore, control measures could be targeted accurately. Further, it is interesting to note that low flying drones have also been used to surveillance the fields for prevalence of pathogens in the air. Aerobiological samples collected by these drones could be periodically tested for pathogen propagules. This system helps in adopting prophylactic measures targeted at areas with higher probability of crop disease. Schmale (2012) has further stated that collecting aerobiological samples above soybean fields and analyzing them using DNA (Restriction Fragment Length Analysis), shows that Fusarial spores detected above the crop were not those just residing in the soil/crop surface below. Those blown into the area via wind drift were also noticed. So, air samples above crop fields periodically may help the agricultural agency/companies to detect presence of pathogens well ahead of it causing devastation. The threshold population of propagules of pathogens need to be understood in order to decide, if prophylactic sprays are required. Studying microbiological and chemical environment in the atmosphere just close to and above the crop could be useful in judging disposition of crops to disease/pollution, if any. We have been doing such tests for soil and water, now with drone technology we may extend it to atmosphere. Public health departments of towns/cities in agricultural belts could also scan the atmospheric samples periodically drawn by drones, for disease causing agents/chemical toxicants.

Drones have been adopted to provide aerial imagery of wheat crop that is afflicted with fungal diseases. Periodic flights over wheat farms have helped farmers in visualizing disease progression and take appropriate spray schedules. For example, in the Southern Plains of USA, wheat crop diseases have been tracked using drones (Basso and Rush, 2013). Early detection of diseases helps in reducing areas that need to be sprayed. Therefore, costs on chemicals and sprays get reduced. Actually, crop damage gets thwarted at an early stage, resulting in better grain yield productivity. Crop disease epidemics could be effectively halted. The agricultural extension department of Texas A and M states that in the High Plains, over 1.1 million ha is periodically affected by mite-vectored viral diseases and water deficiency that appears to be due to improper root growth. So far, Texan farmers have been relying on aerial imagery from different satellite agencies, pictures from low flying piloted air crafts, hyperspectral images captured by human scouts and reports by regular scouts. However, researchers at Texas A and M are trying to use 6 copter drones and standardize digitized data with ground-based observations of viral disease and water stress (Texas A and M Agrilife, 2014). Drones are programed to take flights that make a grid over farmers' fields. There are also reports that late blight of potato could be monitored using drones. Aylor et al. (2011) state that sporangia of *Phytophthora sp* could be detected and quantified using drones. They used at least three different methods to capture and quantify *Phytophthora* propagules in the atmosphere above the potato crop. They suggest that preparing software and appropriate decision support systems could be helpful to spray fungicides on the crop. It has to be based on propagule density detected by the drones. There are also reports that propagules of Asian Rust fungus could be detected using drones flown above crops such as maize, wheat and soybean (Schmale, 2012).

Ehsani et al. (2012) and Lee et al. (2009) have reported that light weighted, low flying drones fitted with HD cameras that operate at visual and NIR band width have been useful in detecting disease afflicted citrus tree and localized areas within the grove. The drones provide detailed imagery of almost each tree if required and display, the intensity of disease and tree growth retardation. Drones supply digitized information instantaneously to farmer's computers, so that they can decide on spray schedules or culling or other suitable crop protection procedures. Drones actually lessen possibility of disease because farmers can identify disease affected trees much ahead and reduce on cost incurred for chemicals and spray schedules. Drones fitted with pesticide/disease control chemicals can spray them accurately on trees. A more recent report by Garcia-Ruiz et al. (2013) states that low-cost, low-altitude, remote sensing multi-rotor drones fitted with visual and NIR cameras have been effective in detecting HLB disease and healthy citrus trees (Lee et al., 2009). They have examined citrus canopies using six spectral band widths between 530 nm and 930 nm. Crop reflectance at 710 nm gave best correlation with HLB disease incidence, when compared with ground data. The accuracy of identification of HLB disease on citrus canopy was 67–85%, if low altitude drones were used. False negatives were low at 7%. However, if a piloted aircraft was used to obtain reflectance, accuracy was 68% and false negatives were higher at 28%. Thus, Garcia-Ruiz et al. (2013) state that in future we may be using drones fitted with multispectral imagery to distinguish areas of citrus groves affected by diseases and those with healthy trees.

In North American Apple producing regions, 'apple scab' is rampant. It affects quality and economic value of fruits. Researchers at the University of New Hampshire have opted to use very low cost drones that fly frequently above the canopies of apple orchards and collect imagery showing apple scab infestation if any. It seems visual and infra-red imagery is enough to suggest apple scab infestation, right at the early stages of infection and spread on apple fruits (Kara, 2013)

3.4.4 DRONES TO SCOUT FOR PEST INFESTATION AND ITS CONTROL

Pest management involves timely identification of insect attack, its spread and a matching response that is accurate and successful in controlling crop deterioration. Initially, pest control measure procedures took care of crop, field by field. Farmers were asked to make blanket application of pesticide considering the insect species and its intensity. Later, particularly since past decade, farmers are exposed to site-specific techniques. They actually scout the crop in field for exact locations that are attacked by pests. Pest control measures are highly focused to cover only the spots with pests or at best a management zone. The pest attacked zones are usually mapped using skilled human scouts or aerial imagery from satellites, if they are easily available. Remote sensing is a method adopted when large expanses of crops are to be watched for pest attack. Remote sensing often provides data about distribution of pest in large geographic areas or agro-ecosystem per se. Generally, it requires satellites or piloted aircrafts to accomplish mapping the large scale pest attack. Area-wide pest attack management decisions such as integrated pest management could also be adopted as a matching response to control pests. Often, remote sensing, GPS, GIS and variable-rate technology are adopted in conjunction. More recent technique is to adopt small, light-weight drones to identify and map pest affected zones. Then, direct the pesticide spray at variable-rates, exactly at each location. Drones are becoming increasingly useful in precision or area-wide control of insect pests (Huang et al., 2008).

Drones are not just used to obtain data regarding occurrence and spread of pests in crop fields. The spectral data of course provides details of pest attack in different places within a crop field. However, farmers owning large areas are prone to use drones to spray pesticides. Drones are quicker and they accomplish tasks with greater ease and human contact with pesticides is least. Drones may at times spray larger quantities of pesticides, if they are not guided by GPS and digitized aerial imagery. Pesticide spray may also overlap leading to detrimental effects on crops and quality of product. Pesticide distribution may not be uniform if wind speeds and directions vary during the drone operation. It is said feed-back regarding exact areas to be sprayed and avoiding overlaps is necessary. Hence, a wireless sensor network that allows farmers to operate the drone and to spray in areas where pests are conspicuous is required. Wireless network is actually used to guide the drone flight accurately into areas that need spray and avoid overlaps. Accurate communications to the drone operator is crucial. Generally, adjustment of drone path based on wireless feed-back and prompting helps in reducing quantity of pesticide used, rather drastically (Costa, 2012). Ruen (2012a) states that copter flight for 20 minute, covers an area of 200 ac field with soybean/maize and it provides imagery with 3.5 inch pxl^{-1}. This degree of resolution of cameras was good enough to identify crop plants attacked by insects. However, if cameras with ability for greater resolution are fitted, drones can help us get clear pictures of actual insects either hovering above or sitting on crops.

Researchers at Minnesota Agricultural Experimental Station have examined the feasibility of using drones to collect data on insect attack and its spread in soybean farms. Drones have been used to scout and pinpoint areas afflicted with aphid. Spectral reflectance of aphid attacked and resistant soybean zones were compared and studied. The idea is to estimate the area and intensity of aphid attack on soybean canopy/leaves. This data is useful when directing UAVs' flight path and guiding it to only spots that need pesticide spray. The insecticide release and spray intensity using variable–rate spray nozzles on octocopters depends on spectral details of soybean crop.

In the Texas High Plains, so far, aerial imagery from piloted aircraft and satellites has proved useful to detect and judge general vegetation. Crop production zones are also studied using such aerial surveys.

However, crux of the situation in farming zones is that aerial imagery that detects and differentiates cotton farms and natural vegetation is necessary. The cotton farms are affected by volunteers and boll weevil. Therefore, imagery using drones that fly low above the crop, that helps to detect insect attacked areas and identifies weeds/volunteers is very useful. Both ground-based and drone supplied multi-spectral reflectance data could be processed to arrive at accurate spray schedules and volunteer culling programs. Hoffman (2013) states that if multi-spectral data from drones or other sources are not used, cotton plants affected with Boll weevil may actually be overlooked. The spectral signatures of healthy cotton plants, and boll weevil affected plants has to be identified correctly.

Farmers have used drones in the Great Plains of North America. For example, a CropCam, was used to study the infestation and spread of root worm attack on corn fields. They divided the 200 acre crop field with soybean and corn into management strips and studied the effect of spraying 'Aztec' insecticide. Drone derived imagery helped in identification of crops that lodge due to insect attack and resistant genotypes. Lodging could be easily detected using visual and NIR imagery of corn crop. It is important to note that drone-based surveillance for insect attack is rapid and imagery is accurate at 3.5 inch pixel^{-1}. About 200 acres of maize was surveyed in a matter half hour flight over the crop (AUVSIAdmin, 2012).

Regarding spray of pesticides/fungicides using drones, Huang et al. (2013) state that it took a few years to develop RMAX helicopter and then to fit it with ultra-low volume spray nozzles and developing remote control systems to guide the drones to spray into exact locations, within a crop field. Commercial experiments to test efficiency of R-MAX began when it was fitted with both liquid and granular pesticide dispersal systems. Firstly, a low volume spray system with a tank that holds 22.7 kg pesticide payload was developed and used. Further, integration of digital maps of insect attack of crops with GPS and pesticide spray system has given good control of pests. Drones could be useful in Site-Specific Management of pests and diseases. Several procedures are accomplished at the touch of computer screen. We can mark the locations in a crop field and decide quantity of pesticide to be applied at each location by the drone. In addition, drones avoid excessive costs on human scouts/labor.

The agricultural zone in China, like many other nations, is now poised to experience perhaps a conspicuous change in the way farmers scout the crops. Crops are scouted for health, pest and disease incidence, using UAVs. For example, 'Minghe UAV' is a small drone that is being deployed in many locations within the cropping belts of China. This drone is low level surveillance and crop dusting vehicle. It can carry a payload of 12 kg that includes insecticide (dust or fluid) for spray (Minghe, 2014).

During past two years, Beijing Agricultural Bureau has conducted field trials with home-made multi-copter drones that are flown over field crops such as wheat, groundnut and vegetables. These home-made drones are used to spread pesticides on wheat crop during mid and late stages of growth. Reports suggest that use of drones to spray pesticides has effectively controlled aphids and powdery mildew in Shunyi district. During past two years, experimental demonstrations of pest detection and sprays using home-made drones has clearly shown effective control of pests/diseases and economic benefits. In some plots, grain yield gain of 10.1% over fields not exposed to drone mediated sprays were observed (Ministry of Agriculture, 2013).

Over all, there are several advantages of using drones to spray pesticides/fungicides. It is said that spraying crops using drones is basically low cost operation compared to traditional manned aircrafts or ground-based skilled human labor using spray cans. Spraying using drones is easier. It can be guided using computers/remote control tablets. Drones have short takeoff and do not require runway unlike a piloted larger aircraft. Drones can negotiate and fly low at 10 m to 50 m above crop and are highly flexible, even if adopted in undulated or hilly terrain. Precise and accurate spraying is possible using digitized maps of insect/pest attack. Fertilizer or pesticide distribution can be exact in areas and as required or decided by computers. Pesticide or fungicide spraying using drones is highly efficient. About 1300 m^2 area is covered per minute with uniform spray using atomizers/nozzles. Drones could spray over 10, 300 m^2 land per flight in a matter of 8–10 minutes of uninterrupted flight. The flight endurance could be enhanced and pesticide tanks made larger, if the aim is to cover larger area at one stretch. At 35 kg pay load and 10 kg pesticide tank a drone runs for 8–10 minutes. Reports suggest that spray of pesticide and uniformity achieved is excellent. Some copter drones are extra efficient in pesticide spray, since the air-flow and disturbance of crop plants that it generates makes pesticide reach even under the canopy effectively. Depending on nozzles, a fine mist of pesticide that settles on the crop leaf can be created. Most importantly, pesticide fluid/granules required gets lessened, if drones are directed to spray the chemical, only in the areas afflicted with disease/pest.

3.4.5 DRONES, IRRIGATION AND INTERCULTURAL OPERATIONS

Remote sensing using satellite imagery and piloted aircrafts has offered insight into soil water storage, moisture distribution in agricultural fields, crop water status and water losses due to surface erosion of soil, if any. These procedures have certain problems related to limits to resolution of imagery, cloud cover and cost of procuring digitized information. However, during recent years, imagery derived from low-flying drones has been used to assess crop water status and requirements. Drones provide more accurate data. Imagery is derived from close range of just few feet above soil/crop surface and it is of high resolution. Most importantly, information on crop water status, need for irrigation and water resources could be obtained easily at any time by flying the drones fitted with visual and thermal imagery cameras.

In Oregon, drones have been used to detect water deficiency and its effect on crop growth. They have used aerial imagery and general crop scouting. Walker (2014) states that, drones help farmers to detect spots with water deficiency and retarded crop growth rather quickly. Drones rapidly conduct field survey of say 1000 ac day^{-1}. In comparison, human scouts have to walk all through the farm for several days. Drones actually help farmers in channeling water (irrigation) exactly and in accurate quantities to locations affected by drought. So, drones are economically advantageous. They reduce on cost of irrigation. Blanket prescriptions for irrigation are really not needed.

A collaborative effort between Scientists from Andalusian Institute for Sustainable Agriculture and University of California is focusing on use of drones and multispectral thermal images, to study water status of crops and its relevance to regulation of irrigation. Drones have been flown over different horticultural crops such as nuts, vineyards, oranges and field crops such as wheat, maize and barley. The multispectral imagery helps in tracing water status of crops at each location, rather accurately (Innova, 2009). Scientists suggest that sharp thermal imagery helps in detecting water deficit zones, excessive use of water and even water leaks from the canopy. A combination of data sets on canopy temperature, crop transpiration rates and water status helps in detecting, if there is water deficit. The information on water status of crops can be directed to irrigation systems, so that water is applied accurately and only as required. One of the arguments about use of drones to detect water need of crop is that, aerial imagery captures information on water that has entered roots and into crop plants. It is this fraction of water that has direct relevance to crop growth and productivity. On the contrary, soil scientists who study soils in depth and use variety of equipment and equations understanding dynamics of water, after all get confused a bit more than necessary. It is water in crop plants that drones measure. Farmers could decide on water supply based on this data. A few farmers growing peach and grapes have expressed satisfaction at use of drone-based thermal imagery for determination of water supply to crops. Infrared technology and sensors have offered precise recommendations for water supply (Innova, 2009). Recent report about usage of drone named 'VIPtero' in Central Italy suggests that such small robots could obtain pictures of soil type, grape vine and its vigor. Multispectral images could be used to decide on nutrient and irrigation inputs (Primecerio et al., 2012). Light weight drones fitted with L-band radiometer have been used to study the soil moisture retrieval and crop water status (Acevo-Herrera et al., 2010).

Grape vineyards in Tasmania have been exposed to use of drones. Drones have accrued data on surface characteristics, particularly, canopy, leaf area and soil moisture. Based on it, farmers prioritized irrigation automatically. Turner et al. (2012) report that drones fitted with thermal imagery cameras capable of making digital pictures at 875–930 nm wavelength width are useful in providing information, about soil moisture distribution in the vineyards. Drip irrigation system could then be triggered using computer controls. This system allows farmers to decide on irrigation rapidly using digital crop surface models and yield goals envisaged.

Further, we may note that periodic drone flights will allow farmers to monitor grapevine vigor, health and fruit bearing trends in response to decision they took regarding irrigation. They can revise quantity and timing of irrigation mid-way, if needed. At this juncture, we have to note that drones could be programed to take definite flight path, assess soil moisture and crop growth. Next, digital data could be used to electronically trigger drip irrigation. These are clear steps towards the concept- 'Push Button Agriculture.' Human drudgery during scouting for crop growth variation, canopy traits, soil moisture dearth and its variability is removed. Similarly, skilled human labor required to irrigate crops gets enormously reduced. Farmers gain by reducing drudgery and labor costs. Inaccuracies related to human labor also gets removed. Hard tasks are after all accomplished at the push of a button!

Scientists in Cordoba, Spain have utilized UAVs to surveillance olive orchards using visual and thermal imagery. The aim was to study the canopy conductance (CC), its temperature, air movement, net radiation, aerodynamic resistance, stomatal conductance and most importantly water status of crops (WSC) and supporting soil profile. Together, such data gets analyzed using appropriate programs to establish maps showing water stress, if any. It is interesting to note that olive trees with different water status show proportionate changes in canopy conductance. There are models that allow calculation of water stress index of each olive tree. Crop water stress index could be utilized to trigger drip irrigation. Over all, imagery accrued by UAVs can aid development of maps of canopy conductance and Crop water stress index (CWSI) (Berni et al., 2009b). The digitized thermal imagery maps can then be utilized to irrigate the olive orchards appropriately.

Bellvert et al. (2013) have characterized water stress across vineyards found in Cordoba in Spain. It is because, knowledge about soil moisture is essential during precision farming. Farmers should have accurate maps depicting variations in soil moisture and water deficits. They estimated water deficits in vineyards using CWSI. The CWSI was calculated based on thermal imagery obtained using low-flying drones. Firstly, infra-red and thermal imagery correlates with leaf water potential that indicates water status. Next the CWSI is correlated to leaf water potential. The timing of measurement of thermal imagery by drones seems to affect the correlation between CWSI and leaf water potential. The correlation (R^2) between CWSI and leaf water potential was 0.46 at 9.30 hours but quite high at 0.77 if measured at 12.30 hr. Further, sensitivity of thermal imagery measured at 0.3 m pixel^{-1} was needed to precisely map CWSI. Lower resolutions at 1.0 or 2.0 m pixel^{-1} did not provide good correlation. Such CWSI maps could then be used to supply water at variable rates under precision farming systems. Drones have been used to monitor leaf growth, canopy size and in estimating leaf carotenoid content in vine yards found in Cordoba, Spain (Zarco-Tejada et al., 2013).

3.4.6 DRONES TO SCOUT FOR WEEDS AND WEEDICIDE APPLICATION

Drones have the ability to scout crop fields accurately and entirely in a relatively short period. Hence, they have also been tested for identifying weeds and weed infestation intensity in fields. Discrimination of crops and weeds is a crucial issue. Satellite imagery and photographs derived using piloted aircrafts have also been utilized to identify weed infestation in crops fields. Satellite imagery provides only low resolution imagery using which weeds and crops have to be located. Conventional aircraft derived pictures could turn out to be costly and it is not easy to cover large field rapidly. Therefore, Pefia et al. (2013) suggest that drones with robust cameras that distinguish weeds of various species, trace their location and present a clear view of their intensity is required. Further, they have stated that researchers at Institute of Sustainable Agriculture, CSIC, Spain have developed drones fitted with automatic object-based image analysis systems. The imaging system has visible and NIR cameras. These drones are capable of discriminating weeds from crops on the basis of spectral reflectance and position in crop rows. The computers are endowed with spectral information of large number of weed species and classified groups. This information helps them to identify weeds (Jones et al., 2006). The drone imagery also generates weed infestation map on a grid. It suggests the intensity of different weeds found on the ground. It is interesting to know that studies that compared weed infestation data derived from drone's imagery and actually measured on ground by skilled farm technicians correlated ($R^2 = 0.89$). Further, drone imagery groups weed infestation maps. The accuracy of weed maps showing intensity of infestation was found to be high. Therefore, spray schedules and type of weedicide that needs to be sprayed could be selected appropriately. Drones can help farmers to organize weed spray schedules. Since, farmers are provided with weed maps ahead, they can concentrate and apply weedicides only where required. Most interesting, in the present context, is the ability of drones to guide a ground robot with ability to spray herbicides exactly to locations with weed infestation. Robots could release herbicides in small quantities in location with few weeds and more in those with higher intensity of infestation. Matching weed intensity maps with herbicide application rates by ground robots economizes on chemical usage, avoids undue contamination and gives greater economic advantage to farmers. Most of the above techniques and agronomic procedures involve just a push of button! or a touch on computer/android's screen!.

Drones or Unmanned Aerial Vehicles fitted with multispectral cameras have been used to map the weed infestation in cropping zones of North America. For example, Torres-Sanchez et al. (2013) have used a MD$-1000 VTOL quadcopter produced by Micro-drones Gmbh. It was equipped with GPS, software for weed detection, telemetry and two cameras (Olympus Pen E mp1 and Tetracam Mini-MCA6) to judge weed intensity. Such drones provide site-specific data about weed distribution, diversity and possible eradication methods. High resolution spectral data about weed species is useful, while operating a variable-rate herbicide applicator in the field.

Drones have been utilized in the crop fields of Denmark to identify weed infestation. Drones flying past the crop just a few meters above it, detect weeds using variations in reflectance patterns of weed and crop plants. Drones pick up the spectral signatures of weeds and crops and discriminate them using appropriate computer programs. Drones fitted with GPS actually relay exact coordinates of areas that are affected by weed infestation. The drones identify dense weed growth, moderate levels and lighter or scattered infestation by weeds. This information is highly useful to farmers since they can channel herbicides in quantities that match the intensity of infestation. Regions without weeds will not receive herbicides. This is unlike blanket applications of herbicides that consume higher quantity of chemicals. Herbicides reach even spots not needing them. Agricultural scientists at the Aarlog University, in Denmark state that usage of drone for weed control is easy, yet it needs a skilled farm worker to control flight path of the drone and apply computer programs so that herbicide spray by ground robots, if any, is accurate (New Scientist Tech, 2013).

Reports from China suggest that drones have been useful in identifying weed infestation in pastures and turfs in the Northeast Plains. Xiang and Tian (2011a) have utilized a drone helicopter of < 14 kg in weight, fitted with multi-spectral cameras to detect weed infested zones within pastures and turfs. A ground station with facility for computer that controls flight path or programs flight path was erected. The computers decode the digital images relayed by the low-flying drone and suggest exact location and quantity of glyphosate (weedicide) to be applied at variable rates (Xiang and Tian, 2011a, b). A skilled farm worker controls the *'push button techniques'* that help in spraying weedicides to vast stretches of turfs and pastures. In the absence of drone-based aerial imagery, scouting for weed spread and accurate application of weedicide is a tedious task and it is costly by many folds.

Rice culture in the tropical regions of Asia is affected by weeds that interfere with crop growth and yield formation. Weeds divert soil nutrients and water meant for rice crop. Weed detection at an early stage of crop and adopting appropriate precision herbicide sprays is essential. There are ground robots that detect and remove weeds physically or spray weedicide with a degree of accuracy at each spot or even a single weed plant. These are time consuming methods. The robots traverse slowly across the large field. Low altitude remote sensing methods using drones have been excellent in detecting weed infestation and mapping their occurrence and intensity. Okamoto et al. (2007) have used hyperspectral cameras to detect weeds based on their spectral signatures. They have used a Specim ImSpector V9 imaging spectrograph mounted on a tractor to control the weeds on-the-go. Often, on-the-go spectral analysis is slow. As an alternative, drone-based digital imaging using multispectral images could be used to map weeds. Digitized maps of weed type, infestation and intensity could then be used to direct variable-rate applicators.

Drones, mainly copters are in vogue in cropping zones of Australia to accomplish weeding (Cornett, 2013). Weed control is achieved first by identifying areas with infestation, then by spraying using low flying pilot-less copters. Such aerial vehicles that act as both weed searchers and spray drones are utilized effectively in few agrarian regions (Meier, 2014)

3.4.7 DRONES IN CROP YIELD FORECASTING

Crop yield forecasting is an important aspect of any farming enterprise. Often, several inputs such as seeds, planting density, fertilizers, irrigation, pesticides and herbicides are dependent on forage/grain yield prediction. Yield predictions could be achieved using tedious and detailed analysis of soil fertility and moisture parameters, weather reports, crop genotype and its phenomic attributes at intervals during the crop season. There are several reports suggesting that crop growth, canopy reflectance and NDVI measurements obtained using stationary *in situ* spectro-photographers could be used to forecast crop yield. Drones fitted with visual, NIR and Red-edge bandwidth cameras also provide data about NDVI. The NDVI data from aerial imagery of drones and those obtained by ground-based measurements correlate. This relationship could be used to forecast for-age/grain yield. Prediction of crop yield using remote sensing has been achieved for several field and fruit crops (Alvaro et al., 2007). For exam-ple, in case of rice, satellite imagery has been used for wide scale yield prediction (Chang et al., 2005; see Table 3.4). Drone derived NDVI val-ues have been used to predict forage/grain yield of corn (Chang et al., 2003; Kahabka et al., 2004), cotton (Thomasson et al., 2000); wheat

TABLE 3.4 Relationship Between Normalized Difference Vegetative Index and Rice biomass, Grain Yield and Grain Protein Content

Relationship	Equation	Correlation ($R^{2)}$)
NDVI$_{LARS}$ vs Biomass	y = 31.851x-23.837	0.7598
NDVI$_{LARS}$ vs Grain yield	y = 22.753–18.342	0.7283
NDVI$_{LARS}$ vs Grain Protein content	y = 0.1593x-0.0277	0.5919

Source: Swain and Zaman (2010);

Note: LARS = Low Altitude Remote Sensing using drones; NDVI $_{LARS}$ based prediction are correlated significantly with crop biomass and grain yield but less so with grain protein.

(Doraiswamy et al., 2003), citrus (Zaman et al., 2006), blue berry (Zaman et al., 2010) and tea (Rama Rao et al., 2007).

No doubt, aerial imagery of crops obtained using drones has found application in many ways. Yield forecasting is most common utility of NDVI. However, NDVI values derived from drones have also been correlated with ground-based actual measurements of yield. This has helped in preparing yield maps of crops rapidly. Drones have supplied us with useful data about NDVI and its relation to crop response to inherent soil-N and fertilizer-N rates. We may note that yield maps of different genotypes of a crop required during experimental evaluation of crop genotypes has also been accomplished using drones. Drones are used to identify crop genotypes based on their spectral reflectance and their NDVI is mapped, then, yield is forecasted based on aerial data. It is said, we can also forecast crop growth yield by developing mathematical relationships between Electrical Conductivity (EC) values of soil and NDVI maps derived from drones. Incidentally, EC is correlated to crop growth and grain yield of many crop species. With regard to commercial crop production, we may note that currently there are several drone companies that regularly obtain aerial imagery of farms and provide forecasts regarding grain productivity.

3.4.8 DRONES IN HIGH INPUT SEED PRODUCTION FARMS

A point to note during high quality seed production is that, it is not just sufficient if seeds of good pedigree and quality are sown. To obtain excellent crop stand, growth and biomass accumulation, it is necessary to strictly follow the agronomic procedures, time them accurately and finish the task rapidly, in time. Drones and robotics are said to be very helpful in obtaining aerial images rapidly, so that decisions on soil and crop management are accurate within a field. Robotics could help in finishing the task with ease and in time. For example, Funk Seeds Inc., in USA is currently using drones to scout their seed production fields, rather extensively. The funk seeds also supplies notes on how to read digitized aerial imagery and convert them into proper agronomic actions (Funk Seeds Inc., 2014). Further, enrolling some or all of your acres in 'Yield Launch' will give farmers the ability to manage both the field and the crop throughout its entire life cycle. Drones launched by the seed company, keep the farmers abreast of happenings in his farm through digitized aerial imagery that are transmitted to farmer's computers. Suggestions on important farm operations are made based on analysis of digitized pictures. Farmers can also take advantage of drone surveillance of their farms.

Drones are very effective tools in experimental stations that evaluate large number of crop genotypes, prior to their release to farmers. Researchers monitor growth pattern, leaf area, canopy characters, seed set, and maturity using drones. Drone based crop evaluations are rapid and provide an overall picture to evaluators. Drones are excellent in acquiring data about crop phenomics rapidly, routinely and at definite intervals during a season. Drones, actually replace thousands of hours of hard work, drudgery and tedious collection of data by skilled farm technicians. The UAV photo system is very easy to use to compare genotypes. Farmers could select crop genotypes for next season rather accurately, after considering several of their own preferences (Editor, 2014).

Crop phenotyping is an essential aspect of evaluation of crop genotypes by geneticists. Plant breeders select genotypes with best phenotypic expression under different stress conditions, such as water deficit, disease or pest pressure. Phenotyping is actually an indirect way of assessing and authenticating the genetic constitution of the crop. Seed farms will rely greatly on drones to rapidly collect data on phenotypic characters periodically. So, in due course, we may expect drones in most of the agricultural experimental farms that evaluate crop genotypes and conduct location trails, prior to release of seeds for open cultivation. Drones collect data, say, twice weekly about morpho-genetics (Dreiling, 2012). Crop genetic evaluation will be that much efficient and rapid, if drones with high resolution cameras are used in place of human scouts and skilled crop evaluators. As a corollary, in a farm show, farmers could easily pinpoint genotype of their choice on computer screen or tablet by using their connectivity to drone relayed surveillance pictures. Stored imagery of crop's progress helps farmers to make better selection after due comparison. Crop breeders will be better placed while deciding on crosses, if they are able to get a clear overall view of different germplasm lines, simultaneously.

In South Africa, drones were introduced into high input seed farms to monitor crop growth, scout for diseases/pests and seed set. Drones were in fact deployed daily to surveillance the crop to decide on irrigation and chemical sprays. Prior to it, farming agencies were chartering aircraft or satellite agencies to provide aerial imagery. Drones were easy to deploy and relatively less costly (Hetterick, 2013).

3.4.9 DRONES IN PRECISION FARMING

Precision Farming is getting due attention from farmers across in several agrarian zones, because of a series of advantages attributed to procedures adopted. Forecasts suggest that in future, success or failure of Precision Farming techniques may largely be governed by the extent of use of drones in performing several of the agronomic tasks. Farmers are already using satellite derived data to accomplish variable-rate techniques. Yet, satellite techniques lack the same extent of accuracy and rapidity with which UAVs can operate in a large farm and perform similar procedures. Precision Farming, as the name suggests, depends to great extent on accurate data collections. Most soil and crop management procedures followed are in response to data collected on ground, through satellites, UAVs, hand-held gadgets, tractors mounted with sensors or stationery instruments perched at vantage points in the field. Data collection meant for precision farming actually begins with yield maps of the previous crop grown in the same area. It tells the farmer about variations in grain harvests, in response to soil fertility and agronomic procedures adopted. It also suggests about the areas that have been depleted of soil nutrients to a greater extent than others. High yield means more nutrients extracted from soil. Satellite imagery provides imagery of soil/crops at visual and NIR wavelength. It provides normalized difference vegetative index. Aerial Imagery using UAVs provides very accurate and close-up shots of soil, crop and insect/disease attack, if any. It is very useful and could be used on-the-go, if integrated or used as digitized data on tractors. Data about crops could also be collected using tractors mounted with sensors. Sensors could measure soil pH, electrical conductivity (EC), organic matter distribution, etc. Skilled farm workers do collect data using hand held gadgets such as leaf chlorophyll meters, photosynthetic light interception meters, soil moisture meters, etc. No doubt, collecting accurate, reliable and geo-referenced data, that gets channeled to tractors (manned or robotic), planters, variable-rate applicators and sprinklers, form the crux of 'Push Button Agriculture' touted in this book.

Drones have the ability to surveillance and store digitized pictures of crop growth. Those fitted with tanks to hold dissolved nutrients or pesticides and connected to GPS and computer-based decision system could also be programed to spray exact quantities to crop canopies. Huang et al. (2010a) state that drones could be fitted with cameras/sensors with super-resolution and used in site-specific management of crops. Drones have the ability to efficiently cover large areas of crop. These drones are fitted with sharp cameras plus low volume accurate spray nozzles. The integration of knowledge about crop mapping, spectral reflectance, crop nutrient deficiency and spray systems using a computer-based decision support allows for rapid accomplishment of nutrient application. Drones with computer decision support systems will swiftly work out foliar application of nutrients (mainly top dress-N) at different rates, at appropriate time and location using precision farming principles. This is not an easy task for skilled human scouts. It is actually a strenuous and time-consuming one. In addition, human errors creep into farm operations. *Most important, in view of context of this book is that, drones help farmers in accomplishing a difficult task at the push of button plus remove enormous amounts of human drudgery.* In fact, Ehmke (2013) has stated that drones (UAVs) and precision agriculture are a natural fit, if farmer's aim is to maximize crop productivity and still keep inputs at optimum. Drones are becoming popular in farms adopting Precision Agricultural Techniques (Pates, 2014). Drones are perfect technology to accomplish agricultural tasks that are dirty, tedious and require long hours of drudgery, dangerous, difficult and dull. Since drones fly very close to the crop they offer better accuracy that is hall mark of precision farming approaches.

Tokekar et al. (2013) have discussed a novel method of symbiotic action between the drone (UAV) and ground-based robot (UGV). The communication between UGV and UAV about soil-N distribution and possible accurate application of fertilizer-N is the center piece of this system. The ground and aerial measurements is used to guide the drone to fly and map the crop growth pattern, chlorophyll and plant-N status. The ground robot moves to zones and spots to apply exact quantities of fertilizer-N as directed by the computer-based decision support system.

Precision farming using such a combination of UAV and UGV with inter-vehicle communication helps farmers in applying nutrients only at spots showing deficiency. This is a good example for 'Push Button Agriculture.'

One of the earliest demonstrations of UAVs capable of performing variable-rate applications on fields kept under site-specific farming techniques occurred at Decateur, in Illinois, USA. Drones fitted with cameras, fertilizer tanks, spray nozzles, computer software to integrate with GPS and flying very low above crops could spray nutrients to foliage. Murray (2013) predicts that drones will be increasingly used to prepare digitized versions of close-ups, so that they could be used on-the-go. They may be used with computer-decision support systems to issue commands to variable-rate applicators. Actually, drones allow higher accuracy compared to remote sensed imagery from satellites.

Regarding 'Big Data and Drones'; it is a massive and complex database that is useful to farmers adopting precision farming procedures. The database is actually built up using ground sensors, robots and UAVs. It includes multi-dimensional digitized imagery of soils, crops and environmental data. This 'Big Data' is a repository of knowledge for making decisions during site-specific or precision farming. Computer-based decision support systems on drones utilize 'Big Data' to arrive at accurate dosages of fertilizers or pesticides.

For specific farms and purposes, there are companies that produce agricultural drones such as Trimble Inc., (California, USA) that specializes in aerial imagery and delivery software. Monsanto's 'Earth map solutions' provides remote sensed data about chlorophyll content of cropping expanses at any time. Autocopter Corporation flies drones and captures detailed data, high resolution imagery of crop fields and even video of crop development. Maps of fields with crops are converted into distortion-free maps. These are supplied to farmers to manage their fields accordingly (Ruen, 2012b).

Big data collection, it seems, is necessary in many of the agriculturally developing countries and those depending predominantly on efficient and large scale crop production. In Brazil, researchers from IBM computers suggest that food grain production companies may heavily depend on accuracy and rapidity with which they collect detailed data about crops growing in the field. Precision techniques require large data sets and matching computer programs that aid accurate dispensation of fertilizers, irrigation and pesticides. In this regard, drones may take over major portion of task of collecting periodic data sets from fields, wherein precision farming techniques are adopted. In fact, it has been opined that for an enlarged population that Brazil is expected to harbor by 2050, efficient and highly productive precision techniques supported by large scale use of drones seems almost necessary.

3.4.10 DRONES IN FIELD CROP PRODUCTION

During rice production, like any other crop, periodic scouting, measuring plant characteristics such as height, leaf area index, leaf number, tillering, panicle number, etc. is important. Many of the agronomic procedures such as fertilizer supply, fixing rates of N top-dressings and timing, irrigation scheduling and harvesting, all depend on farmer's ability to collect accurate data about crop growth pattern. Recently, Bending et al. (2013) have reported use of 'Crop Surface Models' and comparing them with data derived from UAVs flown over rice crop periodically. In Northeast China, rice crop is surveyed and data about crop is collected using small or mini-UAVs of less than 5 kg in weight. These drones collect data about environment, crop height, biomass; areas of nutrient deficiency and healthy growth (see Thenkabail et al., 2000, 2002; Hansen and Schjoerring, 2003; Oryza, 2014). These UAVs make non-destructive measurements about crop chlorophyll and leaf-N status. For example, an Okto-Copter fitted with Panasonic Lumix GF 3–12 Megapixel camera can provide excellent digitized data to computers for decision making regarding in-season fertilizer-N supply. Top dressing using liquid or granular fertilizer-N is accomplished using drones. The variable-rate applicators are provided directions by comparing 'Crop Surface Models' with data accrued by the drones. Farmers are able to decide flight path that can be modified midway, if needed.

Drone usage during rice production is perhaps more pronounced in Japan than in any other location of rice cropping zone. Bennett (2013) states that, in 2010, 30% of rice fields in Japan were observed aerially and sprayed with pesticides using copters. Tadasi et al. (2010) have opined that deploying UAVs is apt when dealing with large rice production zones. In addition, drones could be excellent in accruing data about experimental evaluation of rice genotypes. Drones are used to obtain detailed data on plant growth, leaf area, canopy, leaf nitrogen and disease affliction if any. Since drones could be operated on larger area, we can identify rice genotypes that perform better and offer good quality grains. In fact, entire field could be monitored all through the season and then data could be pooled and analyzed to pick best rice genotype. Tadasi et al. (2010) further state that in Japan, they have examined over 55 individual fields using drones just to choose a rice genotype with good quality grain. Drones fitted with Red-Green-Blue and NIR band width cameras were used for assessing the paddy crop. Drones have actually helped researchers to decrease use of fertilizers, yet reach same yield goal. Further, it has been stated that use of drones is cheaper. They are more flexible and offer high resolution imagery compared with piloted aircrafts or satellite-mediated imagery. Forecasts, suggest that in addition to evaluating rice genotypes to select the best one, we can use drones to assess different rice-based cropping systems. The biomass, grain yield and economic advantages of the entire cropping system could be assessed using drones.

Italy is not an important rice producing country. Yet we have a good example on the use of drones in the rice production on farms as large as 160 ha. Drones have proved useful in reducing cost on fertilizers and chemicals. Fertilizer inputs have been reduced by 15%, about 6 tons less than the usual 41 tons applied on the entire 160 ha. Drones fly over the field on pre-determined flight paths. They study the crop using visual, NIR and thermal imagery. Drones actually transmit information on vigor of the rice crop to a computer that decodes the data and releases only correct amount of fertilizers that need to be applied at different points in the field. Fertilizer supply is made based on precision farming techniques-, for example, variable-rate application of fertilizers. This ensures uniform rice crop productivity.

A more interesting application of drones in rice culture is the fact that recently researchers in Georgia, USA have adopted them to investigate historical rice culture trends in United States of America. They have used Terrestrial laser scanning, LiDAR and UAVs with visual, NIR, red-edge and thermal imagery accessories to scrutinize the Georgian swamps, uplands and coastal plains to ascertain, if rice was grown in the area. Results indicate, that early African slave migration brought with them rice cultivation, mainly upland varieties and those adapted to swamps. Drone technology has produced 3-D imagery of erstwhile rice production zones now abandoned as swamps (Pasqua, 2013). Drones easily move and hover around thick swamps where skilled human labor may not venture at all. Plus, drones provide images from vantage locations above the swamps not generally possible for human scouts. Drones detect rice crop using their spectral signatures.

Drones are currently providing useful photographs produced using NIR-Green-Blue wave length bands. Aerial imagery about field lay-out, crop canopy, leaf chlorophyll and leaf N status are helpful to wheat farmers (Raymond Hunt et al., 2010). Maize researchers from International Maize and Wheat Centre (CIMMYT, Mexico) state that, currently, drones are serving useful purpose in assessing genotypes and in maximizing grain productivity and economic advantages to farmers and commercial farms in Zimbabwe. Mortimer (2013), in fact, prefers to call drone project as 'Skywalker- an aeronautical technology.' Skywalker is a complex phenotyping drone that is fitted with advanced flight control systems and computers that guide it to automatically trace programed flight paths. Skywalker could be programed to fly over each maize genotype of interest and those offering high quality grains at enhanced productivity. Drones are fitted with multi-spectral imaging cameras, mainly visible, NIR and thermal band widths. Soil moisture status too could be assessed using Red-edge band width. Researchers at CIMMYT forecast that drone technology, firstly, allows them to accrue large amount of detailed data on maize phenomics, then it can hasten decoding them using appropriate computer programs and finally decision support systems can enhance maize productivity. A 'Skywalker' flying at 45 km h^{-1} speed for 45–60 minutes over maize fields can cover a very large area in short time. It can accrue accurate data about crop phenology. The greatest advantage it seems pertains to efficient use of soil-N, fertilizer-N and irrigation water. Fertilizer and pesticide requirement of maize grown using drone technology is markedly lesser than traditional agronomic methods. Skywalker may also be used to identify maize genotypes that withstand drought effects better than others. Over all, airborne remote sensing platforms such as helium balloon, parachutes and UAVs may all hold a good promise to evaluate maize genotypes and conduct agronomic procedures with greater ease and accuracy. Lumpkin (2012) points out that drones are fast and non-destructive, while accruing data on maize phenomics. Drones avoid excessive ground-based measurement by human skilled personnel. The aerial imagery from drones is of higher resolution than satellite imagery. Drones could be rapidly deployed and data derived could be almost instantaneously processed by computer decision-support systems. Drone technology could also be employed in conjunction with ground-based leaf color and chlorophyll meters that help in assessing crop-N requirements. In Illinois, corn producers have now started using drone derived imagery to detect regions that are healthy and those afflicted with root worm. It seems farmers can survey about 200 ac corn field for insect attack in a matter of 20 min flight over the crop (AUVSIAdmin, 2012).

Wheat production techniques, particularly fertilizer-N and its effects on crop growth, foliage-N, canopy characteristics and grain productivity has been studied using drones and aerial imagery. Jensen et al. (2007) have reported that digitized data collected using visual, NIR and thermal imagery correlates excellently with crop's response to fertilizer-N, in terms of forage and grain production. Wheat production, in general, could be managed using drones to surveillance the crop for growth rate, assess leaf area, photosynthesis, biomass accumulation pattern, disease/pest incidence and finally grain yield. Many of the agronomic procedures could be organized based on in-season data derived using aerial imagery by drones.

A report from Spanish wheat belt suggests that it is not feasible to use piloted aircraft or satellite derived imagery to evaluate wheat crop and devise agronomic procedures based on data derived from these sources. The spatial resolution, timing and costs may not be congenial for farmers. Wheat farmers are better placed if they use low-flying copters/fixed-wing drones that are fitted with high resolution cameras that pick images at visual, NIR and Red-edge ranges (Torres-Sanchez et al., 2014). This helps to study the wheat crop, its growth, canopy size, leaf area and leaf chlorophyll, leaf-N status and soil/crop moisture status. For ready reference, farmers may preserve maps of NDVI of wheat fields. The copter drones are well suited when wheat farmers intend to obtain in-season data on crop phenomics at periodic intervals say at seedling, tillering, anthesis and grain fill stages (Torres-Sanchez et al., 2014). Finally, data from aerial imagery can be effectively used in precision farming. This allows farmers to save on inputs and obtain uniform crop grain yield. Above examples clearly indicate a trend that leads to 'Push Button Agriculture.' Drone technology progressively reduces drudgery and need for skilled human labor in farms.

3.4.11 DRONES IN HORTICULTURAL CROP PRODUCTION: A FEW EXAMPLES

Drones are now accepted as part of vehicles and implements necessary to manage fruit orchards. They serve the planters in accomplishing variety of tasks, beginning with clearing a natural growth, providing farmers with map depicting topography and contours, so that preparation of land for planting is accurate. Agricultural drones are revolutionizing grape vine and orchard maintenance in California (Paskulin, 2013). Decision regarding timing and intensity of agronomic procedures are being guided increasingly by referring to aerial imagery of orchards. Aerial images of orchards, although available from agencies with access to satellite, owning a drone allows farmers to obtain imagery as many times at different intervals and at very low cost. For example, Kunde (2013) opines that aerial imagery of grape vines indirectly suggests regarding soil depth, fertility and water distribution in the entire field. It helps trace disease or pest afflicted zones. Further, the 3D imagery taken using a Precision Hawk, a drone, using GPS connectivity and at varying altitudes and latitudes, informs farmers about growth of grape vines. Drone derived imagery that provides a bird's eye view takes less than one hr to scout 50–80 ac. However, interesting is the fact that a 3D picture of farm helps farmer in judging on the ripeness of grape bunches and so the area that needs to be harvested can be demarcated on a computer with touch of the screen. There are ground robots (e.g., Wall-Ye) that harvest specific grape vines based on GPS signals and decision-support system. An autonomous ground robot such as Wall-Ye has to be interconnected to take signals from drone. Clearly, drones aid farmers to reach a step further towards the concept of 'Push Button Agriculture.'

Several field trials have been conducted in the Californian vineyards using RMAX pilotless copters. They have aimed at ascertaining pest attack and weed infestation in the vineyards first. After locating the infested zones, RMAX copters with tanks filled with pesticides/herbicides are operated or programed to apply the chemicals. A RMAX copter is said to carry two tanks on either side of the fuselage. The tanks carry 4 gallons of pesticides that could be released selectively at points infested with pests. At full spray, a RMAX copter can unload pesticide for 15 minutes and cover an area of 12 acres. It is indeed very rapid compared to human labor operating pesticide sprayers all over the field and applying based on blanket recommendations over the entire field. It is much faster than a tractor attached spray equipment (Cornett, 2013). In Bordeaux, France, farmers are adopting drone technology to monitor vineyards. Magrez (2014) states that drone fitted with visual and Infra-Red Cameras is currently used as a measurement and management tool. They are used to diagnose plant growth characteristics through aerial photography. Drones are used primarily to study soil deterioration, if any, erosion and loss of top soil and fertility that results in poor grape vine growth. Periodic surveillance helps farmers to apply fertilizers exactly in zones that need improvement in soil fertility. Soil management and replanting of vines is easily accomplished using aerial photography by drones compared to scouting by skilled farm workers.

Turner et al. (2012) have described a novel and rapid method of using UAVs to decide various agronomic procedures within the grape vineyards of Tasmania in Australia. They have used UAVs fitted with hyper-resolution visual, multispectral, NIR and thermal imagery cameras. Technically, drones fly over the grapevines and collect data related to crop growth, leaf area, chlorophyll content, soil moisture pattern and biomass accumulation. The NDVIs using six different wavelength bands were used to compare growth pattern. Next, they compared the digitized data accrued using drones with some of the established grapevine imagery patterns (Digital Surface Models) and ground-data rapidly, using large stored data pool and computer programs. Farmers can actually taylor their agronomic procedures to reach definite fruit yield goals. This system is rapid and easy to adopt, since known models could be used as standard situations (see Hall and Louis, 2008; Hall et al., 2011; Lamb et al., 2001; Lamb et al., 2013; Turner et al., 2012). Grapevine yards have also been examined simultaneously using both low-flying drone and space-borne satellites. Such imagery may be helpful in measuring and mapping grapevine vigor and its variability across the whole orchard (Lamb et al., 2013).

Citrus production in Florida and other regions of North America is an intensive farming enterprise. It involves tedious scouting of trees for general health, nutritional status, water deficits, diseases and pests. Ehsani et al. (2012) state that, so far, scouting citrus grove in Florida has been a time consuming and costly procedure. Further, manual scouting and recording data without accurate GPS coordinates is indeed error prone. Aerial scouting using remote sensing satellites and/or low-flying drones is gaining in popularity within the citrus industry of Florida. Remote sensing can provide hyperspectral or multiband imagery. The digital imagery can be analyzed using appropriate computer programs and decision support systems could be later activated. The spectral images provide detailed information on general vegetation, vegetation in the citrus grove and NDVI values that enable farmers to judge biomass accumulation and fruit bearing trends. Aerial imagery is also useful in detecting water stress and disease incidence in the citrus grove. Ehsani et al. (2012) state that accuracy of aerial imagery and detection of disease/insect incidence can be enhanced using low flying drones with high resolution cameras. Cameras capable of imagery at visual and Near Infra-Red ranges are used in the drones. The drones being used in Florida citrus belt are small, at best 2–5 pounds in weight and are capable of multi-band imagery. The drones fly at low altitudes of 10 m above the citrus grove and so provide excellent images of high resolution of each citrus tree in the grove. Currently, there are several types of drones that are well suited to study and surveillance citrus groves. Ehsani et al. (2012) further report that it is possible to develop images and analyze each single tree, using HD cameras (2 inch pixel^{-1}). A single citrus tree can be represented by 5000 pixels, if the resolution is 2 inch pixel^{-1}. Drones attached with multi-band sensing facility have been able to detect HLB infected trees. Farmers can then direct their disease control measures to areas in the grove that are affected by HLB or individual trees affect by the disease. Forecast by citrus growers suggest that soon, citrus groves may encounter large number of light weight drones all through the day and even night. Citrus grove surveillance may largely become an activity accomplished by drones. In a different study reported from Citrus belt of Florida, Garcia-Ruiz et al. (2013) have reported that citrus greening disease (Huanglongbing) infected plants could be distinguished from healthy one using multi-image sensors placed on UAVs. They forecast that high resolution imagery using drones could become a common method to identify and later adopt control measures to control HLB disease.

Reports suggest that coffee plantations in Hawaii are being managed using drone technology. The UAVs are providing digitized color pictures of the coffee plantations to the farmers. It seems both small and large coffee plantations are reaping advantages of drone technology. Farmers are helped in finding zones that have ripe coffee beans. In contrast, a standard procedure without aerial imagery just adopts mass harvest of all beans. Farmers have to sort the beans later and grade them. Ripe coffee beans are identified using spectral signatures. Actually, spectral data from two different drones are compared and analyzed to arrive at decision regarding harvest of coffee beans. Ripe fruits are yellow in color and they are picked using spectral signature. Coffee plantations are periodically surveyed for soil moisture and nutrient deficiency if any (Herwitz, 2002). In addition, state agencies in Hawaii provide coffee producers with weekly/fortnightly imagery of entire coffee plantation, so that they can decide on a variety of agronomic procedures that suit best. The digitized pictures are posted to individual coffee farmers via internet. A more recent report by Herwitz et al. (2004) states that UAVs were used to surveillance and take decisions on a 1500 ha coffee plantation. The payloads consisted of visual and NIR imagery. A local area network link was used to control the drone. Images from drone were available instantaneously for interpretation. The imagery was used to locate weed infestation. Photo prints were also used to decide fertilization and irrigation. The drones could be precisely navigated using radio controls to spray liquid fertilizers or herbicides. It has been suggested that periodic flights by drones will be helpful in managing coffee plantations with greater ease and accuracy.

Malaysian peninsula is a major oil palm producing region. The palm plantations require constant surveillance and upkeep with regard to fertilizer supply, irrigation and diseases such as *Ganoderma*. Scouting using human labor is in vogue in many regions. However, there are field trials reported about use of Hyperspectral imagery to scout for water stress if any and disease mapping. The multispectral cameras provide excellent images of palm disease and its extent. Airborne measurements of NDVI have also been employed to detect palm plant vigor and water requirements. Shafri and Hamdan (2009) state that, multispectral imagery conducted at Red-Edge band width, provides excellent data on *Ganoderma* infestation and palm vigor. Airborne measurements using UAVs were accurate (84%) and correlated with disease identification.

3.4.12   DRONES IN FORESTRY, PASTURES RANGELAND AND WASTE LAND MANAGEMENT

During containerized tree seedling production of several types of conifers and other tree species, application of chemicals and water has to match the growth rate, expected enlargement of canopy of individual tree and growth anticipated in the next few weeks or months. Blanket recommendations may either over or under estimate need for nutrients, water and pesticide. Nurseries with larger trees in containers have used light interception techniques, and satellite pictures of trees kept in line in open. Satellite images have helped farmers to estimate average canopy size of large tree saplings, so that, actual quantities of nutrients, water and pesticide could be determined (Jeon et al., 2013). Drones flying close to tree saplings kept in a line could be immensely useful. Drones provide accurate measurements of each and every tree sapling (4–5 year old), provide its GPS coordinates. Nursery men could then decide and supply accurate quantities of chemicals and water. Drones, definitely hasten the aerial survey. They also provide very accurate data. Experimental evaluations and nursery trials are needed to estimate input efficiency and economic gains due to usage of drones.

Reports suggest that Light Detecting and Ranging (LiDAR) technology applied using satellite, airborne vehicles or terrestrial platforms have gained in popularity during recent years. The LiDAR is an excellent and useful technique that helps in managing forests. Currently, LiDAR has been used to estimate forest biomass, canopy cover, leaf area index, tree height, etc. These parameters have been effectively used in computing and forecasting stock volume and timber value in commercial forests. In some cases, low-flying drones have also collected information about biodiversity of forest under story. Drones, it seems have offered cost effective method to monitor and estimate forest productivity. Periodic flights and data collection is necessary. Actually, both mini and micro drones fitted with Micro-Electronic Mechanical Systems (MEMS) have enabled them to capture high resolution data. UAV-borne LiDAR systems are useful in full scale forest management. They are able to monitor forest growth, health, defoliation trends, canopy closure patterns, disease/insect attack, etc. Field trials in Tasmania, Australia, have shown that UAV-borne LiDAR can study *Eucalyptus globulus* forests in great detail. Trees could be monitored even at early growth stages. Aerial imagery could be used to trace growth rate of Eucalyptus, its canopy closure patterns, biomass accumulation pattern, etc. Individual trees could also be observed, depending on its growth and size. Foresters could also study the effects of pruning of individual or group of trees on biomass accumulation, leaf growth, etc. (Terraluma, 2014).

Knowledge about forest biomass increase is essential to planters who wish to adopt different agronomic practices. Tree growth scouting and regular measurements of various growth traits is again tedious, time con-suming and it also costs, since skilled human labor has to be employed. Techniques based on drone usage that may enhance accuracy and has-ten estimations will be sought by foresters. In this regard, Jaakkola et al. (2010) have described a method based on UAVs fitted with laser scanning system, in addition to the usual visual and NIR cameras. They have used a drone with two laser scanners, a CCD camera, a spectrometer and thermal camera. The aim is to measure tree growth, its canopy and leaf area using the multispectral imagery and laser scanners. Tree height of conifer stands have been measured accurately using the drone. The tree height and can-opy characters measured using drones, it seems correlates with biomass ($R^2 = 0.92$). Therefore, foresters could use UAV derived data to estimate increments in tree/forest biomass.

Martinsanz (2011) states that sensors placed on drones have an impor-tant role in forest surveillance and production procedures, in general. Sensors have been adopted to obtain data that help foresters to assess soil resources, supply of nutrients, pesticides and herbicides. Drones have been used to measure tree height, crown height, bark thickness, and other variables such as canopy size, leaf area and chlorophyll content. Drones have also been used to aid several aspects of post-harvest processing and transport of forest wood.

Pastures and feed stock production zones in many countries occur on large expanses. Pasture growth and its productivity are monitored periodically. Farmers usually, revise agronomic procedures based on data about growth rate, biomass and nutritional quality of the pasture grass/legumes. Pasture surveillance using satellite-mediated remote sensing is already in vogue in many regions of the world (Ahamed et al., 2011; KSU, 2013). It saves time and cost on scouting the pastures for soil erosion, loss of fertility, retarded growth and disease/pests. Farmers have used the satellite data to amend pastures with fertilizers, water or chemical sprays. They have adopted site-specific nutrient and water management techniques. As stated earlier, satellite imagery is constrained by the clouds and it only provides images of relatively low resolution. Drones equipped with visual and NIR imagery are excellently suited to scout the pastures for various traits such as crop mixtures, growth pattern, biomass and nutrient accumulation, water stress effects and diseases/pests. It seems, in Europe and North America drones are also used to keep a count of cattle heads, their movement and in general management of herd.

Turf grass management in North America involves series of agronomic procedures that need to be administered as accurately as possible. It has also to be economically viable and profitable. The human labor involvement, plus chemical input has to be least. So far, turfgrass managers have adopted ground-based sensors to surveillance large areas of turf. Sometimes, obtaining satellite imagery is difficult because of cloud cover and squally weather and it is also low in resolution. The interpretation of geo-referenced and digitized information from satellites could be cumbersome and costly. Farmers cannot obtain satellite imagery too frequently, because it is expensive. Hence, during past 5 years, turf grass managers in USA have explored the possibility of using Phantom copters (drones) fitted with visual, NIR and thermal imagery (GoPRO Hero 3, 12 Mpxl) to obtain digitized data that can be processed easily by appropriate computer software (e.g., AgPixel) (Stowell and Gelerntr, 2013). Drones were used to assess damage to trufgrass by animals, weeds, insects, microbial diseases and soil erosion. It is said that low cost of light weight drone fitted with cameras, ease with which aerial photographs could be obtained and equally cheaper processing of imagery using computer software makes drone technology popular with turf grass mangers. Turf specialists can also maintain a detailed archived data about the turf grass, using computer programs. Periodic changes caused by natural and man-made factors could be easily documented for ready reference at any time. At present, extension agencies of different states in USA, it seems, are persuading farmers to adopt drone technology to manage pastures and turfs.

3.5 AGRICULTURAL DRONES AND ECONOMIC ASPECTS

Let us consider a few generalized uses of drones in different parts of the world and their economic feasibility and advantages, if any. Drones are among highly useful and economically profitable options in many aspects of daily life of human beings, survey of natural resources and terrain, industry, transport, and surveillance of major installations. For example, Alaskan Oil Pipe line surveillance using traditional ground vehicles and human security personnel costs 3000 US$ hr^{-1}. Compared with it, a common drone helicopter programed to automatically move above the pipe line and conduct aerial surveillance costs 85 US$ hr^{-1}. It means that the cost of a Robotic Drone breaks even within 30 hr (Chakravorty, 2013). Drones are highly recommended and apt in areas that are almost inhospitable for humans. Many of the oil companies conduct aerial vigilance using manned helicopters. However, as stated above, in this ice clad zone of Alaska, a drone is perhaps best option compared to human personnel. It seems drones are more pertinent and profitable, if the area or stretch to be surveillance is long. Human transit on ground-based vehicles is not easy on rugged terrain, while aerial flights are smooth and tidy.

Reports by agricultural drone companies operating in North America suggest that using drones for surveillance has economic advantages. This is in addition to various other positive gains. In the normal course, a walking human scout would require US$ 2/acre^{-1} of cereal crop for visual inspection. An aerial survey using copters can perform the same duties with greater rapidity and provide excellent pictures of happenings in the crop field. The cost of surveillance per acre reduces to 30 cents or even further to negligible level, when a drone is used. Farmers owning drones tend to reduce on cost, still further compared to those hiring drones or drone services from companies (Precision Drone LLC, 2014).

An educated guess suggests that UAVs have great potential in making farmers to spend less on fertilizers, insecticides, fungicides and irrigation, because it avoids blanket applications of these items to crops. Actually, only few spots spread randomly within a farm may have been affected by moisture deficit or pests. Expending large quantities of chemicals and irrigation water under such circumstances is futile and economically inefficient. An evaluation by Rosenstock (2013) states that in USA alone, loss of exchequer due to improper application of pesticides and irrigation reaches US\$ 25 billion yr^{-1}. Drones could thwart such loss.

Crop production specialists at Kansas State University's Experimental Farm state that drones are spreading into farm land rapidly. They are being used efficiently during adoption of precision farming methods. Fixed-wing drones with sophisticated cameras for Visual and NIR imagery cost about 12,000 US\$ and copters about 7000 US\$. However, in the long run, price of a drone is expected to decrease (Huting, 2013). Farmers could break-even costs incurred on drones in a space of 2–3 crop seasons. Therefore, economic potential of drones during cereal production in the Central Great Plains seems immense (Johanssen, 2014)

Drones are being used to accomplish a range farm tasks. Farmers tend to use sophisticated models with accessories that provide highly accurate information. There are established models sold at 25,000 US\$ per unit. Others, mainly local models, that cost just 1000 US\$ for the copter and 300 US\$ for cameras are also available. They are apt, if the purpose is to scout the cattle range, surveillance cattle and monitor cattle heads. Lamb (2014a, b) reports that, in Missouri, farms with wheat, corn and soybean fields plus cattle (300 heads), have been using low cost copters with visual and NIR cameras. Such an option is said to save about 75% of time actually required, if human scouts were used. Monitoring cattle using a computer or a tablet is that much easier. Drones are efficient in inclement weather but may have to be grounded, if winds reach beyond 20 km hr^{-1}.

Reports from Charlotte, in North Carolina suggest that an 'Autocopter' which is a drone helicopter has been improvised enormously to conduct series of tasks relevant to precision farming. It is fitted with excellent DSLR cameras, computer decision-support and GPS connectivity. Effren (2014) has reported that on a farm of 1250 acres, cost for purchase of an

'Autocopter' (25,000 US$) is easily payed-off by the reduced inputs, due to drone mediated variable-rate application of fertilizer-N and pesticides, and enhanced crop yield created. The breakeven on cost of copter occurs within a period of one year. These drones are of tough material and last for several years (Plate 3.8A and 3.8B). There are companies (e.g., 3D Robotics Inc.) that produce drones of cheaper foam material, use computers/processors and cameras of lower costs. Over all, such drones are cost effective, and serve the small farmers in obtaining sharp imagery and processing them (Anderson, 2013)

Green (2013) opines that although drones are perceived as controversial tools in the hands of military and national security teams, in a few years, drones are expected to swarm the skies in big numbers in many of the world's top agrarian regions. The 'Agro-Drone Revolution' may eventually flourish in all farming belts, along with satellite guided GPS techniques. Farming is seen as the most promising, economical and profitable zone to deploy drones. The profitability arises from its ability to fetch excellent aerial imagery of farms, crops and the daily happenings in the entire farm, right to the computer desk or tablet held by the farmer sitting at home (Paul, 2014). In USA alone, drones could create over 100,000 jobs in the manufacturing aspects. There are other opportunities as drone technicians, mainly in its maintenance, computer programming, etc. The forecasts suggest that during next decade, for example, 2015–2025, the drone related monetary turn over could be 82 billion US$. A different way to look at the advantages from use of drones in agrarian zones across different continents is to first perform an economic analysis at different locations on small farms and extrapolate. Paul (2014) further states that even at 1–5% savings in herbicides, fungicides, bactericides and pesticides; it is multi-billion dollar advantage to global agriculture. However, we should that note that use of drone, literally reduces chemical use to very low levels and advantages reported are generally over 80% of usual chemical use.

Canadian farmers are enthusiastic about use of multi-spectral signatures of different pests and diseases (Redmond, 2014). Farmers spray pesticides and fungicides using GPS and an auto-steer system that houses the sprayer. There is also variable-rate sprayers used in some farms. Most importantly, pesticide/fungicide spray is restricted to locations shown in the aerial imagery. The digitized spectral data supplied to farmer shows pest/fungus afflicted zones accurately. The quantity of pesticides applied to early insect instars is drastically low. Even if two sprays are taken, the extent of reduction in pesticide/fungicide requirement is good enough to offer profits to the farmer. Similarly, in case of white mold, focusing on areas affected, instead of blanket sprays reduces quantity of fungicide required. There is also reduction in human labor needed. Further, drones used for aerial survey and robotic sprays avoid any detriment to health of farm workers.

Dobberstein (2014) states that, drones with ability for crop scouting, 3-D mapping, pathogen/pest detection and spot spraying are economically efficient. In the No-till fields, weed infestation is the chief factor that increases cost of production. Weeds could be a severe grain/forage yield retardant, if they are not detected and removed right at early stages of the crop. Drones cut costs on daily surveillance of crops for weeds, reduce input costs because of spot application of herbicides. Drones improve decision making and help in timely control of weeds.

A report by Dobberstein et al. (2014) relates to several aspects of deployment and economics of introduction of drone technology in agriculture, particularly in soil and crop management. They opine that after a fairly long period of its use in military, drones seem to have attracted farmers, to whom it is an economically useful technology, particularly for large farms. Drones seem to have garnered a strong foot hold in 'Precision Farming,' since it allows farmers lessen input costs. Some of the concerns mentioned are farmers may be in possession of drone, perhaps a light weight copter or fixed-wing version, but cost of accessories and purposes that should serve also affect the economic advantage. Currently, cost of a light weight copter or a fixed-wing drone may range from 500 US$ to 2500 US$ but this may reduce eventually as mass production occurs (see Anderson, 2013; Huting, 2013; Heacox, 2014; Dobbs, 2013; Ehmke, 2013). There are copter and fixed-wing drones that cost 17,500 to 22,000 US$ (Vanac, 2014). A slightly larger copter such as RMAX with facility for imagery, spraying and granule application may cost from 40,000 to 60,000 US$ (see Table 3.2). Actually costs of drones vary widely. Yet, we ought to know that drone with high resolution cameras cost relatively more. Drones with a resolution of 3 inch have just to make a few passes over crop fields to map the entire 200 acre stretch in just 20 min. The computer programs and decision support also costs based on sophistication. Reduction in scouting cost and lessening inputs through variable-rate applications are among major factors that offer better economic advantages to farmers. Mapping soil and water resources is an economically profitable function of drones. The 3-D imagery offers excellent opportunity to farmers to reduce on inputs such as fertilizers, water and pesticides. Field maps have often helped farmers to reduce on herbicide sprays. In a few European nations, to make drones and precision technology viable, they have been priced appropriately. Drones with accessories are priced in such a way that it is highly viable, if used by farm owners with over 10,000 ha (Grassi, 2013). Over all, Dobberstein et al. (2014) suggest that success of drone-based enterprises depends much on after sales servicing, particularly in providing latest and sophisticated multi-spectral imaging systems and computer accessories. Heacox (2014) has listed a series of questions asked to farmers and their responses about how they accrued information, knowledge, and details about use and advantages of drones in agriculture. It looks that drone technology was initially perceived as a method to photograph their farms and show physical effects and standing crops with perhaps minor details. However, in due course, economic gains of using drones were explained to farmers, particularly regarding crop scouting, phenomics, lessening of fertilizer and pesticide needs, and accurate decisions on harvesting grains, etc. Drones electronically synchronized to ground robots and steer-less tractors now seem to offer the greatest advantage to farmers. It literally leads farmers to economically highly efficient 'Push Button Technology.' Human drudgery is avoided to a very great extent.

Farmers in North America were exposed to precision farming technique, a few years ahead of the recently popular drone technology (Lyseng, 2006). A sizeable share of advantage attributed to drone technology actually emanates from detailed sampling, survey of soil fertility and crop health plus the adoption of variable-rate technology. The variable rate technique reduces input requirement and allows farmers reach the same yield goal at lowered input costs. In fact, drone technology hinges on the precision farming approaches to a great extent with regard to economizing on inputs such as fertilizers, irrigation, herbicides and pesticides.

In addition, economic advantage due to drones arises from reduced human skilled labor requirement. Rapid completion of crop surveillance, survey for soil fertility and water deficits results in low energy costs. Farm techniques become more accurate and fool proof when drones and precision techniques are adopted in conjunction.

Hetterick (2013) states that drone technology adopted to monitor seed production farms cost few cents acre^{-1} (25–30 cents acre^{-1}) compared with 3.0 US$ acre^{-1}, if agricultural companies chartered aircrafts and pilots to obtain pictures of seed plots. No doubt, seed farms utilizing drones stand to gain compared to those using human labor for scouting the large patches of cereal/legume crops. It seems, quad-copters that are low-cost equipment (<7000 US$) are efficient in terms of economics, even if a swarm of them are hired. Bennett (2013) considers that it is matter of time before drone technology becomes wide spread in the agrarian zones of North America. Here, farmers and commercial crop production companies alike are waiting for the rules and regulations that are to be finalized by late 2015. Whatever is the net gain due to accurate control of fertilizer, irrigation, pesticide supply achieved using drones, the adoption of drone technology, depends on cost of drone and cameras fitted. Drone production could be initially subsidized, if need be, and cost of purchase by farmers kept within reach. Right now, drone models, accessories for imagery and computer programs for decision-support are all available within manageable cost for farmers (see Huting, 2013; Keller, 2014).

Drones used in crop fields of Southern England have been proving profitable. Drones have also induced certain changes such as introduction of Precision Farming methods. Impey (2014) reported that drones were introduced into farming zones in conjunction with steer-less tractors, variable-rate planters, variable-rate fertilizer applicators and GPS-RTK systems. Drone derived soil fertility maps were used to apply N, P and K to the crops. The need for nutrients has reduced because of precision techniques and variable-rate inputs. The fertilizer-N supply has reduced by a clear 5% if drone derived maps and precision techniques are used. Expenditure on P, K and lime has reduced by 50%. If drones are used to spray foliar fertilizer-N, then, need for fertilizer gets reduced enormously.

3.5.1 DRONES MAY AFFECT FARM SIZE, FARM LABOR REQUIREMENT AND FARM WORKER MIGRATION

According to International Association for Unmanned Aerial Vehicles, within next decade, we could encounter a situation where in 80% of the drone production is aimed at satisfying needs of farm scouting and spraying. Farming enterprises are expected to garner most of the drone market. Currently, laws for its commercialization and use in farms and standard guidelines are still in preparation. In 10 years, all farms in Iowa could be using drones. We can easily guess loss of farm jobs once much of scouting and spraying is accomplished through drones. However, agricultural economists and aviation engineers do express that drone production trends and need for skilled human technicians to control or program drones should generate over 10,000 jobs by 2025, in just the Iowa cereal belt. The drone production business, excluding its daily usage in farms is said to generate a revenue of 0.5 billion US$ per year (Doering, 2013). Recent forecasts by soybean growers in Iowa suggests that drone related business, not related to its use in farms, just production and marketing the drone machine could generate 950 million US$ and 1200 jobs in next couple of years. To quote an example, Roboflight, a company in Iowa has posted 3-fold increase in drone production during 2013. The drones sold could easily cover about 40,000 acres in the cereal belt (Ellerbroek, 2014a). A few other forecasts suggest that by 2025 annual drone sales could reach 160,000 yr^{-1} in USA alone (Stutman, 2013)

Regarding individual farms in Iowa, if a farmer with 900 acres purchases a drone at 30,000 US$, he covers about 80 acres of cereal/soybean fields hr^{-1} using it. The drone provides him with digitized imagery depicting areas that need water, nutrients and chemical spraying. The drone could also find weed infested zones. Drones are rapid compared to human scouts. Human labor availability is often seasonal and fluctuations can cause drastic escalations in labor costs or could be disastrous to farms, if they cannot attend to agronomic procedures in time.

A recent report about farming zones of Australia, points out that, farmers are not found toiling hard on soil management and agronomic procedures that need drudgery and constant vigil. For example, farm workers, who were traceable easily in the center of wheat crop that has grown to knee-height is not common today. Farmers more often discuss about uses of drone technology during wheat production. They learn about fixing accessories and computer programs into the light weight drones. Robots and drones together keep out farm laborer. Farm labor needs have depreciated enormously. Hence, farm workers in Australia are asked to depend less on regular recruitment by farm companies. Instead, they are asked to migrate or change professions (Townsend, 2013).

3.6   DRONES IN THE AGRARIAN REGIONS OF DIFFERENT CONTINENTS: PRESENT SITUATION

Currently, drones have been tested, tried and are in use in agricultural farms of over 50 nations (see Table 3.1). Agricultural drones have entered different agrarian regions, irrespective of geographic and weather conditions, farming systems adopted, and economic disposition of farmers. However, they are yet to make a mark as a routine implement or farm vehicle, just the way an animal drawn plow, tractor or harvester.

3.6.1   DRONES IN NORTH AMERICAN AGRARIAN REGIONS

Reports suggest that soon, about 80% of drone usage in North America may be localized into farming and rest for surveillance of transport vehicles, military and aerial imagery of natural resources. Agriculture is expected to be the major user of low flying aerial drones. They may be deployed to aid several farm equipment/vehicles that are sophisticated and auto-piloted (self-steered) robots programed using a computer-based decision support systems (Precision Farm Dealer 2014a,b; DMZ Aerial Autonomous Scouting Robotics, 2013).

Reports from Canadian Institutions suggest that drones are remarkable devices in the hands of personnel involved in several aspects of Military, Industry, Agriculture and general maintenance. Drones are versatile instruments that are gaining ground in Canadian farming zones. Drones being tested and used in Canadian farming zones include both Fixed-wing and Roto-copters. Drones equipped with a few different types of cameras such as visual, NIR, IR, thermal and LiDAR are very useful in monitoring crops, deciding on fertilizer and irrigation input. There are now several drone companies that produce a range of models with accessories meant for use in land and crop management. For example, SkySquirrel Technologies Inc., produces autonomous drones meant for precision agriculture. These drones are built to help farmers engaged in fruit and vegetable production. They monitor crops and measure NDVI (SkySquirrel Technologies Inc., 2013).

Reports by researchers from MIT, Massachusetts, USA suggests that grape vine farmers are able to purchase and utilize advanced UAVs that were once used only in military. For example, drones currently mean low priced small winged airplane, fitted with cameras or small copters with multi-blades and excellent controls and maneuverability above the grape-vines. The low altitude view of the entire orchard and an electronic/digitized map that it generates is of immense use to farmers. It allows them to ascertain areas with low or better soil fertility, areas afflicted with diseases, areas with variations in soil moisture, etc. Such copters it seems are significantly efficient and cheaper in terms of expenditure incurred. A digitized picture or close-up shot of grape vines procured from manned aircrafts costs over 1000 US$. Anderson (2014) has remarked that advent of small copters with 4 rotor-blades or winged drones has also induced excellent improvements in fixtures such as sensors, cameras, gyros, GPS modules, development of digitized pictures of soil or fields. These improvements allow greater resolution of soils/crop maps and make them available to farmers immediately. Farmers have obtained excellent pictures showing spread of fungal diseases on grape vines. Spectral images from drones could also be transmitted instantaneously to stationary computers or those on vehicles (tractors) so that immediate and appropriate actions could be effected. Farmers have also obtained accurate estimates of crop nutrient status and forecasted grape yields accurately using drones.

In New York State, drones and steer-less tractors have started helping farmers in precision farming. Functions such as seed planting, fertilizer application and pesticide sprays have been directed based on information that is available for each square ft. Then, drones that fly periodically keep a vigil on seedling growth, its health and phonemics in general (Dobbs, 2013). Farmers adopting drones and precision farming techniques state that their profits increased since past 2 years, although this technology was adopted some 4 years ago. Precise field data and yield monitoring shows that crop productivity has increased. Steer-less tractors that do not sway off the tracks are a major advantage, since farmers can work both day and night with great accuracy. These tractors are provided information on seeding, fertilizer application and spraying based on aerial imagery derived from drones. Forecasts suggest that within next decade, several large farms in the New York state could be adopting precision farming methods utilizing robots and drones. Dobbs (2013) further states that, in future, farms could be swarmed with smaller and nimble robots, replacing the large tractors. The small robots would depend on data derived from aerial drones. The reduction in inputs and elimination of overlaps of farm operation is an advantage with drones and robots. The proliferation of drones and small robots seems inevitable, yet there are many farmers who are reluctant, because initial costs and complexity of performing all tasks by linking the drones, their images and digitized data to robots that operate driverless.

In Pennsylvania, drone scouting of agrarian region is in vogue since few years. The drones have helped farmers with sharp imagery of terrain, soil types and vegetation of thousands of acres in a span of short period. This data has been used to decide on soil tillage practices, cropping systems and it has also aided in accurate supply of fertilizers, irrigation and pesticides based on yield goals. There are currently, many 'farm shows' that display drones and field days where drones are demonstrated (Noble, 2014).

Reports suggest that farmers in Iowa, particularly those who operate large farms of 1000 ha are using drones for imagery of topography of entire farm. They study the topography of the farm carefully for aspects such as locations prone to soil erosion, those with low soil fertility or inherently poor soil and try to avoid them for re-planting. Farmers are able to focus on areas with high fertility and grain productivity (The Des Moines Register, 2014). In Iowa, the maize/soybean cropping belt is experiencing a kind of drone based revolution in crop management. There are crop consulting companies that prescribe based on aerial imagery derived from drones. Farmers could hire the services of the drone for a definite time and purpose or buy rapid pictures from the agency through internet facility. For example, Labre Crop consulting at Manson in Iowa offers agricultural drone services. The flight plans to suit the field is drawn based on '*eMotion*' software attached to an 'eBee' drone developed by SenseFly Inc., Switzerland (Plate 3.4). Farmers can simulate the flight plan on a mobile or tablet and suitably guide the drone that is powered by a lithium battery. The consultants use a '*Postflight Terra*' software to decode and develop pictures of crops. This software provides visual details at layers and 3D pictures. For Iowa farmers, the real advantage is when consultancies deliver pictures of their farms on a daily basis and suggest them routine procedures and remedies (Labre Crop Consulting Inc., 2014). Additionally, GPS tagging helps farmers to use autonomous robots, if they opt for it.

Drones have been deployed in the wheat, maize and soybean expanses found in the Central Great Plains. It is a few years, may be 4–5, since they have been attracting attention of farmers and food grain production companies. Researchers at the Kansas State University, opine that farmers may find too many, rather endless, advantages and situations with drones. Basically, they exhibit better maneuverability and economics compared to buying satellite imagery of comparatively lower resolution, or manned airplane or human skilled scouts who detect growth retardation, soil moisture deficits, nutrient deficiencies, pests and disease. Farmers have found warnings (alerts) from drone derived imagery very useful, timely and have lessened risks. Farmers have bought drones for as low as 2000 US$. They are fitted with cameras that produce images at visible and near-infra red range of wavelength. Above it, these drones are also amenable to be fitted with tanks that carry pesticides or liquid fertilizer. This facility is highly advantageous, since farmers can restrict application to areas that afflicted. It reduces chemical usage, finishes tasks quickly and lessens burden of hiring large number skilled farm laborers (Doering, 2013; Price, 2013).

In Ohio, the Department of Food and Agriculture has set up demonstrations of UAVS for farmers to learn several possible advantages during wheat and soybean production. Farmers are specifically shown how to obtain detailed imagery of crop stand, plant counts, and NDVI which is a plant health indicator. The chlorophyll estimations allow farmers to decide on fertilizer-N supply to crops (OSU, 2014). Bowman (2014) forecasts that farmers and ranchers in Ohio, could be using UAVs, to assess crop health and nutrient need. Drones could also be used to assess water requirement of crops and insecticide spray schedules. However, in USA, drone based farming has not yet received FAA approval. They are hopeful of regular use by 2015.

Drones have taken to sky over farms in the Mid-West states of Indiana and Ohio. Popular models such as Precision Hawk and several other local versions, with wide range of accessories such as visual cameras, NIR and thermal sensors are used to monitor watershed and measure NDVI. There are also completely computer controlled models such as 'Hawkeye Lancaster' that costs less than US$ 25,000. These drones are used to scout for crop growth, detect pests, diseases, and water deficiency, if any, in the large cereal farms of Ohio (Reese and Higgins, 2013).

Corn and soybean cropping system is prominent in Illinois. Farmers here regularly obtain aerial imagery from satellite companies. Satellite imagery reveals crop growth, NDVI and chlorophyll content. Farmers try to take decisions on planting, fertilizer supply and irrigation based on satellite imagery, although it is of low resolution and may not identify small areas with low intensity scattered insect attack. Hence, corn farmers in Illinois have now adopted drones to collect images at any time of crop growth. Drone derived imagery is of high resolution and provides detailed information on root worm attack (AUVSIAdmin, 2012). Based on digital data, drones could then be guided to apply pesticides only at spots that are attacked by insects, instead of the usual blanket applications.

Recently, farmers in Virginia, were shown the advantages of using drones in crop production. They were exhibited the small 'MikroKopter,' a drone that produces immaculate imagery of crops such as wheat and soybean and their growth status. These drones can also pick close-up video or stills showing details of insect/disease attack on crops (Kimberlin, 2013).

During past two years, farmers in North Carolina are being exposed to a new drone known as 'Autocopter.' It has flight endurance of over 2 hr. at a stretch, and carries a small payload of 3 cameras. The sensors include a visual camera, DSLR, multi-spectral camera and a HD videographer. It is being used to scout grape vines, apple orchards, maize and cotton fields. Reports suggest that use of data from Autocopter that is timely and agronomic procedures conducted based on it, are economically beneficial. For example, on a 1250 acres farm, reduced inputs, and increased productivity due to use of Autocopter was perceptible (Unmanned Systems Technology, 2014).

In Oregon, Hamm (2013) is researching on use of UAVs in Precision farming. He states that drones with ability to fly at low altitudes over crops saves a lot of guess work, time and resources, in addition to reducing the elaborate walking exercise that crop scouting requires. The soybean fields were efficiently scouted for crop growth, nitrogen deficiencies, and insects. It is believed that, sooner or later, drones may be common sight above cereal/soybean fields. Gonzalez (2013) reports that farmers in Oregon are using drones to monitor potato crop for growth, biomass accumulation and diseases.

Agricultural drones have invaded wheat farming belt of Montana since long. Drones are used to study natural resources, estimate forest biomass (NDVI) and crop production zones. Recently, drones have been used to supply aerial imagery of wheat crop highlighting the zones that are afflicted with fungal, bacterial and viral diseases. Wheat affected by impaired water uptake has also been mapped using drones (Basso and Rush, 2013). A report from Idaho suggests that a light weight (< 10 lb), 5 ft. long drone of the size of a large hen or turkey is working wonders for farmers engrossed in production of potato, wheat, barley, peas and alfalfa. Wozniacka (2014) states that, drones are very useful instruments in the hands of farmers. The imagery helps to decide on various soil and crop management procedures and revise as many times at short intervals. In-season data accruals are easy, rapid and less costly than hiring skilled farm workers. Drones seem to be perfect for large farms of size more than 500 acres, where scouting and variable-rate supply of inputs is rather difficult. Further, in Idaho, farmers have generally paid to buy satellite imagery of their crop fields/plantation. Whenever, imagery of greater accuracy and resolution were needed they relied on imagery from piloted airplanes. These were costly and prone to difficulties, if clouds were interfering. However, recent trend with farmers producing field crops and plantations alike is to deploy drones to get close-up images. It has helped them to monitor and obtain disease free crop and produce. Similarly, in Colorado, skies over farms have been taken over by drones that are used to surveillance fields, scout crops and suggest farmers about various agronomic measures that should follow. These drones have GPS connectivity and hence are very accurate, in guiding the farms to exact locations within large farms that need their attention. In Missouri, large wheat farms, cattle ranches and pastures are being monitored using drones (Lamb, 2014a,b).

The agrarian zones of Michigan have been exposed to use of drones since mid-1990s. Drones have been used to collect detailed data about weather and soils, mainly to prepare maps for use by farmers, agricultural companies and other natural resource related agencies. Drones have also provided data on natural vegetation, cropping systems and productivity. Currently, farmers in Michigan have been demonstrated the various uses of drones in crop production. Drones have been integrated with crop models to help in accurate decisions and appropriate fertilizer and pesticide spray schedules. Drones have also been used to accomplish various tasks related to precision farming (Azorobotics, 2014).

Reports from Arizona indicate that production of drones meant for farming sector has gained in popularity during recent years. Drone producers expect agricultural farming to be the major user of drones after 2015, by when FAA regulations get standardized. Drones could actually be flying in thousands over crops grown in Arizona (Shinn, 2014).

It seems sugarcane and corn producers in Louisiana are finding drones useful to monitor the crop. UAVs capable of flying programed routes have been adopted to pick aerial imagery of crops. Drones detect herbicide-resistant weeds, also insects and fungi affected zones of crops efficiently. Autonomous copters powered by batteries have also been used to monitor experimental rice plots at LSU AgCenter, Baton Rouge, USA. Drones collect data about performance of rice genotypes. It is much easier compared to skilled technicians making several still pictures that need to be arranged to obtain a good assessment of a rice genotype (Schultz, 2013).

In the Southern Plains region of Texas, cotton and sorghum production zones are conspicuous. Cotton, in particular, is affected by water and nutrient deficits, boll weevil attack, weeds and diseases. Therefore, current research with drones is focused to solve any or all of these problems, to the extent possible (Hoffman, 2013). Drones with multi-spectral sensors and ability for high resolution pictures of 0.5–2 cm are being tested to detect healthy cotton plants, boll weevil infested zones, soil moisture deficit and disease afflicted plants. Accurate sprays using variable-rate techniques are adopted based on high resolution digitized data that are supplied to aerial/ground spray equipment. Such an amalgamation of drone-based crop survey, use of digitized data and maps, and sprays using variable technique will be efficient in many ways. It reduces use of chemical and soil pollution problems related to it. It reduces cost of production and enhances profitability. Timely detection avoids build-up of insect/disease problems and ensures better yield of larger areas within the cotton belt.

Citrus Research and Education Centre, situated at Lake Alfred in Florida is aiming at introducing low cost drones to conduct precision farming. The lithium battery powered, six or eight copter drones may cost around 7000 US$ a piece, but it is highly useful in scouting long stretches of citrus groves that are so prominent in Central Florida.

In Arkansas, drones with 6 copters are being examined for usefulness and profitability during soybean production. The Soybean Board in Arkansas is trying to recommend drones to scout soybean farms for canopy, leaf area index, chlorophyll and nitrogen content of crop. These drones could also alert the farmer about disease and pest incidence through periodic aerial imagery of the crop.

In the Caribbean, drones are in use in agrarian regions. In addition, drones are used to study other aspects such as natural resource monitoring, border security, fire-fighting, flood monitoring, pollution monitoring, pipeline and infrastructure surveillance, natural disaster assessment, etc. (UAV-Belize Ltd, 2013). Within the Caribbean agricultural regions, drones are used for crop disease detection, fungicide spray or dusting, soil moisture monitoring, crop growth monitoring, fertilizer management (aerial surveying), detection of crop maturity and harvesting. These drones are being used to monitor crops like sugarcane, coffee, jathropa and paddy. The UAV agencies in the Caribbean provide aerial imagery of commercial farms for a fee. The drones fly past a pre-determined path that engulfs several farms and pick up images of crop stand and other details. Farmers have the option of buying 3D and layered pictures of their farms.

3.6.2 DRONES IN SOUTH AMERICAN FARMS

The South American nations, like those of other continents began deployment and research on development of drones during 1990s. It was mostly meant for military surveillance, surveying natural resources and tracing zones with disasters and natural calamity. In addition, Latin Americans used drones to trace and smoke out drug smuggling rings. Drones programs in Latin America got initiated as early as 1980s in Brazil, and 1990s in Argentina (Glickhouse, 2013). For example, VT-15 series and Orbis series in Brazil, and Lipan M3 drones in Argentina were developed for different civilian and military purposes. However, during recent years, predominantly agricultural nations such as Brazil and Argentina, that possess large farm companies has embarked on use of drone in farms. They are using drones to monitor crops such as coffee, cocoa, wheat and soybean. Several other nations such as Uruguay, Venezuela, Colombia, Equador and Mexico are utilizing drones mainly to map natural resources, floods, natural disasters, etc.

Forecasts suggest that Brazilian Agriculture that thrives on crops such as soybean, maize, wheat, sugarcane and coffee is to experience fairly large scale usage of drones to accomplish several of agronomic procedures with greater ease. Reports from EMBRAPA, Sao Paolo, indicate that drones and software for decoding and interpreting digital imagery are being standardized. Drones are being examined for use in imaging rural topography, soil types, soil moisture distribution and possible cropping systems, etc. Further, aerial imagery is adopted to monitor crop growth, chlorophyll content, leaf area and N status, water requirements, etc. In future, many of the large agricultural companies that produce major cereals (maize and wheat) and soybean may utilize visual and NIR imagery to keep a vigil on diseases/insects and their spread. Drones could also be used to spray pesticides and weedicides. The anticipated cost of drones to be developed in Brazil may range from 3200 to 6500 Euros (EMBRAPA, 2014).

In Peru, drones are being used to monitor natural resources, agricultural cropping expanses and to trace or reach difficult archaeological locations. A drone model designed by Peru's Catholic University's engineers is small, but equipped with high resolution and precision video-graphic instruments. The drone is attached with visible range and near-infra red camera. These agricultural drones help agronomists to assess crop growth, occurrence of pests and diseases. It also helps in tracing soil erosion if any. The high quality images of soil and crop stand helps in detecting occurrence of drought, soil moisture and nutrient deficiency. Perhaps, in Peru, farm companies, co-operatives and groups of farmers together may opt for usage of drones during for crop production.

3.6.3 DRONE USAGE IN EUROPEAN FARMS

European plains support a vast stretch of crops. Crop production trends include both intensive and expansive. Drones have been used in Europe to surveillance and produce imagery of crop fields periodically. The Drones fitted with visual and NIR cameras are also connected to GPS and computers that allow them to depict crop growth status, chlorophyll and plant-N status, and most importantly weed and pest incidence. The computers can read the spectral signatures of crops and weeds and discriminate them. Herbicide usage gets reduced because chemicals are applied only at spots that require it and based on weed intensity (Jones et al., 2006; New Scientist Tech 2013).

In France, drones have been in vogue in agricultural farms for a few years now. French companies started producing drones for crop scouting some 5 years ago. Drones have gained in popularity since then. Drones with GPS connectivity and predetermined flight path are being used over field crops and grapevines. Drones are fitted with cameras to take chlorophyll and nitrogen status measurements. The crop growth and nutrient status maps are later used to distribute fertilizers accurately. Aimov (2014) forecasts that, a large number of drones say over 5000 units could be sold to farmers in next few years. They currently cover farms of over 1000 farmers in Southern France. The usage of drones is being extended into surveying water resources for farms, pesticide application based on maps showing insect attacks. In the North, wheat belt is finding drones useful in obtaining maps prior to fertilizer-N application. Magrez (2014) has reported that drones are being used in French grape production zones, mainly to study soil deterioration if any and in replanting of grape vines. The drones provide detailed aerial images of entire grapevine yard. This helps farmers to direct soil alleviation programs such as contouring, soil erosion control and fertilizer supply at points that need attention most. Soil management cost is reduced since blanket treatment of entire field is avoided.

In Spain, researchers believe that drones have potentially initiated an agricultural revolution *not* because that they are flying above the crops without pilots or that they could be programed to take a definite flight path over the crops or not even the fact that they could be controlled using remote controllers or GPS connectivity. The revolution is indeed dependent on the *sensors* that move very close and just above the crop. The miniaturized sensors, cameras and their connectivity to GPS and computers is the portion crucial to transforming Spanish Farming systems (Zarco, 2014). In fact, Grassi (2013) also expresses that farmers in Europe actually perceive and realize the utility of both visual and NIR cameras (sensors) on the drones. They actually see the magic whenever they switch between the two, as the drone flies over the fields picturizing crops. A recent opinion about agricultural drones suggests that it is like having a satellite that can produce a soil/crop map of the field, at sharp focus and high resolution of 3–10 cm compared with 4–5 m resolution of satellite pictures. Farmers need not weight for 3–4 days for the satellite to move above their fields to obtain aerial images. Over all, information and status of farm/crop is at finger tips of farmers, if they owned a drone.

Olive production is pronounced in certain provinces of Spain. They are well established tree plantations that yield oil bearing fruits. The productivity of orchards is highly dependent on soil water distribution and fertilizer supply at crucial stages. Olive orchards experience heterogeneous distribution of soil moisture and are prone to drought. Researchers at Institute for Sustainable Agriculture in Cordoba, Spain have devised methods based on drones, visual and thermal imagery to obtain tree water status and soil moisture distribution (Berni et al., 2009b). This information is contained in digitized map that is processed by computer-based decision systems. The irrigation to olive orchards is then regulated by electronic triggers. This is a step towards 'Push Button Farming' of oil bearing olive orchards.

Para-gliders have been used as drones that fly at low altitudes over crops such as wheat, barley and legumes. Field experiments in the Polish plains suggest that visual and NIR cameras placed on para-gliders that hovered above the crops at 15 km hr^{-1} speed provided high resolution photographs about fungal diseases, weed infestation and crop growth variability (Pudelko et al., 2008; 2012). It seems use of para-gliders is efficient and economically profitable.

3.6.4 AGRICULTURAL DRONES IN AFRICAN CONTINENT

Natural resource monitoring in Africa is an important frontier where drone usage could be maximized. Drones are excellent in drawing digitized imagery of natural resources, deserts, the fluctuations of fringes of Arid desert, Sudano-Sahelian, Sahelian and Guinean regions. Drones could be used to study the natural vegetation, its biomass potential, rivers, rivulets and their flow rates, cropping patterns, etc. Thermal imagery is useful in detecting water resources. Aerial observation of natural vegetation and cropping systems could be done better using drones. Satellite imagery provides only low resolution images, while drones with HD cameras and close-up shots can offer high resolution pictures. Soil erosion due to gully erosion, sheet erosion and surface loss of fertile soil is rampant in many regions of Africa, particularly, arid and semi-arid belts. The soil erosion could be monitored periodically using drones and information conveyed to farmers, so that they could take appropriate measures. To quote an example, soil erosion has been monitored periodically in the dry land cropping regions of Morocco using small drones. A fixed wing drone takes only 20–25 minutes to fly past 200 acres of cropping zone and provide aerial view of the land degradation, extent of erosion, loss of vegetation and crop, etc. Farmer could then take remedial efforts quickly, and restrict remedies to exact spots (D'Oleire-Oltmanns et al., 2012). Further, as a routine, drones have been utilized in some parts of Africa to prepare maps of topography, ground cover, water resources, cropping expanses, etc. They have actually integrated and developed aerial imaging using both satellite imagery and UAV derived digitized information. Drones are apt when small scale aerial imagery of cropping districts is needed. Satellite imagery and drones have been used in conjunction to study the riverine water resources in West Africa. The fluctuations in soil moisture, crops and biomass accumulation could be observed using satellite imagery. However, drones with high resolution cameras provided local details of crop fields as a function of rain fall pattern better. Drones are also used to study local changes in water flow on surface, local floods and erosion problems, if any. In the Sahel, impact of drought and land use changes on soil, water, vegetation, cropping pattern and crop productivity has been studied using satellite (Descroix et al., 2011, 2012). Drones could be better alternative or additional methods to assess such effects of land use change and drought. Drones could be operated at low cost and at any time.

Drones are perhaps best alternatives to ground-based assessment and satellite derived imagery about land degradation, soil fertility loss and depreciation of crop productivity. There are indeed innumerable reports and treatises that deal with detailed causes for soil fertility deterioration in Sub-Saharan Africa. The rampant loss of fertile surface soil due to sheet and gully erosion, percolation of nutrients to lower horizons, nutrient leaching, gaseous emissions have after all reduced crop productivity. These deleterious factors affect crop stand, seedling growth rate, leaf area index, canopy size, biomass accumulation pattern and forage/grain yield. Soil fertility decline has been actually attributed to variations in soil pH, organic-C, N, P, CEC (cation exchange capacity) etc. Soil moisture fluctuations further accentuate nutrient related effects, because, all nutrients are absorbed in dissolved state through water. Drones could be periodically flown over large expanses during cropping season, to ascertain the impact of soil fertility deterioration on crop growth and productivity. Drones collect aerial data about crop growth and productivity rapidly and matching this data with ground realities (data) will hasten identification of soil fertility loss and land degradation.

Large seed farms specializing in production of maize, wheat and legumes began using copter to monitor their plant breeding programs and seed production farms. They were also using aerial imagery from piloted air planes. Traditional aircraft produced photographs were of much less utility compared with those from low-flying drones operated using radio control equipment. Seed production agencies in South Africa have reported that, drone technology is efficient, useful to geneticist and seed certifiers and costs are significantly low. If aerial imagery from piloted aircraft costs 3 US$ ac^{-1}, those derived from drone technology cost < 25 cents ac^{-1} (Hetterick, 2013). The initial cost of drone with controller is <10,000 US$ per unit.

Maize production is an important agrarian enterprise in Zimbabwe. Commercial farms adopt high input intensive production technology. Fertilizer-based nutrient and irrigation are kept at higher rates in order to achieve higher grain yield goal. Drones have been used to assess maize crop growth, vigor and leaf-N status. Drones have also been used to evaluate several maize hybrids and composites grown in large fields. Drones keep track of crop phenology and grain production traits of several genotypes at regular intervals. Drones actually lessen costs on human labor required to scout and apply fertilizers (Mortimer, 2013).

In South Africa, drone services have been offered to farmers at a cost. Drone services include weather data and forecast for the individual farms, observing fields for soil erosion, crop health and growth pattern, data about phenomics along with suggestions for split applications of fertilizer-N, instructions for irrigation, weeding and pesticide application. The private agricultural consultancy agency such as SGS-South Africa is helping farmers to adopt precision farming techniques using drones, to accumulate data about soil fertility and moisture variation. Farmers are advised about precision methods based on digitized pictures derived from drones (SGS South Africa, 2014).

3.6.5 ASIAN AGRARIAN ZONES AND DRONE USAGE

China has large agrarian stretches that support cereals, legumes, oilseeds and plantations. The terrain includes vast plains, undulated and hill country farm land. Drones are apt for use in several different agrarian regions of China. Drones that are versatile in operation are best suited to survey, scout and conduct accurate aerial imagery of crops. Farms in China are not yet swarmed with drones or even ground robots, but it seems imminent. Several types and models of small aerial drones are being tested and released for use during crop production.

Let us consider a typical home-made unmanned aerial vehicle in vogue in the crop production zones of China. This drone has been deployed during vegetable and wheat production (Ministry of Agriculture, 2013). It seems Beijing Agricultural Bureau has spearheaded this project that induces farmers to use home-made small copters with high resolution cameras. Such drones are in use to detect pests and diseases occurring on field crops. The vast expanse of pastures and turfs in North China is currently being exposed to drone technology, mainly to manage weed and pest infestation. Drones fitted with multi-spectral cameras are used to obtain data regarding weed spread, its intensity and species that dominate.

Then, computer-based decision-support systems are used to decide on spraying glyphosate at variable-rates using the same drone that has a tank to hold weedicide/pesticide (Xiang and Tian, 2011a).

The drone helicopters are already in use rather routinely in Japan. In this country, they are being used to spray pesticides and distribute fertilizer (granules) uniformly across rice fields. Currently, it seems, 35% of rice fields in Japan are supplied pesticides using 'Pesticide Spraying Yamaha RMAX helicopter.' These drones are remote controlled and powered by 2 stroke engine. Totally the equipment weighs 218 lbs. While picking aerial photographs they fly about 16–100 ft. above the rice field. Currently, drones may cost 150,000–230,000 US$ depending on the engines fitted, cameras used and multiple purposes that they accomplish. It seems companies such as Yamaha Inc., for example, sell entire set of drone (helicopter), ground station components, antennae, computers, monitors and decision-support systems for rice production at about 10,00,000 US$ (Precision Farming Dealer, 2014b).

Japan is among the most advanced nations regarding use of agricultural drones. Both copters and fixed- wing types are currently popular in paddy zones. Green (2013) believes that agricultural drones have found strong foot hold in the rice belt. The Japanese rice farmers have been using drones to scout rice paddies and spray pesticides almost since 1990s. Historically, agricultural drones took to flight above the Japanese farms some 15 years ago. The unmanned helicopters were used to spray liquid and dust formulations of pesticides, foliar sprays of fertilizers dissolved in water and to obtain pictures of fields. Currently, drones are employed for a variety of other activities such as seeding (broadcasting) in forestry zones, prophylactic eradication of pests, observations of fields after floods or natural calamities, etc. In Japan, drones have also been used to study the geographical changes, natural resources such as water, vegetation and soil types. Rice production is an important aspect of Agriculture in Northeast China (Miao et al., 2012; Peng et al., 2011) and Japan. Studying rice phenology periodically is necessary to prescribe fertilizer and irrigation. Currently, drones are being used to obtain stereo-images of the crop at various critical stages of growth. These images are compared with established 'Crop Surface Models (CSM).' Since this is a non-invasive method of estimating rice crop phenology, it is gaining in popularity

(Bending et al., 2013). Plus, drone technology is rapid and less cumbersome compared to tedious soil sampling and chemical analysis.

Japan UAV Association (2014) states that, at present, there are about 2000 UAV helicopters and large number of fixed-wing drones operative in the agricultural zones. However, Cornett (2013) reports there are more than 2500 Yamaha RMAX autonomous copters covering an area of 2.0 Million ha of rice belt in Japan. The RMAX drone copters are being used to survey weeds and spray exactly at locations affected by weed growth. Drones are also in vogue in other Southeast Asian countries. Drones are used to surveillance the several thousand small islands, in addition to forests and cropping zones in Indonesia.

3.6.6   DRONES IN AUSTRALIAN FARMS

Drones have gained in popularity in accomplishing a range of tasks. The Australian drone fliers have used them to deliver small goods such as books, newspaper, to conduct aerial survey and report with pictures about spread of fires, floods and other types of disasters. However, within the context of this book, we may note that drones are taking up many of the farm jobs. There is currently a project on popularization of UAVs in farming zones and it is called 'Eye in the Sky.' This project aims at informing farmers about land care, such as plowing, irrigation need, pest and disease occurrence, timing of sprays, etc. (Chester, 2014). The drones are useful in providing pictures that allow farmers to pinpoint areas where a group of plants or even individual plant needs to be treated. Based on cameras fitted, Australian farmers have obtained images with resolution of 1.7 to 3 cm. The aerial imagery is highly accurate. Agricultural drones such as eBee have been demonstrated in the sorghum growing regions of Queensland (Plate 3.4). There are also private agencies that conduct aerial surveys of farms using drones and provide imagery, help the farmer in post-imagery processing using different computer programs and of course suitably suggest regarding agronomic procedures that should follow soon. Drone companies forecast that, in due course, cameras with greater resolution, lighter and costing much below the current price line will be available.

According to Wilson (2014), drones being deployed actively in Australian farms are connected closely to advances in software technology. Drones technology has been touted by groups such as Minegnew Irwin Group. They are spearheading adoption of UAV technology during precision farming. The drones currently produced are directed towards site-specific nutrient management, fertilizer sprays, pesticide sprays, weed detection and control. Most of the recent models of drones, particularly copters are fitted with pesticide/liquid fertilizer tank and GPS connectivity. Drones help farmers to provide digitized maps to decision-support system and therefore in precision application of inputs. Data acquired by drones through aerial photography could also be used to direct robotic tractors and deep tillage equipment, robotic seeders and irrigation equipment.

In the grapevine yards of Tasmania, Australia, farmers are being exposed to use of drones to derive digitized data about surface features of grape vines, collect information on canopy, leaf area index, soil moisture and biomass accumulation pattern. They are trained to compare the data collected periodically with established digital surface models (DSM) and then decide on soil and crop management strategies. In-season management decisions are easier, if farmers have to just compare the visual imagery with previously known trends (DSM). Computer programs that match the data obtained with established DSM are available. Farmers are able to revise crop management schedules and yield goals (see Turner et al., 2012; Hall et al., 2011).

Banana plantations in Queensland, Australia have been examined for use of aerial imagery using remote sensed data from SPOT-5. Banana plantations exhibit heterogeneous growth characteristics and water distribution. Each banana plant may exhibit different water status and requirements that needs to be met using irrigation water. Satellite imagery at a pixel ≤ 2.5 m was required to distinguish banana plant rows and each banana plant (Johansen et al., 2009). It is possible to classify the entire heterogeneous plantation into segments with homogeneous growth (management zones), canopy structure and water distribution. Such demarcation helps in channeling water and fertilizers more accurately to plantations. Input efficiency increases, since only areas with need for water and nutrients are supplied and at accurate rates. In the present context, it is possible to use drones and collect aerial imagery at visual, NIR and IR ranges to decide on irrigation. Drones accrue multi-spectral data from very close range and at high resolution. Each banana plant could be analyzed accurately using hyper-spectral imagery and HD video if needed. Fruit yield prediction should also be possible using appropriate data on canopy growth rates, leaf area, photosynthetic radiation interception, chlorophyll content and N status.

3.6.7   REGULATIONS FOR USE OF DRONES IN AGRICULTURE— FEDERAL AVIATION ADMINISTRATION (FAA) OF USA AND OTHER NATIONS

At present drones are largely confined to military programs of different nations. Next, it is often used to assess atmospheric parameters while forecasting weather. Drones are also common with agencies that keep vigil and surveillance of natural resources, like river flow, snow caps, forest cover changes, etc. The law enforcement agencies of each of different nations have their stipulations about commercial use of drones. For example, in USA, FAA has to permit use of drones in the civil airspace. There are indeed several aspects of drones, such as its size, model, flight endurance, altitude, power source, its' fixtures such as cameras of visible and near infrared range, fittings for carrying payload, for example, fluid spray, etc. Many more characteristics of drones and purposes they serve are considered by FAA, prior to offering permission. We may note that, mostly drones < 55 lbs are in use in agricultural farms in North America and other zones. Currently, it seems, most farmers have been procuring drones and using them just like any other farm implement or vehicle. It's flight is restricted to the farm and at close range above the crop. Agricultural drones are confined to regions with crop production and away from populated cities or aerodromes. They do not transgress into airspace of commercial airways or military zones. Agricultural drones are restricted to heights less than 400 ft. Drones are no doubt autonomous and safe. However, during operation of agricultural drones, damages may occur to vehicle, if it strays and drops on crop, others' properties or into water ways, etc. Next, chemical sprays done by drone could drift and reach neighboring farms. Such chemical drifts may affect unintended fields. FAA still has to discuss and derive a code of conduct and penalties for drones that stray or affect crops/properties of others. There are also concerns about who can effectively deploy, operate drones using remote controllers or computer programs. The personnel have to be trained firstly and proper certification may become necessary in due course. However, right now, there are suggestions that farmers in North America, Europe and in other regions should become more conversant with different types of drones, understand intricacies of several drone models and perfect their skills to use drones during crop production. Meanwhile, FAA may get a standard set of regulations to apply in USA and other regions. There are several aspects of drone usage in military, civil administration and agriculture that need detailed discussions prior to developing rules and regulations. Aspects such as licensing, privacy of drone usage, intrusions, etc., need to be addressed while developing rules (EPIC, 2014; Ehmke, 2013). Obviously, aviation regulations listed by each nation may vary based on specificities related to geography, topography, cropping systems, type of drones in use, and economic aspects of drones, etc.

Reports from Massachusetts suggest that by 2020, there would be over 30,000 drones in use. Federal laws allow only few agencies to fly drones. Farmers and drone operators are informed about privacy laws. Drones are restricted from flying over crop fields of other farmers.

Reports from Missouri suggest that FAA rules for operating drones in farms need to identify between violation of privacy of neighboring farms and their personnel. Repeated flying over others' farms and home could be deemed violation of privacy. However, chance intrusion onto others airspace may not be construed as violation. The problem gets conspicuous when drones are applied to regularly stalk other farmer's crops and livestock (Hetterick, 2013). According to Green (2013), about 30 states in USA have tried to develop a set of rules and regulations for drone use for commercial purposes. There are other arguments about use of drones in free air space of a farming community and that of a particular farm. We know that several low flying piloted civilian aircrafts and even passenger aircraft often fly over farms. This is true in most parts of the world (see Kimberlin, 2013)

In Canada, there are regular trainers who help farmers to learn flying drones for agricultural and other tasks. There are a few rules and permits necessary to be obtained prior to use of drones outside private farms.

Special flight operation certificates from Canadian Transport Authority are necessary for those who handle drones. Two other requirements are; first the drones should be safe regarding mechanics and second pilot/controller should show proficiency in handling drones (Epp, 2013; Redmond, 2014; Privacy Commissioner of Canada, 2013; Fitzpatrick and Burnett, 2013). The drones are to be flown well below the altitudes utilized by commercial and other aircrafts. Drones, if flown very close to airports, ground controller has to file a notice to air traffic control systems in place in that area. Person operating drone should be conversant with radio broadcasting methods. Drones should also keep a safe distance of 5–30 m from farm buildings and other installations.

In Latin America, at least 14 nations have purchased, developed and used drones since long for variety of purposes. It includes surveillance of forest cover, agricultural crop scouting, mapping coffee, wheat and soybean fields and ascertaining crop health. Yet, there are no government legislations regarding use of drones. However, Brazil has set of laws regulating drone use in their sky (RT New Team, 2013).

As stated earlier, drones are used more frequently in the agricultural zones of Japan. They are deployed to scout the crop, detect pest/disease affliction and take appropriate pesticide sprays. Liquid formulations of fertilizers are applied using small drones with pay load of 5–8 kg pesticides. There are now standard regulations deliberated and promulgated by the Japan Agricultural Aviation Association for drone use on field crops, plantations and forests. These guidelines are meant for low altitude drones that hover over farms and accomplish tasks related to crop production. The Japan UAV Association certifies and allows memberships to farmers trained to fly these vehicles (Japan UAV Association, 2014)

In Australia, there are drone users who help the natural resources department by providing them digitized pictures of local areas. They focus on geographical details, landscape, natural vegetation and cropping pattern. These drone users do stray into areas outside their limits. They may affect the privacy of neighboring farmers by taking imagery of their farms, or show up others private and public buildings in that area. Hence, drone users are generally advised to restrict to their own farms and not intrude into those of others. Drone filled skies with each of them moving in different directions could be distracting and may be an unacceptable situation (Ansley, 2014).

KEYWORDS

- **Aerial Spraying**
- **Agricultural Drones**
- **Copters**
- **Crop Growth Monitoring**
- **Flat-Winged Drones**
- **Multispectral Cameras**
- **Normalized Difference Vegetative Index**
- **Ortho-Mosaics**
- **Scouting**
- **Sensors**
- **Yield Forecasting**

REFERENCES

1. Abuzar, M., O'Leary, Fitzgerald, G. F. (2009). Measuring water stress in a wheat crop on a spatial scale using Airborne thermal and multispectral imagery. Field Crops Research 112: 55–65.
2. Acevo-Herrera, R., Auasca, A., Xavier, B. L., Jose, M. F., Nilda, S. M., Carlos, P. G. (2010). Design and first results of an UAV borne L-band Radiometer for Multiple monitoring purposes. Remote Sensing 2: 1662–164.
3. Adamchuck, V. I. (2005). Characterizing Soil Variability Using On-the-Go Sensing Technology. International Plant Nutrition Institute, Site-Specific Management Guidelines, SSMG-44. Norcross, Ga. www.ipni.net/ppiweb/ppibase.nsf /b369c6db e705dd13852568e3000de93d/8bfeba411afe85e28525718000690a10/$FILE/SSMG-44.pdf, pp. 1–8.
4. Adamchuck, V. I., P. Jasa. 2014. On-the-Go Vehicle-Based Soil Sensors. University of Nebraska Lincoln Extension ECO_2–178. www.ianrpubs.unl. edu/epublic/live/ec178/build/ec178.pdf, pp. 1–4.
5. Aguera, F., Carvajal, F., Perez, M. (2011). Measuring sunflower nitrogen status from an Unmanned Aerial Vehicle based system and an on the ground device. International Archives of the Photogrammetry. Remote Sensing and Spatial Information Science. Conference on Unmanned Aerial Vehicles in Geometrics, Zurich, Switzerland, pp. 1–12.
6. Ahamed, T., Tian L., Zhang, Y., Ting, K. C. (2011). A Review of Remote Sensing methods for Biomass feedstock production. Biomass and Bioenergy 35:2455–2469.

7. Aimov, S. (2014). More and More farmers using drones in France. http://www. freshplaza.com/ article/118379/More-and-more-farmers-using-drones-in-France 1–3 (July 27th, 2014).

8. Alvaro, F., Garcia del Moral, L. F., Royo. C. (2007). Usefulness of Remote sensing for the assessment of growth traits in individual cereal plants grown in the field. International Journal of Remote Sensing 28: 2497–2512.

9. Anderson, C. (2013). In: Drones and Agriculture: Unmanned Aircraft my revolutionize Farming, experts say. Ghose, T. (Ed.). Huffington Post Science http://www. huffingtonpost.com/2013/05/20/drones-agriculture-unmanned-aircraft-farming_n_ 3308164. html?ir=India., pp. 1–4 (January 22nd, 2015).

10. Anderson C. (2014). Lower priced UAVs give Farmers a new way to improve crop yields. MIT Technology Reviews. Lessiter Publications and Farm Equipment, Brookfield, Wisconsin, USA, pp. 1–3.

11. Ansley, G. (2014). Drone filled skies an emerging headache. The New Zealand Herald. http://www.nzherald.co.nz/world/news/article.cfm?c_id=z&objectid=11231127, pp. 1–3 (September, 2014).

12. Arjomandi, M. (2013). Classification of Unmanned Aerial Vehicles. The University of Adelaide, Australia. http://www. personal.mecheng.adelaide.edu/major/arjomandi/ aeronautical%20engineering%20/projects/2006/groupd9.pdf, pp. 1–45 (July 23rd, 2014).

13. ASME 2012 Unmanned Aerial Vehicles Soar High. American Society of mechanical Engineers.htps://asme.org/engineering-topics/articles/robotics/unmanned-aerial-vehicles-soar-high.htm, pp. 1–3 (March 20th, 2014).

14. AUVSIAdmin, 2012 Corn and Soybean Digest: Put Crop Scouting on Auto-Pilot. Corn and Soybean Digest, http:// http://increasinghumanpotential.org/corn-and-soybean-digest-put-crop-scouting-on-auto-pilot/, pp. 1–3 (September 8th, 2013.

15. Axel, B., Sandro, L. (2014). Development of Micro UAV swarms. In: Autonome Mobile System. Springer Verlag Berlin Heidelberg, Germany, pp. 217.

16. Aylor, D. E., Schmate, D. G., Sheilds, E. J., Newcombe, M., Nappo, C. J. (2011). Tracking the potato late bight pathogen in the atmosphere using Unmanned Aerial Vehicles and Lagrangian modeling. Agriculture and Forest meteorology 151:251–260.

17. Azorobotics, 2014 Explore use of Drones, UAVs and Crop Models at growing Michigan Agriculture Conference. Azorobotics.com, pp. 1–3 (August 12th, 2014).

18. Ball, M. (2012). Let the Drone mapping Race Begin. http://www.sensysmag.com/ spatialsustain/let-the-drone-mapping-race-begin.html, pp. 1–2 (August 6th, 2014).

19. Barnard Microsystems Ltd 2013 Yamaha R-MAX Type-II-G unmanned helicopter. http://www.barnardmicrosystems.com/UAV/uav_list/yamaha_rmax.html, pp. 1–5 (July, 26th, 2014).

20. Basso, B., Rush, C. (2013). Montana State researchers wheat crop disease from the air. http://diydrones.com/profiles/blogs/montana-state-researches-wheat-crop-disease-from-the-air-1, pp. 1–6 (August 12th, 2014).

21. Bellvert, J., Zarco-Tajeda, Girona, J., Ferreres, E. (2013). Mapping crop water stress index in a 'Pinot-noir' vineyard: Comparing ground measurements with thermal remote sensing imagery from an unmanned aerial vehicle. Precision Agriculture DOI 01.1007/x11119–013–9334–5, pp. 1–22 (September 9th, 2014).

22. Bending, J., Wilkomm, M., Tilly, N., Guyp, M. L., Bennertz, S., Qiang, C., Miao, Y., Lenz-Weidman, J. S., Bareth, G. (2013). Very high resolution Crop surface Models (CSMs) from UAV-based images for rice growth monitoring in Northeast China. International Archives of the Photogrammetry, Remote Sensing and Spatial Information Sciences, 22:45–50.

23. Bennett, C. (2013). Drones begin descent on US Agriculture. Western Farm Press, http://westernfarmpress.com/blog/drones-begin-descent-us-agriculture, pp. 1–3 (September 6th, 2014).

24. Berni, J. A. J., Zarco-Tejada, P. J., Suarez, L., Feres, E. 2009a Thermal and Narrowband Multispectral Remote Sensing for Vegetation Monitoring from an Unmanned Aerial Vehicle. IEEE Transactions on Geoscience and Remote Sensing. DOI: 10.1109/TGRS.2008.2010457, pp. 1–3 (January 12th, 2015).

25. Berni, J. A. J., Zarco-Tajeda, P. J., Sepulcre-Canto, G., Feres, E., Villalobos, F. 2009b Mapping Canopy conductance and Crop Water Stress Index in Olive orchards using High Resolution thermal sensing imagery. Remote Sensing of Environment 113: 1380–1388.

26. Biermacher, J. T., Epplin, F. M., Brorsen, J. B., Solie, J. B., Raun, W. R. (2006). Maximum benefit of a Precise Nitrogen Application System for wheat. Precision Agriculture 7: 193–204.

27. Blackmer, T. M., Schepers, J. S. (1996). Aerial Photography to detect Nitrogen stress in corn. Journal of Plant Physiology 148:440–448.

28. Boike, J., Yoshikawa, K. (2003). Mapping of preiglacial geomorphology using kite/ balloon aerial photography. Permafrost Periglacial Process 14:81–85.

29. Bowman, L. (2014). UAV drones for farmers and Ranchers. Ohio State University Extension Service. Farm Science Review, http://fsr.osu.edu, pp. 1 (August 20th, 2014).

30. Buerkert, A., Mahler, F., Marchner, H. (1996). Soil productivity management and plant growth in the Sahel: Potential of an aerial monitoring technique. Plant and soil 180: 29–38.

31. Caldwell, J. (2014). Trimble rolls out crop scouting drone. http://www.agriculure. com/news/technology/trimble-rolls-out-cropscouting-drone-6-ar41300.html, pp. 1–2 (July 22nd 2014).

32. Carr, E. B. (2004). Unmanned Arial Vehicles: Examining the Safety, Security and Regulatory issues of Integration into US Air Space. http://www.ncpa.org/pdfs/sp_ drones-long-papers.pdf, pp. 1–55 (September 20th, 2014).

33. Chang, J., Clay, D. E., Dasted, S., Clay, S., O'Niel, M. (2003). Corn (Zea mays) yield prediction using multispectral and multidate reflectance. Agronomy Journal 95: 1447–1453.

34. Chang, K., Shen, W. Y., Lo, J. C. (2005). Predicting rice yield using canopy reflectance measured at booting stage. Agronomy Journal 97: 872–878.

35. Chakravorty, S. (2013). Oil Field Drones: Monitoring Oil pipelines with Drones. Energy and Capital http://www.energyand capital.com/articles/oil-field-drones/3490, pp. 1–3 (February 15th, 2015).

36. Cheng, S. W. (2008). Rapid deployment UAV Academic Search 10.1109/ AERO.2008.4526564, pp. 1–2 (August 23rd, 2014).

37. Chester, S. (2014). UAV: Growing for Australia. http://www.spatialsource.com.au, pp. 102 (July 23rd, 2014).

38. Cornett, R. (2013). Drones and pesticide spraying a promising partnership. Western Plant Health Association, Western Farm Press, http://westernfarmpress.com/grapes/drones-and-pesticide-spraying-promising-partnership, pp. 1–3 (August 30th, 2014).

39. Costa, F. G., 2012 The use of unmanned aerial vehicles and wireless sensor network in agricultural applications. Proceedings of Geoscience and Remote Sensing Symposium (IGARSS)., pp. 5045–5048.

40. DARC, 2013 About the Drones and Aerial Robotics Conference. New York city, New York, USA, https://droneconference.org, pp. 1–8 (March 20th, 2014).

41. DeAngelis, S. (2014). High Tech Agriculture. http://www.enterresolutions.com/author/bradd.htm, pp. 1–12 (August 3rd, 2014).

42. Descroix, L., Bouzou, I., Genthon, P. Sighmnou, D., Mahe, G., Mamadou, I. (2011). Impact of Drought and land-use changes on surface-water Quality and Quantity: The Sahel Paradox. INTECH http://dx.doi.org/10.5772/54536, pp. 243–258 (September 9th, 2014).

43. Descroix, l., Genthon, P., Amogu, O., Rajot, J. Sighomnou, D., Vauchin, M. (2012). Hydrograph: The case of recent red the Niamey region. Global and Planetary Change 99: 18–30.

44. Doraiswamy, P. C., Moulin, S., Cook, P. W., Stern, A. (2003). Crop yield assessment from remote sensing. Photogrammetry Engineering and Remote Sensing 69: 665–674.

45. Dreiling, L. (2012). New Drone aircraft act as scout. http;//www.hpj.com/archives/2012/apr12/apr16/SpringPlanting MACOLDsr.cfm, pp. 1–3 (March 8th, 2014).

46. DMZ. Aerial Autonomous Scouting Robotics 2013 Unmanned Aerial Vehicles and scouting. http://www.dmzaerial.com/uavscouting.html., pp. 10 (August 15th, 2014).

47. Dobberstein, J. (2014). Drones could change face of No-Tilling. No-Till Farmer http://www.no-tillfarmer.com/pages/Spre/SPRE-Drones-Could-Change-Face-of-No-Tilling-May-1, −2013.php, pp. 1–7 (July 26th, 2014).

48. Dobberstein, J., Kanicki, D., Zemlicka, J. (2014). The Drones are coming: Where are the dealers. Farm Equipment. http://www.farm-equipment.com/pages/From-the-October-2013-Issue-The-Drones-are-Coming-Where-are-the-Dealers.php, pp. 1–16 (September 6th, 2014).

49. Dobbs, T. (2013). Farms of the future will run on Robots and Drones. http://pbs.org/wgbh/nova/next/author/taylor-dobbs/, pp. 1–22 (July 3rd, 2014).

50. Doering C 2013 Growing use of Drones is poised to transform Agriculture. USA Today http://www.usatoday.com/story/money/business/2014/03/23/drones-agriculture-growth/6665561/.html, pp. 104 (August 2nd, 2014).

51. D'Oleire-Oltmanns, S., Marzolff, I, Peter, K. D., Ries, J. B. (2012). Unmanned Aerial Vehicle (UAV) for Monitoring soil Erosion in Morocco. Remote Sensing 4: 3390–3416.

52. Doward, J. (2012). Rise of drones in United Kingdom Airspace prompts civil liberties warning. The Guardian. http://www.theguardian.com/world/2012/oct/07/drones-uk-civil-liberty-fears, pp. 1–6 (September 9th, 2014).

53. Dreiling, L. (2012). New Drone aircraft to act as Crop Scout. http://www.hpj.com/ archives/2012/apr12/apr16/SpringPlantingMACOLDsr.cfm, pp. 1–2 (February 15th, 2015).

54. DroneLife 2014 Precision Hawk: Designing the future of Agricultural UAS. http:// www.dronelife.cm/2014/04/29/precisionhawk.ariculture-drone/, pp. 1–4 (August 5th, 2014).

55. Duan, S., Li, Z. L., Wu, H., Ma, L., Zhao, E., Li, C. (2014). Inversion of the PRO-SAIL. Model to estimate leaf area index of maize, potato and sunflower fields from Unmanned Aerial Vehicle hyperspectral data. International Journal of Applied Earth Observation and Geoinformation. 26: 12–20.

56. Editor (2014). Up and Coming Precision Agriculture Technology: Farm Drones. Cropmetrics. http://cropmetrics.com/2014/01/up-and-coming-precision-agriculture-technology-farm-drones/, pp. 1–8 (March 20th, 2014).

57. Effren, D. (2014). AutoCopter-The Precision Ag solution. PR Newswire. http:// ireach.prnewswire.com, pp. 1–2 (August 25th, 2014).

58. Ehmke, T. (2013). Unmanned Aerial Systems for Field Scouting and Spraying. CSA News, pp. 4–9.

59. Ehsani, R., Sankaran, S, Maja, J and Garcia, F. (2012). Advanced-stress-detection technologies for citrus. Citrus Industry http://www.crec.ifas.ufl.edu/extension/trade_ journals/2012/2012_May_tree_stress.pdf, pp. 1–7 (September 2nd, 2014).

60. Ellerbroek, P 2014a In: Growing use of Drones is poised to transform Agriculture. Doering, C. (Ed.). USA Today, http://www.usatoday.com/story/money/business/2014/03/23/drones-agriculture-growth/6665561/.html, pp. 1–4 (August 2nd, 2014).

61. Ellerbroek, P 2014b In: Roboflight seeks partnership with Wal-Mart for precision Ag project. Industry News http://www.farm-equipment.com/pages/Spre/PFD-Precision-News-RoboFlight-Seeks-Partnership-with-Wal-Mart-for-Precision-Ag-Project-March-27, –2014.php, pp. 1–3 (April 9th, 2015).

62. EMBRAPA, 2014 The Drones land in Brazil. Fresh Plaza. http://www.freshplaza. com/article/120401/The-drones-land-in-Brazilian-agriculture, pp. 1–3 (September 8th, 2014).

63. EPIC 2014 Domestic Unmanned Aerial Vehicles (UAVs) and Drones. Electronic and Privacy Information Centre. https://epic.org/privacy/drones.htm, pp. 1–14 (March 14th, 2014).

64. Epp,M.(2013).UAVsTakingAgronomytoNewHeights.CropLifehttp://www.croplife. com/equipment/uavs-taking-agronomy-to-new-heights/, pp. 1–2 (March, 20th, 2014).

65. Farm Aerial Drones 2014 Farm aerial drone provides you with precision. http:// farmaerial.co.uk, pp. 1–5 (March 20th, 2014).

66. Filella, I., Serrano, L., Penuclas, J. (1995). Evaluating wheat nitrogen status with canopy reflectance indices and discrimination analysis. Crop Science 35: 1400–1405.

67. Fitzgerald, G. J., Rodgrigues, D., O'Leary. (2010). Measuring and predicting canopy nitrogen nutrition in wheat using a spectral index-the canopy chlorophyll content index. Field Crops Research 118: 324–328.

68. Fitzpatrick, S., Burnett, K. (2013). Regulation and use of Drones in Canada. The Canadian Bar Association, http://www.cba.org/CBA/section_airandspace/newsletter/2013/drones.aspx. html, pp. 1–2 (September 21st, 2014).

69. Funk Seeds Inc., 2014 Funk's yield launch is the secret to increasing production. http://billfunkinc.com/fuav-program/, pp. 1–2 (August 4th, 2014).

70. Gale Encyclopedia of Espionage and Intelligence, 2014 Unmanned Aerial Vehicles. http://www.answers.com/ topic/unmanned-aerial-vehicle, pp. 1–32 (July 24th, 2014).

71. Galimberti, K. 2014a Can Drones offer new ways to predict Storms, Save lives. http://www.accuweather.com/en/weather-news/drone-weather-safety-prediction/27739091, pp. 1–4 (September, 20th, 2014).

72. Galimberti, K. 2014b Drones offer new horizon, solutions for weather modification. http://adam.curry.com/art/1402371079_bmc3E96z.html, pp. 1–4 (September 20th, 2014).

73. Garcia-Ruiz, F., Sankaran, S., Maje J. M., Lee, W. S., Rasmussen, J., Ehsani, R. (2013). Comparison of Two imaging platforms for identification of Huanglongbing-infected Citrus disease. Computers and Electronics in Agriculture 91: 106–115.

74. Ghose, T. (2013). Drones and Agriculture: Unmanned Aircraft may Revolutionize Farming, Experts Say. Hust Post Science. http://www.huffingtonpost.com/2013/05/20/drones-agriculture-unmanned-aircraft-farming_n_3308164.html, pp. 8–14 (September 3rd, 2014).

75. Gil-Perez, B., Zarco-Tajeda, P. J., Correa-Guimaraes, A., Relea-Gangas, E., Navas-Garcia, L. M., Hernandez-Navarro, S., Sanz-Requena, J. F., Berjon, A., Martin-Gil, J. (2010). Remote sensing detection of Nutrient uptake in Vineyards using narrow-Band Hyperspectral Imagery. Vitis 49:167–173.

76. Gogarty, B., Robinson, I. (2012). Unmanned Vehicles: A History, Background and Current State of the Art. Journal of Law, Information and Science 21: 1–14.

77. Gonzalez, L. (2013). Agricultural groups experiment with Unmanned Vehicles to Manage crops. http://.psfk.com/ 2013/05/farmrs-drones-crops.html#yR79h.htm, pp. 1–5 (July 3rd, 2014).

78. Glickhouse, R. (2013). Explainer: Drones in Latin America. http://www.as-coa.org/articles/explainer-drones-latin-america., pp. 1–5 (August 3rd, 2014).

79. Grassi, M. J. (2013). Rise of the Ag Drones. Kawak Technologies Inc., http://kawak-aviation.com/what-people-say/rise-of-the-ag-drones/, pp. 1–4 (August 6th, 2014).

80. Grenzdorffer, G. J., Engelb, A., Teichertc, B. (2008). The photogrammetric potential of low-cost UAVs in Forestry and Agriculture. http://www.isprs.org/proceedings/XXXVII/congress/1_pdf/206.pdf, pp. 1–12 (August, 2014).

81. Green, M. (2013). Unmanned Drones may have their greatest impact on Agriculture. http://www.thedailybeast.com/articles/2013/03/26unmnned-drones-may-have-their-greatest-impact-on-agriculture.html#stash.c36uDpsT.dpuf., pp. 1–4 (August 3rd, 2013).

82. Hall, A., Lamb, D. Holzapel, B., Louis, J. (2011). Optical Remote sensing applications in Viticulture—A Review. Australian Journal of Grape and Wine Research 8: 36–47.

83. Hall, A., Louis, J. P. (2008). Low resolution remotely sensed images of wine grape vineyards map spatial variability in planimetric canopy area instead of leaf area index. Australian Journal of Grape and Wine Research 14: 9–17.

84. Hamm, P. (2013). Farming's Newest Precision Agriculture Tool Takes Data to New Heights. http://unitedsoybean.org/article/new-precision-agriculture-could-revolutionize-farming/#sthash.Leri7bSa.dpuf, pp. 1–4 (August 12th, 2014).

85. Han, S. Hendrickson, L., Ni, B. (2001). Comparison of Satellite Remote sensing and aerial photography for ability to detect in-season nitrogen stress in corn. An ASAE Meeting presentation paper No. 01–1142, ASAE, St Joseph, Michigan, USA, pp. 1–3.

86. Hansen, P. M., Schjoerring, J. K. (2003). Reflectance measurement of Canopy biomass and Nitrogen status in Wheat crops using Normalized difference Vegetation Indices and partial least squares regression. Remote Sensing of Environment 86: 542–553.

87. Heacox, L. (2014). Real-World UAV experience in Agriculture. CropLife. http://www. croplife.com/equipment/real-world-uav-experience-in-agriculture.htm, pp. 1–4 (March 20th, 2014).

88. Herwitz, R. (2002). Coffee harvest optimization using Pathfinder-Plus (solar powered aircraft). National Aeronautics and Space Agency, USA, http://www.nasa.gov/missions/research/FS-2002–9-01-ARC.html. (September 2nd, 2014).

89. Herwitz, S. R., Johnson, L. F., Dunagan, S. E., Higgins, R. G., Sullivan, D. V., Zheng, J., Lobitz, B. M., Leung, J. G., Gallmeyer, B. A., Aoyagi, M., Bass, J. A. (2004). Imaging from an unmanned Aerial Vehicle: Agricultural Surveillance and Decision Support. Electronics in Agriculture 44:49–61.

90. Hetterick, H. (2013). Drones can Positive and Negative for the Agricultural Industry. Ohio's Country Journal. Ag Net. http://ocj.com/2013/05/drones-can-be-positive-and-negative-for-the-ag-industry/pp 1–2 (August 14th, 2014).

91. Hoffman, W. C. (2013). Aerial application research for efficient crop production. Agricultural Research Service, College Station, Texas, USA, http://www.ars.usda.gov/pandp/people/people.htm?personid=2539, pp. 1–7 (September 22nd, 2014).

92. Horcher and Visser, 2005 Unmanned Aerial Vehicles: Applications for natural Resources Management and Monitoring. http://uavm.com/images/Forest_Mangement-UAV_for_resources_Management.pdf, pp. 1–5 (July 20th, 2014).

93. Huang, Y., Hoffmann, W. C., Lan, Y., Thomson, S. J., Fritz, B. K. 2010a Development of Unmanned Aerial Vehicles for Site-Specific Crop Production Management. Proceedings of 10th International Conference Precision Agriculture, Denver, Colorado, CDROM http://www.ars.usda.gov/research/publictions/publictions.htm?SEQ_NO_115= 253718, pp. 1 (March 20th, 2014).

94. Huang, Y., Lan, Y., Westbrook, J. K., Hoffman, W. C. (2008). Remote Sensing and GIS applications for Precision area-wide Pest Management.: Implications for homeland security. In: Geospatial Technologies and Home Security. Sui, D. Z., Cutter, S. L. (Eds.). Research Frontiers and Challenges, New York, NY, Springer, pp. 242–256.

95. Huang, Y., Steven, J.T, Hoffmann, C., Lan, Y., Fritz, B. K. (2013). Development and Prospect of Unmanned Arial Vehicle Technologies for Agricultural Production Management. Journal of Agricultural and Biological Engineering 6:1–10.

96. Huang Y., Thomson, S. J., Lan, Y., Maas, S. J. 2010b Multispectral Imaging systems for Airborne remote Sensing to support Agricultural production Management. International Journal of Agricultural and Biological Engineering 3: 50–62.

97. Huting, K. (2013). The spectrum of unmanned Arial Vehicles for Agriculture. Farm Industry News. http://www.farmindustrynews.com/precision-farming/spectrum-unmanned-aerial-vehciles-agriculture.htm, pp. 1–4 (March 20th, 2014).

98. Impey, L. (2014). Drones to have a bigger role in mapping arable crops. Farmers Weekly. http://www.fwi.co.uk/articles/22/01/2014/142533/drones-to-have-a-bigger-role-in-mapping-arable-crops.htm, pp. 1–4 (September 21st, 2014).

99. Innova, A. (2009). Estimating crop water needs using Unmanned Aerial Vehicles. Science Daily http://www.sciencedaily.com/releases/2009/07/090707094702.htm, pp. 1–5.

100. I'nen, I. P., Saari, H., Kaivo Saijo, J., Honkavaara, E., Pesonen, L. (2013). Hyperspectral imaging based biomass and Nitrogen content estimations from light-weight UAV. Proceedings of SPIE 8887 Remote sensing for Agriculture, Ecosystems and Hydrology. DOI: 10.1117/12.2028624, pp. 1–14 (January 12th, 2015).

101. Jaakkola, A., Hyyppa, J., Kukko, A., Yu, X., Kaartinen, H., Lahtomaki, M., Lin, Y. (2010). A low cost multispectral mobile mapping system and its feasibility for tree measurements. ISPRS Journal of Photogrammetry and Remote Sensing 65: 514–622.

102. Japan UAV Association 2014 The Japan Unmanned Arial Vehicle (UAV0 Association. http://www.juav.org, pp. 1–2 (August 8th, 2014).

103. Jarvis, B. (2014). Drones are helping meteorologists decipher tropical cyclones. Novnext. http://www.pbs.org/wgbh/nova/next/earth/drone-meteorology/., pp. 1–3 (September 21st, 2014).

104. Jensen, A. M., McKee, M., Chen, Y. (2014). Procedure and processing thermal images using Low cost Microbolometer Cameras for small Unmanned aerial systems. IEE/IGARSS International Geoscience and Remote Sensing Symposium Quebec City, Canada., pp. 34–45.

105. Jensen, T., Apan, A., Young, F., Zeller, L., Cleminson, K. (2007). Detecting the attributes of a wheat crop using digital imagery acquired from a low altitude platform. Computers and Electronics in Agriculture 59:66–77.

106. Jeon, H. Y. Zhu, H. Derksen, R., Ozkan, E., Krause, C. (2013). Evaluation of Ultrasonic Sensor for Variable-rate spray applications. Computers and Electronics in Agriculture 75: 213–221.

107. Johansen, K., Phinn, S., Witte, C., Seasonal, P., Newton, L. (2009). Mapping Banana plantations from object-oriented Classification of SPOT-5 imagery. Photogrammetric Engineering and Remote Sensing 75: 1069–1081.

108. Johanssen, J. (2014). Drones to increase Profitability. http://agwired.com/2014/01/13/drones-to-increase-profitability/, pp. 1–3 (August 15th, 2014).

109. Jones, G., Gee, C., Truchet, F. (2006). Crop/weed discrimination in simulated images. http://www.agrosupdijon.fr/fileadmin/user_upload/pdf/Recherche/UP_GAP/19[th]_SPIE_SanJose07_jones.pdf, pp. 1–12 (August 30th, 2014).

110. Junge, B. (2008). Soil Conservation options in the Savannahs of Africa: New Approaches to assess their potential. An overview of BMZ/GTZ-Project. International Institute for Tropical Agriculture, Ibadan, Nigeria, http://www.slideshare.net/IITA-CO/soil-conservation-options-in-the-savanna-of-west-africa-new-approaches-to-assess- their -potential., pp. 1–18 (March 17th, 2014).

111. Jouzapavicius, J. (2013). How weather Drones will Unravel How tornados are formed. The HuftingtonPost.com., pp. 1–3 (September 20th, 2014).

112. Kahabka, J. E., Es, M. V. McClenahan, J., Cox, W. J. (2004). Spatial analysis of maize response to nitrogen fertilizer in Central New York State. Precision Agriculture 5: 463–476.

113. Kaiser, S. A. (2006). Legal Aspects of Unmanned Aerial Vehicles. Zeitschrift Fur Luft-Und Weltraum- Recht. 55: 345–346.

114. Kara, A. (2013). Low Cost UAV fights disease devastating Apple Crops. https://www.linkedin.com/groups/Low-Cost-UAV-Fights-Disease-53140.S. 5809026007457361924, pp. 109(January 15th, 2015).

115. Keane, J. F., Carr, S. S. (2013). A brief history of Early Unmanned Aircraft. Johns Hopkins Technical Digest 3: 558–567.

116. Keber, P. (2013). Weather Drones: New Technology in forecasting. The Weather Network. http://www.theweathernetwork.com/news/articles/weather-drones-new-technology-in-weather-forecasting/7772/, pp. 1–3 (September 20th, 2014).

117. Keller, R. (2014). Agriimage adds innovative UAVs to the market. AG Professional http://www.agprofessional.com/media-center/Agriimage-adds-innovative-UAVs-to-the-market-249441971.html?source=related, pp. 1–2 (March 12th, 2014).

118. Kennedy, M. W. (1998). The moderate course of United States Air Force for Unmanned Aerial Vehicle Development. Research Report No 0211, pp. 41.

119. Kimberlin, J. (2013). Virginia Beach crop drone gives peek into farming's future. Pilotline.com, pp. 1 (August 7th, 2014).

120. Knoth, C., Prinz, T. (2013). UAV-based high resolution remote sensing as an innovative monitoring tool for effective crop management. Institute for Geo-informatics, University of Munster, Germany. http://www.slideshare.net/CKnoth/uavbased-highresolution-remote-sensing-as-an-innovative-monitoring-tool-for-effective-crop-management-18052317, pp. 1–22 (September 22nd, 2014).

121. Koh, L. P., Wich, S. A. (2012). Dawn of Drone Ecology: Low-cost Autonomous Aerial Vehicles for conservation. Journal of Tropical Conservation Science 5: 121:121–132.

122. Koontz, L. (2013). Drones become a new tool in scouting crop. http://www.brwn-fieldagnews.com/2013/07/03/drones-become-a-new-tool-in-scouting-crop-fields., pp. 1–2 (July 22nd, 2014).

123. Krienka, B., Ward, N. (2013). Unmanned Arial Vehicles (UAVs) for Crop Sensing. Proceedings of West Central Crops and Water Field Day. University of Nebraska, Lincoln, NE, USA. p. 2.

124. KSU 2013 Small Unmanned Aircraft Systems for Crop and Grassland Monitoring. Farms Com http://www.agronomy.k-state.edu/documents/eupdates/eupdate040513. pdf, pp. 1–4 (January 24th, 2015).

125. Kunde, R. (2013). In: Drones plus Wine: How UAVs can help farmers harvest grapes. 3D Robotics Inc., http://www.3drobotics.com/2013/drones-wine-how-uavs-can-help-farmers-harvest-grapes/, pp. 1–4 (August 2nd, 2014).

126. Labre Crop Consulting Inc., 2014 Iowa crop consulting firm offers Ag services from drones. http://www.agprofessional.com/news/dealer-update-articles/Iowa-crop-con-sulting-firm-offers-ag-services-from-drones-245201631.html, pp. 1–5 (August 10th, 2014).

127. Lamb, D., Hall, A., Louis, J. (2001). Airborne remote sensing of Vines for Canopy variability and Productivity. Australian Grape Grower and Winemaker 5: 89–94.

128. Lamb, D., Hall, A., Louis. (2013). Airborne/Spaceborne Remote Sensing for the Grape and Wine Industry. Geospatial Information and Agriculture. http://regional. org.au/au/gia/18/600lamb.htm, pp. 1–5 (September 5th, 2014).

129. Lamb, K. 2014a Missouri Farmers consider using Drones. Washington times http://www.washingtontimes.com/news/2014/jan/17/mo-farmers-consider-using-drones/?page=all, pp. 1–8 (August 13th, 2014).

130. Lamb, K. 2014b Drones could be timesaver for farmers. The Joplin Globe. http://www.joplinglobe.com/topstories/x2011611121/Drones-could-be-timesaver-for-farmers., pp. 1–4 (August 13th, 2014).

131. Lee, W. S., Chichulum, R., Ehsani, R. (2009). Airborne hyperspectral imaging for citrus greening disease detection. Proceedings of the 3rd Asian Conference on Precision Agriculture (ACPA) Beijing, China, pp. 244–149.

132. LeMieux, J. (2012). Unmanned Aerial Vehicles (UAV): How will widespread domestic use of Drones change our lives. http://www.quora.com/How-will-widespread-domestic-use-of-drones-change-our-lives, pp. 1–7 (September 7th, 2014).

133. Leon, C. T., Shaw, D. R. Cox, M. S., Abshire, M. J., Ward, D., Wardlaw, M. C., Watson, C. (2003). Utility of Remote Sensing in predicting Crop and Soil characteristics. Precision Agriculture 4: 359–384.

134. Lerner, K. L. (2004). "Unmanned Aerial Vehicles (UAVs)." Encyclopedia of Espionage, Intelligence, and Security. 2004. *Encyclopedia.com*. (July 24, 2014). http://www.encyclopedia.com/doc/1G2–3403300785.html, pp. 1–7.

135. Linehan, P. (2012). Drones and Natural Resources. White Pine https://sites.psu.edu/whitepine/2012/08/13/drones-and-natural-resources/, pp. 1–2 (September 9th, 2014).

136. Lumpkin, T. (2012). CGIAR Research Programs on Wheat and Maize: Addressing global Hunger. International Centre for Maize and Wheat (CIMMYT), Mexico. DG's Report, pp. 1–8.

137. Lyseng, R. (2006). Ag Drones: Farm tools or Expensive toys. The Western Producer, http://www.producer.com/2006/02/ag-drones-farm-tools-or-expensive-toys/, pp. 1–5 (September 7th, 2014).

138. Magrez, B. (2014). Magrez embraces drone technology in Bordeaux, France. Wine Searcher http://www.wine-searcher.com/m/2013/12/drone-technology-takes-off-in-bordeaux, pp. 1–3 (September 2nd, 2014).

139. Martinsanz, G. P. (2011). Sensors in Agriculture and Forestry. Sensors 11: 8939–8932.

140. Mattern, P. (2013). Crop Scouts in the Sky: Researchers use drones on mission to assess crop stress. Solutions. http:// www.cfans.umn.edu, pp. 1–4 (March 13th, 2014).

141. Meier, W. (2014). The future of drones in Agriculture. Airborne. http://www.airbornedronescom/blogs/news/15025505-thefuture-of-drones-in-agriculture., pp. 1–4 (September 21st, 2014).

142. Miao, Y., Stewart, B. A., Zhang, F. (2012). Long-term experiment for Sustainable Nutrient Management in China. A Review. Agronomy for Sustainable Development 31: 397–414.

143. Minghe, 2014 Minghe UAV. http://dgmingh.1688.com/, pp. 1 (August, 7th, 2014).

144. Ministry of Agriculture, 2013 Beijing applies "Helicopter" in Wheat pest control. Ministry of Agriculture of the Peoples Republic of China-Report http://english.agri.gov.cn/news/dqnf/201306/t20130605_19767.htm, pp. 1–3 (August 10th, 2014).

145. Moritani, S. (2010). Monitoring of Soil Surface under Wind and Water Erosion by Photogrammetry. http://www.intechopen.com, pp. 1–24.

146. Mortimer, G. (2013). 'Skywalker' Aeronautical Technology to improve maize yield in Zimbabwe. DIY Drones http://diydrones.com/profiles/blogs/skywalker-aeronautical-technology-to-improve-maize-yields-in-zimb, pp. 1–5 (September 5th, 2014).

147. MSS Drone Division 2014 Larger drones can help farmers in broadcasting seeds. http://www.midsecsystem.com/drones.htm, pp. 1–3 (September 21st, 2014).

148. Murray, P. (2013). Drones close in on farms, the next step in Precision Agriculture. http://singularityhub.com/2013/05/28/drones-close-in-on-farms-the-next-step-in-precision-agriculture/, pp. 1–5 (March 20th, 2014).

149. New Scientist Tech 2013 Precision Herbicides drones launch strikes on Weeds. http://www.newscientist.com/article/dn23783-precision-herbicide-drones-launch-strikes-onweeds.html#. VACJrvmSx0w, pp. 1–3 (August 30th, 2014).

150. Noble, D. (2014). Drone scouting demonstrations is used at Field day. American-Farm.com, pp. 1 (August 22nd, 2014).

151. Okamoto, H., Murata, T., Kataoka and Hata, S. (2007). Plant classification for weed detection using hyperspectral imaging with wavelength analysis. Weed Biology and Management 7: 31–37.

152. Oregon State University, 2014 Drones to check out Acres of Potatoes. Crops and Soils Oregon State University, Corvallis, OR, USA, http://www.cropandsoil.oregon-state.edu/contents/drones-check-out-acres-potatoes. Html, pp. 1 (July 23rd, 2014).

153. Oryza 2014 Can Rice farmers use drones for fertilizer Management?. http://www.oryza.com/news/rice-news/can-rice-farmers-use-drones-fertilizer-management. p. 1 (September, 4th, 2014).

154. OSU, 2014 Unmanned Aerial Vehicles (UAV) Demo. http://fabe.osu.edu/about-us/multimedia/unmanned-aerial-vehicle-uav-demo.htm, pp. 1–4 (August 9th, 2014).

155. Pasqua, A. (2013). Use of Terrestrial laser Scanning (TLS) and Unmanned Aerial Vehicles (UAV) to investigate Rice Cultivation on the Isle of Hope, Georgia. Centre for Geospatial Research, Department of Geography, The University of Georgia, Athens, Georgia, USA. Internal Report, pp. 1–12.

156. Paskulin, A. (2013). Drones plus Wine: How UAVs can help farmers harvest grapes. #D Robotics Inc., http://www.3drobotics.com/2013/drones-wine-how-uavs-can-help-farmers-harvest-grapes/, pp. 1–4 (August 2nd, 2014).

157. Pates, M. (2014). Drones for farms a challenge, but popular at Precision Ag Summit. AGWEEK http://www.agweek.com /event/article/id/22532/, pp. 1–3 (March 12th, 2014).

158. Paul, R. (2014). In: Drone could change face of No-till. Dobberstein, J. (Ed.). No-Till Farmer http://www.no-tillfarmer.com/pages/Spre/SPRE-Drones-Could-Change-Face-of-No-Tilling-May-1, −2013.php, pp. 1 (July 26th, 2014).

159. Pefia, J. S., Torres-Sanchez, J., Isabel de Castro, A., Kelly, M., Lopez-Granados, F. (2013). Weed mapping in Early-Season maize fields using Object-based analysis of Unmanned Aerial Vehicles (UAV) images. Institute for Sustainable Agriculture. CSIC Spain, DOI: 101371/journal.pone.0077151, pp. 1–16 (August 23rd, 2014).

160. Peng, S., Buresh, R. J., Huang, J., Yang, J., Zou, Y., Zhong, X., Wang, G., Zhang, F. (2011). Strategies for overcoming low agronomic nitrogen use Efficiency in irrigated Rice systems in China. Field Crops Research 96: 37–47.

161. Perry, E. M., Band, J., Kant, S., Fitzgerald, G. J. 2012a Field based rapid phenotyping with Unmanned Aerial Vehicles (UAV). http://www.regional.org.au/au/asa/2012/precision-agriculture/7933_perrym.htm, pp. 1–5 (August 23rd, 2014).

162. Perry, E. M., Fitzgerald, G. J., Nutall, J. G., O'Leary., Schulthess, U., Whitlock, A. 2012b Rapid estimation of Canopy Nitrogen of Cereal crops at Paddock scale using a Canopy Chlorophyll Content index. Field Crops Research 118: 567–578.

163. Precision Farm Dealer 2014a UAV Technology Symposium to demonstrate potential uses in Agriculture. Delta Ag Tech Symposium, Memphis, Tennessee, USA http://www.precisionfarmingdealer.com/content/uav-technology-symposium-demonstrate-potential-uses-agriculture, pp. 1–3 (August 14th, 2–14).

164. Precision Farming Dealer 2014b Unmanned Yamaha RMAX helicopter sprayer displayed at AUVSI conference. http://www.precisionfarmingdealer.com/content/content/unmanned-yamaha-rmax-helicopter-sprayer-displayed-auvsi-conference.htm, pp. 1–3 (May 16th, 2014).

165. Picard, R. (2014). Using Aerial Imagery for Site-Specific Fungicide Application. http://www.umanitoba.ca/ faculties/afs /agronomists-conf/media/Rejean_Picard_fungicide-app-poster.pdf, pp. 1 (July 22nd, 2014).

166. Precision Drone LLC, 2014 Precision Scout Drone. http://www.precisiondrone.com/precision-drone-dealers.html, pp. 1–7 (August 30th, 2014).

167. Precision Hawk 2014a Lancaster Platform. Precisionhawk.com, pp. 1–17 (August 5th, 2014).

168. Precision Hawk 2014b Lancaster Platforms. http://www.precisionhawk.com/index.html#industries., pp. 3–4 August 5th, 2014).

169. Precision LLC, 2014 Drones for Agricultural Crop Surveillance. http://www.precisiondrone.com/drones-for-agriculture. htm, pp. 1–3 (July 23rd, 2014).

170. Price 2013 AgCenter Researchers study use of Drones in Crop Monitoring. Louisiana State University. Baton Rouge, USA http://www.lsuagcenter.com/news_archive/2013/december/headline_news/AgCenter-researchers-study-use-of-drones-in-crop-monitoring.htm, pp. 1–3 (March 8th, 2014).

171. Primecerio, T., Filippo Di Gennaro, S., Fiorillo, J., Generio, L., Lugato, E., Matese, A., Vaccari, F. P. (2012). A flexible unmanned aerial vehicle for precision Agriculture. Precision Agriculture 13: 517–523.

172. Privacy Commissioner of Canada, 2013 Drones in Canada: Will the proliferation of domestic drone use in Canada raise new concerns for privacy. http://www.priv.ge.ca/information/research./drones 201303_e_asp., pp. 1–23 (September 21st, 2014).

173. Pudelko, R., Kozyra, J., Nierobca, P. (2008). Identification of the Intensity of Weeds in Maize plantations based on aerial photography. Zembdirbyste 3:130–134.

174. Pudelko, T., Stuczynski, T., Borzecka-Walker, M. (2012). The suitability of an unmanned Aerial Vehicle (UAV) for the evaluation of Experimental fields and crops. Zemdirbyste-Agriculture 99: 431–436.

175. Rama Rao, Kapoor, M., Shrama, N., Venkateswarulu 2007 Yield prediction and waterlogging assessment for tea plantation land using satellite imaging based techniques. International Journal of remote Sensing 28: 1561–1567.

176. Raymond Hunt, E., Hively, W. D., Fujikawa, S. J., Linden, D. S., Daughtry, S. S. T., McCarty, G. W. (2010). Acquisition of NIR-Green-Blue digital photographs from Unmanned Aircraft for Crop monitoring. Remote Sensing 2: 290–305.

177. Redmond, S. (2014). The future of UAV's for Agriculture. Hensall District Cooperatives. http://www.hdc.on.ca/grain-marketing/hdc-reports/29-grain-marketing/253-hdc-future-ofUAV-ag-steve-redmond.html, pp. 1–3 (July 21st, 2014).

178. Reese, M and Higgins, T. (2013). Cleared for takeoff at the Farm Science Review. Ohio Country Journal http://ocj.com/2013/09/cleared-for-takeoff-at-the-farm-science-review/, pp. 1–2 (August 13th, 2014).

179. Reyniers, M., Vrinsts, E., Baerdemaeker, J. (2004). Fine-scaled optical detection of Nitrogen stress in Grain crops. Optical Engineering. 43: 3119–3129.

180. Reyniers, M., Vrinsts, E. (2006). Measuring wheat Nitrogen status from Space and Ground-based platform. International Journal of Remote Sensing 27: 549–567.

181. Rosenstock, S. (2013). How Drones are helping farmers produce healthier crops. DIY Drones http://diydrones.com/profiles/blogs/aha-how-drones-are-helping-farmers-produce-healthier-crops 1–3 (August, 21st, 2013).

182. RT New Team 2013 Latin American drone use on the rise and unregulated- A Report. http://rt, news/latin-american-drones-unregulated-216, pp. 1–3 (August 3rd, 2014).

183. Ruen, J. 2012a Tiny planes coming to scout crops. Drone planes take Aerial imaging to a New level. Corn and Soybean Digest. http://cornandsoybeandigest.com/corn/tiny-planes-coming-scout-crops., pp. 1–3 (August 16th, 2014).

184. Ruen, J. 2012b Take that Red Baron. AG Professional http://editiondigital.net/article/Take+that, +Red+Baron/1108676/0/article.html, pp. 1–3 (August 23rd, 2014).

185. Sarapatka, B., Netopil, P. (2010). Erosion processes on intensively farmed land in Czech Republic: Comparison of Alternative Research Methods. Proceedings of 19th World Congress of Soil Science, soil Solutions for a Changing World. Brisbane, Australia, pp. 47–50.

186. Scharf, P. C., Lory, J. A. (2002). Calibrating corn color from aerial photographs to predict side dress nitrogen need. Agronomy Journal 94: 397–404.

187. Schmale, D. (2012). In: Tiny planes coming to scout crops. Drone planes take aerial imaging to a new level. Ruen, J. (Ed.). Corn and Soybean Digest http://cornandsoybeandigest.com/corn/tiny-planes-coming-scout-crops, pp. 1–3 (August 16th, 2014).

188. Schwing, R. P. (2007). Unmanned Aerial Vehicles-Revolutionary tools in War and Peace. United States Army War College, Research Project Report. http://www.fas/irp/program/collect/docs/97–6230D.pdf.Au/Acsc/0230D/97–03, pp. 1–3 (July 21st, 2014).

189. Schultz, B. (2013). Ag Centre Researchers study use of drones in Crop Monitoring. LSU College of Agriculture Baton Rouge, LA USA, http://www.lsuagcenter.com/news_archive/2013/december/headline_news/AgCenter-researchers-study-use-of-drones-in-crop-monitoring.htm, pp. 1–3 (August 15th, 2015).

190. Scott, W. (2011). Technology used to collect weather data. Bright Hub http://www.brighthub.com/environment/science-environmental/articles/107783.aspx, pp. 1–6 (September 20th, 2014).

191. SenseFly, 2013 Swinglet CAM-user manual. https://www.sensefly.com/fileadmin/user_upload/images/BROCHURE-swingletCAM.pdf, pp. 1–29 (August 4th, 2014).

192. SGS South Africa 2014 Flying in to increase Crop yields and reduce losses. Seed and Crop Services. http://www.sgs.com/—/media/global/Documents/Technique.Htm, pp. 1–7 (September 21st, 2014).

193. Shafri, H. Z. M., Hamdan, N. (2009). Hyperspectral imagery for Mapping Disease infection in Oil palm Plantation using Vegetation indices and Red-Edge Techniques. American Journal of Applied Sciences 6: 1031–1035.

194. Shinn, M. (2014). Future of Drones: Faming looks to flying tractors.' Arizona Daily Star. www.azstar.com. http://tucson.com/business/local/future-of-drones-farming-looks-to-flying-tractors/article_00cc2dea-22a6–5da1–8d6a-00223d5f5d71.html, pp. 1–3 (August 28th, 2014).

195. Silva, G. (2013). Unmanned Aerial Vehicles for Precision Agriculture. Michigan State University Extension Service. http://msue.edu/news/unmanned_arial_vehicles_for_precision_agriculture., pp. 1–2 (March 20th, 2014).
196. Singer, P.W 2009a A Revolution once more. Unmanned Systems and the Middle East. Brookings Institute, Washington D. C. USA, http://www.pwsinger.com/pdf/Future_Horizons_English.pdf, pp. 1–4.
197. Singer P. W. 2009b How the Military can win the Robotic Revolution. The Brookings Institution, Washington D. C. USA. http://www.brookings.edu/research/articles/2010/05/17-robots-singer, pp. 1–6.
198. SkySquirrel Technologies Inc., 2013 Unmanned Aircraft Solutions. http://skysquirrel.ca/application.html/, pp. 1–5 (September 21st, 2014).
199. Stowell, L. J., Gelerntr, W. D. (2013). Unmanned Aerial Vehicles (Drones) for Remote Sensing in Precision Turf grass Management. Water, Food, Energy and Innovation for a Sustainable World. Proceedings of International Annual Meetings of ASA, CSSA and SSSA, Tampa, Florida, USA, pp. 1–2.
200. Stutman, J. (2013). Agricultural Drone Investing: Unmanned Aircraft Industry Ready to sore. Energy and Capital http://www.energyandcapital.com/articles/agricultural-drone-investing/3526&c=sGhRgHFmZIyvrE4C-ZCQJK_K9aw_hqN7kO1jy5WS14U&mkt=en-us, pp. 1–5 (September 20th, 2014).
201. Swain, K. C., Zaman, Q. U. (2010). Rice crop Monitoring with Unmanned Helicopter Remote Sensing Images. http://www.intechopen.com, pp. 253–272.
202. Tadasi, C., Kiyoshi, M., Shigeto, T., Kengo, Y., Shinichi, L., Masami, F., Kota, M. (2010). Monitoring rice growth over a production region using Unmanned Aerial Vehicle: Preliminary trial for establishing Regional Rice strain. 3:178–183.
203. Terraluma, 2014 Applications Selected Case Studies. http://www.terraluma.net/showcases.html, pp. 1–8 (August 28th, 2014).
204. Tetrault, C. (2014). A short History of Unmanned Aerial Vehicles (UAVs). Dragonfly Innovations Inc., http://www.draganfly.com/news/2009/03/04/a-short-history-of-unmanned-aerial-vehicles-uavs/, pp. 1–4 (July 24th, 2014).
205. Texas A and M Agrilife 2014 Researchers using UAV to track disease Progression in Wheat fields. http:// texasagrilifeextension.edu/, pp. 1–2 (August 10th, 2014).
206. Thamm, H. P. (2011). SUSI A robust and safe parachute UAV with long flight time and good payload. International Archives of Photogrammetry, Remote Sensing and Spatial Information Services 38: 1–6.
207. Thamm, H.P and Judex, M. (2006). The Low Cost Drone- An interesting tool for process monitoring in a high spatial and temporal resolution. In: International Archives of the Photogrammetry, remote Sensing and Spatial Information Science, ISPs commission 7th Midterm Symposium. Remote Sensing: From Pixels to processes. Enschede, The Netherlands, 36: 140–144.
208. Thenkabail, P. S., Smith, R. B., Pauw, E. D. (2000). Hyperspectral vegetation indices and their relationship with Agricultural Crop Characteristics. Remote Sensing of Environment 71: 152–182.
209. Thenkabail, P. S., Smith, R. B., Pauw, E. D. (2002). Evaluation of narrowband and Broadband Vegetation Indices for determining Optimal Hyperspectral Wavebands for Agricultural Crop Production. Photogrammetric Engineering and Remote Sensing 68: 607–621.

210. The Des Moines Register 2014 Backers say drones will prove useful for Farmers. http://www.washingtontimes.com/news/2014/mar/23/backers-say-drones-will-prove-useful-for-farmers/?page=all, pp. 1–2 (August 16th, 2014).

211. Thomasson, J. A. Sui, R., Akins, D. C. (2000). Spectral changes in picked cotton leaves with time. In: Proceedings of 5th International conference on Precision agriculture and other precision Resources management. Madison, Wisconsin, USA, pp. 567–569.

212. Thorp, K. R., Tian, L. F. (2004). A Review on Remote sensing of weeds in Agriculture. Precision Agriculture 5: 477–508.

213. Tokekar, P., Hook, J. V., Mulla, D., Isler, V. (2013). Sensor Planning for a Symbiotic UAV and UGV system for Precision Agriculture. IEEE/RSJ International Conference on Intelligent Robots and Systems (IROS) Tokyo, Japan., pp. 5321–5333.

214. Torres-Sanchez, J., Lopez-Granados, F., De Castro, A. I., Pena-Barragan, J. M. (2013). Configuration and Specifications of an Unmanned Aerial Vehicle (UAV) for early Site-Specific Weed Management. PLOS one DOI: 10.137/ journal.pne.0058210, pp. 1–11 (May 20th, 2014).

215. Torres-Sanchez, J., Pefia, J. M., De Castro, A. I., Lopez-Granados, F. (2014). Multispectral mapping of the vegetation fraction in Early-season wheat fields using images from UAV. Computers and Electronics and Agriculture 103:104–113.

216. Townsend, S. (2013). Drones dull the drudgery of Aussie Farming. Daily Telegraph, Sydney, http:// www. dailytelegraph. com.au/news/nsw/drones-dull-the-drudgery-of-aussie-farming/story-fni0cx12–1226729554517, pp. 1–3 (September 6th, 2014).

217. Trimble Navigation Systems 2014 Trimble UX5 Aerial Imaging Rover. http://www.trimble.com/survey/ux5.aspx, pp. 1–2 (July 27th, 2014).

218. Trout, T. J., Gartung, J. (2006). Use of Crop canopy size to estimate crop coefficient for vegetable crops. Proceedings of World Environmental and Water Resources Congress, Omaha, Nebraska, USA, pp. 233–239.

219. Trout, T. J., Johnson, L. F., Gartung, J. (2008). Remote Sensing of canopy cover in horticultural crops. Horticultural Science 43: 333–337.

220. Turner, D., Lucier, A., Watson, C. (2012). Development of an Unmanned Aerial Vehicle (UAV) for hyper resolution vineyard mapping based on visible, multispectral and thermal imagery. http://citeseerx.ist.psu.edu/viewdoc/summary?doi=10.1.1.368.2491, pp. 1–12 (September 4th, 2014).

221. UAV-Belize Ltd, 2013 The case for UAV helicopters in the Caribbean. http://www.uav-belize.com (August 18th, 2014).

222. UCMERCED, 2014 Unmanned Aerial Systems (UAS). MESA lab-Mechatronics, Embedded systems and Automation. Internal Report, pp. 1–5.

223. Unmanned Systems Technology, 2014 Autocopter UAV launched for Agricultural Applications. http://www.unmannedsystemstechnology.com/2013/06/autocopter-uav-launched-for-agricultural-applications/, pp. 1–3 (August 10th, 2014).

224. Vanac, M. (2014). Drones are the latest idea to improve farm productivity. The Columbus Dispatch http://www.dispatch.com/content/stories/business/2013/09/19/eyes-in-the-skies.html, pp. 1–4 (September 6th, 2014).

225. Volt Aerial Robotics 2010 Venture- Multipurpose-Professional grade. Volt Aerial Robotics, Missouri, USA, http://www.voltaerialrobotics.com/quadcopter.htm, pp. 1–3 (July 26th, 2014).

226. Walker, M. (2014). Oregon Company nabs funding for water saving farm drones. http://www.bizjournals.com/portland/blog/sbo/2014/03/oregon-company-nabs-funding-for.html?page=all., pp. 1–4 (August 14th, 2015).

227. Wharton, C. (2013). Nevada looks at 'drones' for Economic development and Natural Resource efforts. http://www.unce.unr.edu/news/article.asp?ID=1871, pp. 1–4 (September 9th, 2014).

228. Wikipedia, 2014 History of Unmanned Aerial Vehicles. http://en.wikipedia.org/wiki/History_of_unmanned_aerial_vehicles., pp. 1–12 (July 24th, 2014).

229. Wilson, K. (2014). Farmers gear towards robots. Farm Weekly. http://farmweekly.com.au/news/agriculture/machinery/general-news/farmers-gear-towards-robots/2683809.aspx., pp. 1–5 (March 11th, 2014).

230. Wozniacka, G. (2014). Drones could revolutionize Agriculture, Farmers Say. Huff Pos Business http://www.huffingpost.com/2013/12/14/drones-agriculture_n_4446498.html, pp. 1–5 (August 3rd, 2014).

231. Xiang, H., Tian, L. 2011a Development of a low-cost Agricultural Remote Sensing System based on an Autonomous Unmanned Aerial Vehicles (UAV). Biosystems Engineering 108: 174–190.

232. Xiang, H., Tian, L. 2011b Method for Automatic Geo-referencing Aerial Remote Sensing Images from an Unmanned Aerial Vehicle (UAV) Biosystems Engineering 108: 104–113.

233. Yintong Aviation Supplies 2014 Agriculture Crop Protection UAV. http://china-yintong.com/en/productshow.asp?sortid=7&id=57, pp. 1–3 (July 31st, 2014).

234. Zaman, Q., Schumann, W., Hostler, K. H. (2006). Estimation of citrus fruit yield using ultrasonic sensed tree size. Applied Engineering in Agriculture 22:39–44.

235. Zaman, Q. U., Swain, K. C., Schumann, A. W., Percival, D. C. (2010). Automated low cost yield mapping of blue berry fruit. Applied Engineering in Agriculture 26: 225–232.

236. Zarco, P. (2014). Drones in Spain's Agriculture. http://www/freshplaza.com/article/114743/Drones-land-in-Spain-Agriculture. htm/pp. 1–2 (July 2nd, 2014).

237. Zarco-Tejada, P. J., Gonzalez-Dugo, V., Bemi, J. A. J. (2012). Fluorescence, Temperature and Narrow-Band Indices acquired from a UAV platform for water stress detection using micro-hyperspectral imager and a thermal camera. Remote Sensing of Environment 117: 322–337.

238. Zarco-Tejada, P. J., Guillen-Climent, M. L., Hernandez-Clemente, R., Cataline, A., Gonzalez, M. R., Martin, P. (2013). Estimating leaf carotenoid content in Vineyards using High resolution hyperspectral imagery acquired from an Unmanned Aerial Vehicles (UAV). Agricultural and Forest Meteorology 171: 281–294.

239. Zhu, J., Wang, K., Deng, J., Harmon, T. (2009). Quantifying Nitrogen status of Rice using low altitude UAV-mounted system and object-oriented segmentation methodology. Proceedings of the ASME 2009 International Design Engineering Technical Conferences and Computers and Information in Engineering. San Diego, California, USA, pp. 1–8.

CHAPTER 4

SATELLITE GUIDED AGRICULTURE: SOIL FERTILITY AND CROP MANAGEMENT

CONTENTS

4.1   BACKGROUND

4.1.1   HISTORICAL ASPECTS OF SATELLITE AND REMOTE SENSING METHODS IN AGRICULTURE

Sputnik is the first artificial satellite that was launched in October, 1957 by erstwhile Soviet Union. It was a very small satellite that orbited earth once in 98 minutes. Later, in 1958, Explorer-1 also known as Satellite-Alpha was launched by National Aeronautical Space Agency (NASA) of United States of America. This was followed by pioneering effort by NASA to assess and harness satellites in communications and aerial imagery. In 1960s, a series of satellites like Telstar, Relay and Syncom were launched (Launius, 2005; Dickson, 2009; Harford, 1997; USDA-ARS, 2005).

The satellite aided observation of earth's natural resources was initiated in 1966 under the name 'Earth Resources Technology Satellites Program.' Its name was subsequently changed in 1975. In 1970s, Landsat program managed by NASA highlighted the usefulness of satellites in agriculture and remote sensing of natural resources such as land, soil, water, vegetation, etc. Landsat provided large amounts of data and color pictures of earth. Landsat program was used for several aspects of crop production, such as, mapping soils, topography, cropping systems, forest and vegetation cover, monitoring drought, floods, erosion, locust movement, diseases, etc. Landsat was extensively used in compiling weather data for agriculturists and to note effects of global climate change. Reports suggest that first Multi-Spectral Scanners (MSS) was devised and tested in 1969. The era of satellite imagery began with first use of MSS to observe half domes of Yosemite National Park.

Landsat program experienced difficulties with regard to funding and long term objectives. However, its continuation was ensured by efforts of Mr. Daniel Quayle, Vice President of USA as Chairman of National Space Council of USA. It was continued to obtain imagery and archive them for posterity and to conduct usual multi-spectral survey to help different agencies and farmers worldwide. In mid-1980s, long term Landsat Program was envisaged and it was managed by a private company- 'Earth Observation Satellite Company (EOSAT).' This company operated Landsat 4 and 5. Processing of satellite imagery and supply were exclusively done by EOSAT. With the launch of Landsat-7 in 1994 by EOSAT, digital data, maps and images were provided to wide range of users including Agricultural agencies and farmers worldwide. Landsat imagery was provided at lowest cost, hence it attracted greater number of clients (see Landsat Program, 2014). Today, it is said that CORONA series in case of military aspects and LANDSAT series for Natural Resources monitoring, including agriculture are among the earliest and longest serving satellites (Ruffner, 1995; Landsat Program, 2012; Landsat 7 Gateway, 2012).

Agricultural development during past 5 decades has hinged on our ability to understand weather parameters and develop accurate forecast.

Satellite technology geared to collect data about weather, understand the several atmospheric processes and provide farmers with tangible forecasts has been useful. Let us now consider a few salient historical facts about use of satellites to study weather. During mid-1950s, first meteorological satellites were planned in USA. The first payload for meteorological experiments was placed aboard Explorer VII in 1959. First weather satellite was launched from Cape Canaveral in Florida, USA in April, 1960. The satellite weighed 122 kg, orbited at 435 km from earth's surface, carried several sensors, video cameras and power communication systems (Davis, 2011). In 1965, Nimbus-1 satellite was launched. It carried several sophisticated sensors for meteorological observation. In 1970s, Tiros/ESSA series were launched for collecting data about weather. GOES-series satellites 1–15 launched between 1981 and 2011, improved meteorologists' capabilities by continuously profiling temperature, water vapor along with several other atmospheric parameters. During recent past, private satellite agencies are also engaged in launching constellations of satellites and they collect useful agricultural information all through the year each day. The satellite imagery and crop production prescriptions are sold to farmers and companies engaged in large scale farming (Growing Nebraska, 2014).

In West Africa, satellite imagery has been utilized to study natural resources, topography, land cover changes in vegetation, water resources, rivers and cropping trends. Landsat imagery has been used extensively since 1988 to record expansion or shrinkage of crop production zones in West Africa. The agricultural and weather related data and imagery collected has been coordinated and used effectively by researchers at AGRHYMET, Regional Centre in Niamey, Niger (Irons et al., 2014b; Aron, 2013). Nigeria is an important agricultural nation that has adopted satellite technology to monitor crop production zones, spread and shrinkage of Sahelian cropping areas, droughts, floods, soil deterioration, erosion, desertification, etc.

In West Asia, Israel's farming experts have already deployed satellite techniques to accomplish various land use and crop production aspects. A recent report by Smadar (2012), states that Israel's Agricultural Department has offered geo-spatial techniques to farmers. Firstly, farmers are advised and offered maps of topography and elevation of fields, biomass production trends, crop yield maps and sensor data for overlaying with yield maps, etc. In general, farmers could requisition hyperspectral and thermal imagery patterns obtained through satellites. Thermal imaging is helping farmers to judge soil mineral and water resources in the area adopted for farming.

Russia has elaborate satellite systems used for variety of purposes such as defense, transport, cartography (e.g., Karatograph-OEN2, Resurs-P), monitoring natural resources such as forest cover, crop production zones and land use pattern in general. During past 25 years, Russia has over 50 weather satellites operating in space to transmit weather information useful to farmers, and to warn of impending droughts, floods and other natural disasters (Clark, 2009; NASA, 2013). During recent years, Russian agricultural satellites of the Kosmos series have been utilized to direct farming operations and take policy decisions. Precision techniques that require regular satellite services are popular. Remote sensing satellites operate in constellations to provide information to agriculturists. To quote an example, *Kosmas SKa* to be launched in 2015, is an agricultural satellite useful to monitor crops and development of new cropping zones (Zak, 2014). Rockets such as Vostok, Proton and Soyuz series launched from Baikonur in Kazakhstan have helped in placing these agricultural satellites in orbit (Group of Earth Observations, 2014). Several other nations operate agricultural satellites to inform their farmers. For example, Argentina, Brazil, South Africa, India, China, Australia, and many European nations have their own satellites.

4.1.2 SATELLITES RELEVANT TO NATURAL RESOURCES, FORESTRY AND AGRICULTURAL ASPECTS AND THEIR SALIENT FEATURES

During past five decades, satellites launched by different countries have served different purposes such as defense, transport, monitoring global and regional weather phenomena. They have been extensively used for collecting data about natural resources such as land, water and biomass. As time lapsed, some of the satellites were deployed or re-deployed to study agricultural parameters, study cropping systems, monitor and note crop productivity. Currently, there are series of satellites launched by different nations that are capable of monitoring agricultural crop production (see Plates 4.1–4.4). A few of them are exclusively utilized to provide information, imagery and digital data about agricultural field. Let us consider salient features of a few examples of satellites used in agriculture.

PLATE 4.1 Ikonos Satellite (Source: Dr. Davon Libby, Corporate Communications, DigitalGlobe Corporation, Dry Creek, Longmont, Colorado, USA; Note: Ikonos Satellite is a multipurpose facility that offers satellite imagery of natural resources. It aids agricultural monitoring and satellite controlled farm operations. It belongs to DigitalGlobe Inc., Longmont, Colorado, USA).

Salient characteristics of IKONOS are as follows:

Launch Date	24 September 1999 at Vandenberg Air Force Base, California, USA
Operational Life	Over 7 years
Orbit	98.1 degree, sun synchronous
Speed on Orbit	7.5 km S^{-1}
Speed Over the Ground	6.8 km per second
Revolutions Around the Earth	14.7, every 24 hours

Altitude	681 km
Resolution at Nadir	0.82 meters panchromatic; 3.2 m multispectral
Resolution 26° Off-Nadir	1.0 meter panchromatic; 4.0 m multispectral
Image Swath	11.3 km at nadir; 13.8 km at 26° off-nadir
Equator Crossing Time	Nominally 10:30 AM solar time
Revisit Time	Approximately 3 days at 40° latitude
Dynamic Range	11-bits per pixel
Image Bands	Panchromatic, blue, green, red, near IR

PLATE 4.2 QuickBird satellite (Source: Dr. Davon Libby, Corporate Communications, DigitalGlobe Corporation, Dry Creek, Longmont, Colorado, USA).

Note: QuickBird Satellite's salient characteristics are as follows:

Launch Date	October 18, 2001
Launch Vehicle	Boeing Delta II
Launch Location	Vandenberg Air Force Base, California, USA
Orbit Altitude	450 km/482 km – (Early 2013)
Orbit Inclination	97.2°, sun-synchronous

Speed	7.1 km/sec (25,560 km/hour)
Equator Crossing Time	10:30 AM (descending node)
Orbit Time	93.5 minutes
Revisit Time	1–3.5 days, depending on latitude (30° off-nadir)
Swath Width (Nadir)	16.8 km/18 km – (Early 2013)
Metric Accuracy	23 meter horizontal (CE90)
Digitization	11 bits
Resolution	Pan: 65 cm (nadir) to 73 cm (20° off-nadir)
	MS: 2.62 m (nadir) to 2.90 m (20° off-nadir)
	Pan: 450–900 nm
	Blue: 450–520 nm
Image Bands	Green: 520–600 nm
	Red: 630–690 nm
	Near IR: 760–900 nm

PLATE 4.3 Landsat 8—An Agricultural Satellite (Note: This latest of Landsat satellite series has multispectral imaging cameras that operate at 8 different bandwidths; Source: Drs. Jeannette Allen and James Irons, National Aeronautical and Space Agency, USA; http://landsat.gsfc.nasa.gov/wp-content/uploads/2013/01/ldcm_2012_COL.png).

PLATE 4.4 Spot 7 Satellite (Source: Systeme Pour l'Observation de la Terre, Centre national d'Etudes Spatials (CNES), Paris, France http://www.satimagingcorp.com/satellite-sensors/spot-7/).

Note: SPOT-7 Satellite Sensor Specifications

Launch Date	June 30, 2014
Launch Vehicle	PSLV
Launch Location	Satish Dhawan Space Center, Sriharikota, AP., (India)
Multispectral Imagery (4 bands)	Blue (0.455 µm – 0.525 µm)
	Green (0.530 µm – 0.590 µm)
	Red (0.625 µm – 0.695 µm)
	Near-Infrared (0.760 µm – 0.890 µm)
Resolution (GSD)	Panchromatic – 1.5 m
	Multispectral – 6.0 m (B, G, R, NIR)
Imaging Swath	60 km at Nadir

4.1.2.1 Landsat 8

'Landsat 8' is latest satellite of the LANDSAT series. It was developed by National Aeronautical and Space Agency and United States Geological Survey crew and launched by an Atlas rocket in February, 2013 from Vandenberg Air force base, California (see Plate 4.3). The pay load includes evolutionarily advanced instrumentation for imagery. It consists of Operational Land Imager (OLI) and Thermal Infrared Sensor (TIRS). These sensors are used to collect images of global land mass with a resolution of 30 m. The sensors include Visible bandwidth cameras, Near-infrared, Red-Edge, Thermal Infrared (15 m) and Panchromatic (15 m). The multi-spectral sensors (MSS) obviously are of wide bandwidth. Spectral bands operative in Landsat 8 satellite are:

Band-1: 435–451 nm for Coastal/Aerosol observation; *Band-2:* 452–512 for blue; *Band-3:* 533–590 nm for green; *Band-4:* 636–673 nm for red; *Band-5:* 851–879 nm for Near Infrared; *Band-6:* SWIR-1– 1566–1651 nm; *Band-7:* SWIR-2–2107–2294 nm; *Band-8:* 503–676 Panchromatic imagery; *Band-9:* 1363–1384 nm Cirrus; *Band-10:* TIR-1–1060–1119 nm for thermal imagery; *Band-11:* TIR-2–1150–1251 nm for thermal imagery.

The thermal infrared sensors in the Landsat 8 collect data using two narrow bandwidths. Compared with it, previous Landsat versions have one single thermal bandwidth. Landsat 8 has been returning 555 scenes per day. The size of each scene is 185 km. and the imagery is derived from a distance of 705 km. The resolution of imagery is 12 km (Irons et al., 2014a).

As stated earlier, Landsat series satellites harbor several sensors, among them of course, MSS were popular during early phase of the program, but later the Thematic Mapper (TM) was sought by many governments and agricultural agencies. Actually, routine data collection using MSS was reduced and almost halted by 1992. The TM accrued data started with Landsat-4 and it superseded the MSS data. Following are the channels, bands and immediate applications of Thematic Mapper found in Landsat 8 (Natural Resources Canada, 2014; Table 4.1).

4.1.2.2 IKONOS Satellite Series

The IKONOS satellite that belongs to DigitalGlobe Inc., was launched in September 1999 from Vandenberg Air Force Base, California, USA (Satellite Imaging Corporation, 2014b; Plate 4.1). Its orbit is 98.2° inclined, sun-synchronous and is 681 km above earth's surface. It has a 3 day revisit time. IKONOS satellite is capable of images at 3.2 m resolution using multispectral and NIR bands and at 0.82 m resolution

TABLE 4.1 Thematic Mapper Channels, Spectral Resolution, Bandwidths and Applications

Channel	Wavelength (μm)	Applications in Agricultural and Natural Resources
TM1	0.45–0.52 (blue)	Soil, crops and natural vegetation discrimination, coastal mapping, urban feature depiction
TM2	0.52–0.60 (green)	Mapping natural vegetation, crops, urban feature identification
TM3	0.63–0.90 (red)	Detection of soil, crops, plant species, chlorophyll abundance, NDVI and identification of urban installations
TM4	0.760–0.90 (NIR)	Identification of plant and vegetation type, crop health, productivity, soil moisture and drought stress on crops, water body delineation
TM5	1.55–1.75 (SWIR)	Soil moisture and vegetation discrimination and mapping, discrimination of snow and cloud covered areas
TM6	10.4–12.5 (TIRS)	Identification of water stress on crops and natural vegetation, crop productivity, monitoring forest cover and clear cutting, Thermal mapping of cropping zones and moisture storage
TM7	2.08–2.38 (SWIR)	Vegetation mapping, soil moisture, discrimination of mineral and rock types

Source: Natural Resources Canada, 2014.
Note: NIR = Near Infrared; SWIR = Short Wave Infrared; TIRS = Thermal Infra-Red Sensor.

using panchromatic wave lengths at nadir. IKONOS satellites are suited to provide imagery for assessing natural resources, disasters such as drought, dust bowls, floods, etc. It is used effectively in agriculture and forestry to study vegetation, crop growth, collect data on green vegetative index, etc. IKONOS is also used in homeland security and monitoring coastal regions. IKONOS satellite is also capable of stereo images required for agriculture. It provides digital elevation models of farm topography. Major image bands are blue, green, NIR, red-edge and panchromatic.

4.1.2.3   QuickBird Satellite

QuickBird satellite that again belongs to DigitalGlobe Inc., was launched from Vandenberg Air Force Base, California on October 18, 2001, by a Boeing Delta vehicle (Plate 4.2; Satellite Imaging Corporation, 2014c). It orbits earth at 450 km altitude which is sun-synchronous and at an inclination of 97.2°. It transits at a speed of 71 km sec^{-1} (25560 km hr^{-1}). Major imaging bands are blue 450–520 nm, green 520–600 nm, NIR 760–900 nm, and Panchromatic 450–900 nm. The QuickBird satellite provides higher resolution images of 65 to 73 cm, if panchromatic wavelength is used and at 2.62 m to 2.90 m, if multi-spectral bandwidths are used. QuickBird satellite is supposedly excellent for collecting data on environmental and agricultural aspects. Currently, it is used to detect effects of climate change on agricultural and forest productivity. QuickBird satellite imagery is also used to surveillance pipelines, industries, in oil and gas exploration, etc.

4.1.2.4   SPOT (Systeme Pour l'Observation de la Terre)

SPOT is a series of earth observation and imaging satellites designed and launched by Centre National d'Etudes Spatials (CNES) of France (Plate 4.4). This program is supported by Sweden and Belgium. The SPOT satellites observe earth from sun-synchronous orbits at altitudes of 830 km above earth's surface. SPOT repeats its orbit once in 26 days. SPOT satellites use push broom scanning technology. These satellites have two high-resolution visible imaging systems that could be used independently or simultaneously. They also have a panchromatic mode that is coarser and multi-spectral. The resolution of SPOT is 10 m, but the three MSS have 20 m spatial resolution. The high-resolution visible spectral ranges available with SPOT are:

Multi-spectral Scanners Bands and Wavelength (μm)
Band 1: 0.51–0.73 (blue)
Band 2: 0.61–0.59 (green)
Band 3: 0.79–0.89 (near infrared)
Panchromatic 0.51–0.73 (blue-green-red)
Source: Natural Resource Canada (2014); Satellite Imaging Corporation (2014a).

Let us consider some salient features of SPOT-5 satellite and its sensors. This satellite was launched by Arian rocket in May 2002, from Kourou in French Guyana. It is said SPOT-5 is relatively superior in the quality of imagery and resolution of images. It is also cost effective. One of the key features of SPOT is its ability to provide maps of natural resources, vegetation and cropping belt at 1:25,000 to 1:10,000 scale. The resolution of sensors is 2.5 m to 5 m (Satellite Imaging Corporation, 2014a). The viewing angle of sensors found on the SPOT-5 can be adjusted on either side of satellite's vertical tract. This allows off-nadir viewing and allows it to revisit same location several times in a week. Therefore, specific regions could be monitored more accurately using SPOT imagery. Hence, applications such as in agriculture and forestry are served better by SPOT satellite. SPOT-5 is also capable of 3D imagery and terrain modeling. There are sensors that provide stereo image. This facility provides extra advantage to farmers and farming companies studying the topography in greater detail and in designing earth work and land preparation accurately. SPOT is versatile with regard to sending imagery to clients. SPOT provides rapid orders for satellite data and imagery of relatively higher resolution. SPOT, indeed has several other advantages that makes it more popular over other satellite derived imagery/data. It has high spatial resolution pointable sensors.

4.1.2.5   IRS (Indian Remote Sensing) Satellite Series

IRS system is a constellation of satellites of Indian origin. These satellites are meant mainly to assess natural resources, land use and land cover monitoring, forest survey, wetland mapping, biodiversity estimates, flood risk alert and damage assessment, snow melt and water flow, mineral prospecting coastal vegetation monitoring, etc. Initially, during 1980s and 1990s, these IRS satellites were launched through Russian launch vehicles. More recent satellites have been put into orbit using PSLV rockets of India. Agricultural crop monitoring, mapping and guidance to farmers in India and elsewhere in other parts of the world are among most important tasks performed by IRS satellites. Major functions are: delineating agroecosystems, cropping systems analysis, soil erosion inventory, soil carbon dynamics and land productivity assessment, integrated drought management, water resource monitoring and allocation (Bhan et al., 2012).

4.1.2.6   NigeriaSat Series

NigeriaSat is a set of satellites launched for use by various agencies of Nigeria. NigeriaSat-1 was launched in 2003 and it lasted till 2012. Other satellites in the series are NigeriaSat-2 and NigeriaSat-X. These satellites are equipped with high-resolution multispectral cameras. They are tuned to monitor agricultural crop production, fluctuations in the Sahelian zone, Dust bowls, Drought and Soil erosion, etc. These satellites are also geared to monitor disasters, so that, farmers could be alerted ahead (Aron, 2013).

4.1.2.7   What is Resolution of Satellite Imagery?

According to Campbell (2002), there are at least few different types and definitions for resolution of imagery obtained by different satellites. They are appropriate to be used for different purposes. The four major types of resolution commonly recognized are Spatial resolution, Spectral resolution, Radiometric resolution and Geometric resolution.

Spatial resolution is defined as pixel size of the image representing the size of the earth's surface being measured and determined by the sensors' instantaneous field of view.

Spectral resolution is defined by the wavelength interval in the electromagnetic spectrum and number of intervals that sensor is measuring.

Temporal resolution is explained as the amount of time elapsed between image collection events of a given location on earth's surface.

Radiometric resolution refers to effective bit-depth of the sensor. It is generally expressed as 8-bit, 12 bit, 16 bit, etc.

Geometric resolution refers to the satellite sensor's ability to effectively image a portion of the earth's surface in a single pixel. It is expressed as

Ground Sample Distance (GSD). For example, GSD of Landsat is 30 m x 30 m and that of SPOT is 60 m x 60 m.

As stated earlier, regarding drones in Chapter 3, the crux of satellite technology too is dependent on the multispectral cameras, sensors, quality of spectral imagery and resolution of images produced. Currently, among the satellites that could be used for agricultural purposes, GeoEye-1 launched in 2008 has best resolution of imagery. This satellite collects ground images with a resolution of 41 cm (16 inches) in black and white and at 165 cm (64 inches) in panchromatic mode (GeoEye Inc., 2011). DigitalGlobe's Worldview-2 is a civilian commercial satellite that provides images of choice at high-resolution of 46 cm. QuickBird, another satellite by DigitalGlobe Inc., supplies images of 60 cm resolution at nadir using pan chromatic images. SPOT, by CNES, France provides imagery of wide range of resolution ranging from 25 cm to 1 km. A few other satellites such as EROS A and B too supply spectral imagery of high-resolution that is useful various aspects of agricultural cropping such as land preparation, identification of disease/insect attack on large expanses of crops, etc. The resolution of images from EROS-B is 70 cm and that EROS-A is 120 m.

4.1.2.8 What is Satellite Imagery and How Is It Processed?

Satellite imagery, in its raw form is not amenable for use in agricultural production systems. It is processed using suitable computer programs that convert raw images into maps that depict topography, crop growth pattern and grain yield. In addition to computer based technology to convert raw images into digitized and easily readable crop maps, it is necessary to choose apt satellites with resolutions as needed. Computer programs that utilize low, medium and high-resolution imagery obtained using multiple sensors are preferred. Some satellites offer unique imagery at medium resolution obtained from different angles. A few computer programs are equipped with software that remove distortions caused by environmental factors (see GDA Corp, 2014d). Satellite agencies use an array of image processing computer software that suits raw images from particular satellites and to serve customer's needs. For example, raw imagery could be converted into color coded maps, digitized data loaded into a chip or into relief maps that could be studied.

Some examples of computer software, programs and mapping services of utility during satellite imagery guided Precision Farming is as follows:

AgPixel Software: AgPixel transforms raw data from satellite imagery into useful field maps showing variations in vegetation, crop stand and soil traits. This software agency offers services such as automatic image processing and secure storage of data. It also updates software for image processing and detailed analysis of images. In particular, AgPixel allows farmers to obtain NDVI and green vegetation index, Raw data fields with image conversion packages, false color images, cropping maps and atmospheric calibrations (Goldfinch Inc., 2014).

Pixel Mapping: Pixel mapping is another agency that offers software for processing satellite imagery. They offer geo-spatial mapping facility for farmers as and when required during a crop season. Major tasks handled are conversion of satellite data to maps for farmers, calibrating camera systems, balancing aerial imagery, processing images from vertical and oblique images. They provide digital imagery maps of crops (Pixel Mapping Inc., 2014).

PLM Software: The main package of this software relates to keeping field records, mapping soil and crops and analyzing the growth and production trends, using satellite imagery. Satellite imagery is rapidly converted and interferences from soil and other sources in the field are removed, to leave only crop related reflectance for mapping and analysis. The software also deals with layers of multispectral data, a variety of topographical maps, and yield data from different years, prior to analysis of crop performance. The digitized data could be used to feed variable-rate applicators. The software tracks inputs and crop production trend. Average multiple year grain yield maps are useful in calculating quantity of inputs required. The software could also be used to assess performance of crop genotypes during different season/years in response to inputs.

Farm Star: Farm Star was first produced in 1997 in Australia to develop yield maps. It was entirely re-written in 2000–2001, then again in 2006. It is a software that is utilized to: (a) Create sophisticated images of spatially variable data, mainly grain yield; (b) It can automatically generate contour maps and spatially variable precision farming data such as grain yield data from combines, map and locate soil samples and develop maps using on-line data about EC, color, pH, soil moisture, etc.; (c) Leaf-N data could be mapped, so that management zones could be developed; (d) Most importantly, it can develop digitized data for use in the variable-rate applicators for accurate dispensation of seeds, fertilizers and other chemicals.

4.1.3   SATELLITE GUIDED AGRICULTURAL MONITORING

Satellite aided monitoring and imagery is apt when timeliness is not too important. Routine, periodic monitoring once a fortnight or a month by satellites is cost effective and they cover large area in an agrarian zone. Satellite agency controls the timing and resolution of imagery. Satellite imagery can be archived in digitized formats and can be retrieved at any time. Aerial imagery using aircrafts is another mode for observing the farms and crops. Farmers may use small aircraft such as Cessna. Imagery is relatively sharper and of higher resolution compared to that derived from satellite companies. At present, Drones or UAVS are getting ever popular in farms. They are used for monitoring crops at short intervals and obtaining high-resolution close-up shots of entire crops (Plate 4.5 and 4.6). They are easy to purchase and manage. However, monitoring agricultural activities within cropping expanses or even individual farm using satellite technology is the center piece of this chapter. Satellite-guided crop monitoring was accomplished initially only by few satellite agencies. During recent years, there is greater emphasis on agricultural applications of satellite technology. It has been aided by availability of several satellites, launched by different nations and private companies. Therefore, satellite-guided soil management and crop monitoring has become common in many agrarian regions. Satellites actually provide a new, easier and accurate analysis of crop related events in farms. Currently, several governmental agencies and private firms offer to supply satellite imagery derived using spectral bands and with different resolutions, at a cost. They help farmers even in remote regions to judge crop growth and ascertain fertilizer and irrigation prescriptions easily from experts stationed at agricultural call centers. In some agrarian regions, satellite imagery is sent periodically to farmers via internet, so that they could analyze the digital data using several computer-decision support systems, then adopt appropriate agronomic procedures. For example, 'Savepi,' a company situated in Douains, France, provides satellite data related to surveillance of land surface and crop progression in large farms. Satellite agencies offer such data, five to 10 times a season, to farm managers directly through electronic mailing systems. Several farmers are members of such satellite companies. They receive color coded images from agency. The color codes actually are derived after detailed analysis of digital data using a few computer programs that offer top few best options regarding land management, fertilizer, water, and pesticide sprays at that period. The agency literally alerts the farmers with timely best options. Worldwide, currently, France is supposed to be the leader in satellite imagery services (New Economist, 2009). According to Infoterra, a subsidiary of Astrium, France monitors a greater share of its farm land using satellite imagery and surveillance techniques. Researchers in agricultural agencies situated in Toulouse, in Southern France, forecast that demand for satellite imagery and well evaluated prescriptions will increase as climatic vagaries get accentuated. Farmers who have revised their yield goals upwards will need greater flexibility and accurate prescriptions rapidly, to adapt to such changes in environmental parameters. Farmers may not be able to rely on tedious, time consuming soil and plant analysis to commit investment on fertilizers and other inputs. Instead, constant monitoring and versatility in crop management gets preferred. In few other European regions, encompassing German and Spanish farms, satellite imagery has helped immensely in deciding on crop loans and insurance schemes. For example, 'RapidEye' offers detailed data of crop stand to farm insurance companies. It also offers tangible yield forecasts. Farm disasters such as drought or floods are analyzed in greater detail, so that rehabilitation agencies can offer appropriate loans to farmers. It seems farm yield forecasting by satellite companies is becoming popular in North America and Europe. Satellite monitoring and GPS services is also necessary for adopting precision techniques. Again, satellite companies are helping farmers in running tractors, spray equipment and harvesters by providing matching satellite data and support. In developing nations, agricultural monitoring has been supported by subsidy from government agencies. For example, in Africa, satellite imagery has helped farm agencies with data that replaces efforts of several farm workers required for soil digging, sampling, processing and soil chemical analysis. They say in one region itself, a set of detailed digital data and satellite imagery has replaced human drudgery otherwise required to obtain and analyze 100,000 soil samples (New Economist, 2009).

Satellite imagery is used by Agricultural agencies of several nations and FAO, Italy, to monitor and estimate seasonal or yearly crop acreage. For example, United States Department of Agriculture, Washington, utilizes satellite imagery to supplement data about crop acreage collected by regular farm specialists. The National Agricultural statistics is regularly upgraded using crop monitoring techniques to improve accuracy in judging fluctuations in crop production zones. Expansion and shrinkage of area under specific crop species and its genotypes is also monitored (USDA-NASS, 2009). The spread of cultivation of different crop species, a few specific genotypes, land management techniques, water supply systems or fertilizer usage is monitored using satellite images. We may note that there are situations when crop acreage and related data derived from satellites falls short in accuracy or usefulness (USDA-NASS, 2009). Satellite imagery has also been used to estimate annual productivity and light use efficiency of crop land (Lobell et al., 2002).

A different report from USDA-ARS, Beltsville, in Maryland, USA states that, we can use imagery from SPOT 5 obtained at four spectral bandwidths (green, red, NIR and mid-infrared) to detect each crop species. At a resolution of 10 Mpixel, imagery could easily discriminate crops such as maize, cotton, grain sorghum and sugarcane grown in Southern Texas. They have pointed out that SPOT 5 images were accurate to the extent of 78–91% and they were used routinely for monitoring agricultural crops (Chenghai et al., 2008).

Satellite imagery is also used to monitor agricultural weather patterns. They are used to collect spatial and temporal changes in weather parameters over a field or an agrarian region. Satellite imagery about topography and weather changes has allowed modeling and preparation of maps using data sets from several years. Such maps are of utility during monitoring of climate change effects. Satellite monitoring helps farmers with up-to-date geo-spatial information, rather rapidly. Monitoring landscape changes, land use pattern in agrarian zones, surface climatology during crop production and alterations in hydrology are few other applications of satellite imagery (Satellite Imaging Corporation, 2014d). Satellite images showing climate change effects, influence of conservation methods, extent of natural disasters such as floods, erosion, drought, etc. could be obtained periodically and apt measures could be applied. It therefore curbs severe and extended loss. Satellite imagery could also be used to assess changes in plant biodiversity and crop pattern, by adopting high-resolution sensors.

Agricultural expanses are susceptible to different types of diseases that may occur in small patches or spread into large areas and reach epidemic proportions. Agricultural agencies need to be extra careful about fast spreading and devastating diseases that afflict the commercial mono-cropping belts. Mono-cropping stretches may become particularly vulnerable if there is breakdown of resistance. Break down of genetic resistance also occurs, if the disease organism mutates to virulence or new strains invade the crop belts. Satellite based surveillance of cropping expanses is therefore utmost needed. Periodic analysis of satellite imagery helps in curbing disease or insect pest attack, right at early stages of impending epidemic. Satellite-guided disease/insect control could be accomplished by preparing spray schedules based on satellite data. Precision sprays could also be adopted. It helps to reduce pesticide usage.

Weeds could be monitored using satellite images. Sensors and computer software that identify weeds and crop species using spectral signatures are available. Detecting weeds or its patches among crop plants needs high-resolution imagery. Precision techniques could be adopted using digital data from satellites to control weeds. It allows farmers to direct herbicides accurately in apt quantities to control weeds.

Satellite imagery is also used to detect and monitor several geophysical processes that affect cropping zones, their expansion and productivity. Soil erosion is an important natural malady that restricts farming zones.

It potentially reduces top soil fertility and gullies that it creates could reduce farm productivity. Gully erosion is a common phenomenon that could be monitored and imaged using satellites. Satellite imagery and weather forecasts could be used to forecast extent of gully erosion that may affect a particular zone (Okwu-Delunzu et al., 2013). Satellite monitoring can also help agricultural agencies to identify gully erosion at early stages and adopt remedial measures.

Satellite imagery is also used to assess soil moisture resources on a large scale. Soil holds only a small portion of total global water resource compared to oceans, yet it plays vital role in controlling global water cycle plus density and spread of vegetation. Agricultural crops depend immensely on soil water storage. Satellite imagery and estimates of soil water are useful in locations where *in situ* ground estimates of water are sparse or absent. A general impression of soil moisture that is obtained via satellite imagery is highly informative and useful to farmers.

4.1.3.1   Satellites and Agricultural Intelligence

Satellite imagery has been most useful in obtaining intelligence reports about crop production zones across different agrarian zones of the world. Agricultural intelligence through satellite provides early warning about crop disasters if any, and to suspect low or high grain productivity of different crop commodities in a village, zone or nation. Agricultural monitoring also helps in re-mapping a zone based on changes noticed in the cropping pattern. Most commonly, satellite-guided crop monitoring across regions/nations helps in organizing business prospects. For example, deciding on import/export policy for grains, etc. There are several private satellite companies, in addition to space agencies of some nations that regularly monitor agricultural cropping pattern, assess the growth and productivity status, and supply tangible forecasts. Usually Agricultural crop production forecasts are based on sampling at few spots in a region. However, there are agencies that make detailed satellite monitoring of crops, cropping pattern, crop acreage, grain/forage production, etc. prior to forecasting. For example, Geospatial Data-analyzing Corporation offers agricultural intelligence reports after considering basic crop growth and yield data plus ancillary information such as historical trends of a location, crop calendar, conditions, agricultural maps of the area, crop phenology models and yield models (GDA Corp, 2014c). Computer decision supports could be utilized to match the forecasts provided by the satellite company. Some of these satellite companies and National Space agencies also provide regular intelligence reports about crops, crop acreage, production trends on a global scale. Such reports, for example, from LANDSAT or SPOT satellite help in adjusting and readjusting business trends in a given region or even across continents. The USDA Foreign Agricultural Services that utilizes satellite imagery extensively is an example for agency that routinely monitors cropping belts across different continents and offers reports with forecasts pertaining to yield, disasters and bumper crops.

4.1.3.2 Satellite Aided Mapping of Agricultural Zones

Satellite mapping of agrarian regions is not a simple task of just obtaining images from camera and putting a scale on it. While utilizing satellite imagery and digitized data for assessing crop acreage and mapping the crops that occupy a particular region or crop mixtures common to a location, we may have to be particularly careful about fluctuations in spectral signatures of the crops, natural vegetation and weeds (USDA-NASS, 2009). The spectral signatures of crop species or even genotypes of a particular species are not constant. Satellite-based classification of cropping zones and crop species is actually based on energy emitted or reflected. These spectral emissions vary and this fact is used to classify the plants on the ground. The spectral signatures of crops vary based on the wavelength of the sensors found on the satellites. However, even within a season, as the crop grows from succulent seedling through vegetative and reproductive stages, the peak spectral signatures vary. Spectral signatures of healthy and disease affected patches of the crop differ. Crops under stress too show different spectral signatures. Sometimes density of crops and weeds that are common to the region may affect the spectral signatures. Further, spectral measurements, if done immediately after rainfall may show some difference compared to those not irrigated. Hence, while studying satellite imagery and digital data, due care should be taken to decipher crop reflectance measurements accurately. Calibrations for various factors that affect the spectral signatures are needed (USDA-NASS, 2009) while preparing maps of cropping expanses. There are reports that satellite imagery is currently used to assess progress of planting, and spread of maize belt in Northeast China. Annual maize acreage in North China is estimated using satellite imagery (Li et al., 2011). There are methods that allow identification of predominant crops. For example, corn production zones from a satellite (Landsat) imagery that generally depicts wide range of vegetation could help us in classifying and mapping the land cover as 'most likely corn,' 'likely corn' or 'unlikely corn' zones (Maxwell et al., 2004). At present, 3D laser scanning showing crops and elevation, phenotyping and monitoring crop growth using hyper-spectral imagery is routine in North America (Galileo Geo Inc., 2014).

As stated above, satellite imagery is a good option to monitor agricultural crops, map their expanses, rate of spread, and productivity. However, a basic requirement is that the spectral bands used and resolution should be able to distinguish and identify different crops species. The imagery should provide an accurate estimate of cropping area and its fluctuations. Agricultural monitoring and mapping has been attempted using different satellite systems. For example, Chenghai et al. (2008) have shown that high-resolution imagery obtained using SPOT 5 satellite helps in recording crops such as maize, wheat, sorghum and sugarcane and map the areas covered by each of them. These crops grown intensely in the Great Plains of North America could be easily mapped with acceptable accuracy. They have examined the satellite imagery obtained at 20 and 10 Mpixel. They suggest that both coarser images and sharper ones have their utility during studying cropping systems and allocating resources. We should note that these satellite aided crop maps done using computer and satellite connectivity works at 'Push of a Button.' We can browse the entire agrarian belt or large farm on a computer screen. Compared with it, during yester years, several human scouts had to move across the entire cropping expanse or else costly aircraft imagery had to be procured. At this juncture, we may note that such applications of satellites, reduces human drudgery in agricultural zones to very great extent.

The above introductory paragraphs should make it clear that, satellites along with their paraphernalia such as multispectral sensors, thermal imagery, computers and software for decision support, the ground robots and GPS connectivity are all set to make agriculture, physically and economically a more easier enterprise. Several of the tedious agronomic operations could be accomplished most accurately by just adopting few computer programs and pushing appropriate buttons of the farm equipment. Now, in tune with the context of this chapter, let us consider current status of involvement of satellites in 'Push Button Agriculture,' particularly, in soil management and crop production, in greater detail.

4.1.3.3   Evolution of Site Specific Management to Satellite-Guided Precision Farming

Regarding evolution of satellite-guided farming; we should note that during early stages, it began as Site-Specific fertilizer/nutrient management. Farms showing perceptible soil fertility variation within a field were noticed. Grain yield loss due to fertility variation was rampant. Blanket prescription of fertilizer made by State Agricultural Agencies or those suggested under best management practice (BMP) or maximum yield practice (MYP) were either too high or less than sufficient in many areas within a field. Hence, during 1990s, researchers were advising farmers to make a through and detailed assessment of soil fertility variations within a field, then, decide on fertilizer distribution accordingly at each spot within a field. This procedure aimed at obtaining uniformity in terms of soil fertility across entire field. Field maps showing soil fertility and crop growth variations had to be prepared using tedious soil sampling, chemical analysis and mapping. Grid sampling and formation of management zones were adopted to make it easy to assess soil fertility and apply fertilizers accordingly to each grid cell or zone. Precision farming took shape when large scale use of satellite imagery, digitized maps of crop growth and yield pattern were made available. Satellite imagery and yield maps from GPS guided Combine harvesters offered the much needed primary information about soil fertility and crop productivity variation to the farmers. Development of soil sensors and ability to map variations of soil pH, electrical conductivity, soil-NO_3, moisture and organic matter has currently provided a great impetus to those touting satellite-guided precision farming. Development of steer-less farm vehicles and GPS guided robots to accomplish variety of tasks such as tillage, weeding, top dressing of nutrients and harvesting makes satellite-guided farming, a promising concept for the future. Satellite-guided farming along with usage of drones and robots could hasten several of the farm activities, offer greater accuracy to farmers, economize on inputs, stabilize soil fertility and bring in uniformity to grain productivity. As satellite-guided techniques evolve it adds to ease and accuracy with which we can accomplish the farm tasks. Hence, it leads us to a kind 'Push Button Farming.'

4.2   SATELLITE GUIDED SOIL MANAGEMENT

4.2.1   *SATELLITE AND GPS USAGE TO STUDY TOPOGRAPHY AND CLEAR VEGETATION*

Knowledge about land surface and three dimensional features of agricultural farm land is an essential first step, if the intention is to start a crop production enterprise. The expansion and understanding of the feasibility of land for farming has usually involved preparation of detailed maps. Techniques adopted to study topography and map it has evolved gradually. Initially, farm topography was mapped using manual methods. It was followed by aerial surveys involving flights and use of sharp high-resolution cameras, specially developed to conduct aerial photography. Then, topographic data was converted into digitized maps and held in computers (USGA, 2012). It is easy to retrieve and utilize topographic data held in computers. Most common features noted during aerial photography are related to land surface, natural vegetation and diversity of plant species, water resources, soil type, crop species and productivity. According to reports by FAO (2007), obtaining a land resource inventory prior to initiation of farming in a virgin zone is essential. This has been done using manual collection of data on the ground, aerial photography and currently much of this aspect is accomplished using satellite-guided study of farm land. Some of the aspects noted using satellites are general pictures of the farming area, topographic data, soil survey with details on soil type, water resources, drainage data, present vegetation, weeds, crops, etc., soil erosion pattern if any, weather characteristics of the location, etc. The early steps of farming involves use of satellite imagery to clear natural vegetation using GPS guided bull dozers, tree cutting vehicles, clearing of shrub and herbs prior to deep plowing. Aspects such as land clearance, leveling, contouring, plowing and ridging could be accomplished with greater accuracy using a range of GPS guided vehicles.

4.2.2   *GPS GUIDED TRACTORS AND TILLAGE VEHICLES*

Ploughing is one of the agronomic procedures that require high human energy, his time and it leads to definite drudgery. Ploughing methods have evolved enormously from the original human dragged wooden plows to those using animal draft, then to automotive engine fitted vehicles. Earliest evidence for use of steam energy driven tractors is from Nebraska in the Great Plains region (see Krishna, 2002). Currently, we have tractors with IC engines energized by petrol or electric engines. Tractors are one of the important agricultural vehicles or gadgets that have mechanized land and soil preparation, improved accuracy of field operations, reduced human drudgery, improved economic advantages and allowed humans to expand agricultural cropping into hitherto uncultivated areas. Since that event in mid 1800s, great strides have occurred in the sophistication and electronic controls added to tractors. Innumerable types of attachments, hitches and improvisation in operation have occurred. The most recent invention, rather attachment to already relatively highly sophisticated tractors is GPS guidance and steer-less auto guidance based on predetermined navigation maps. These GPS tractors are currently most important aspects of 'Push Button Systems.' Many of them do not require a human pilot (driver). So, tractors are perhaps most glaring of the examples of how 'Push Button Agriculture' is evolving.

The basic idea behind attaching GPS guidance and improving accuracy of soil and crop management is to increase efficiency of inputs. In other words, lessen inputs and enhance productivity. In addition to economic gains, we should note that satellite aided steer-less tractors help farmers to accomplish tasks that hitherto were unsurmountable or impossible, requiring large number of human scouts and farm workers. Satellite-guided steer less tractors could be operated day and night. There are two types of GPS guided tractors, namely those with navigation aids and those with auto-guidance (Hest, 2012; Kinze Inc., 2014; Clemson University Cooperative Extension Service, 2014). Relatively inexpensive GPS connected navigational aids known as parallel tracking devices (light bars) are attached. They allow operators to identify their position in the field and adjust steering systems. Positional accuracy depends on quality and electronic sophistication of DGPS receiver and driver's ability to follow light bars. There are several brands of DGPS equipment, for example, John Deere Starfire, Omnistar, Beacon, Topstar WAAS etc. Most DGPS receivers offer sub-meter (<1.0 m) accuracy in field operations. The sub-meter accuracy equates to 2–4 ft. year-to-year and < 1.0 ft. regarding pass-to-pass errors. These DGPS systems are mounted on tractors and used during tillage, fertilizer supply and pesticide applications. Some of the highly accurate operations require greater accuracy than sub-meter. Accuracy of one decimeter equates to 4–8 cm year-to-year and 3–5 inches pass-to-pass errors and it is feasible when accuracy required is greater. Tractors are guided using local base stations or using dual frequency receivers with satellite-guided corrections to navigational errors. Some examples of the DGPS correction systems are Omnistar-HP, John Deere Starfire 2, etc. Pre-determined auto-guidance can be used if further higher accuracy is required. Usually centimeter level accuracy is obtainable during tractor operations, if a local base station with Real time Kinematic differential correction is used (Clemson University Cooperative Extension, 2014). Tractors with such high accuracy GPS-RTK systems are used during strip tillage, drip tape placement and land leveling. Stiers (2014) expresses that most farmers in North America are positive about using steer-less auto-guidance tractors. Farmers usually pay to get auto-guidance signals. The accuracy of navigation ranges from few inch to less than an inch. The precision reduces overlap during tilling and rarely needs some steering at the corners in a field. The hands-free driving and tilling reduces fatigue in the farm to a great extent. The auto-steer, satellite-guided tillage reduces usage of fuel, seed and herbicides. The auto-steer system helps farmers to concentrate on implements and its performance (see Plates 2.23 and 2.24 of Chapter 2).

4.2.3 SATELLITE AND GROUND ROBOTS: AUTO-GUIDANCE

Auto-guidance in agricultural farming is currently among the most sought after aspects. Satellite mediated guidance and activity of robots seems to hold key for high efficiency farming in the future years. Automation and auto-guidance of farm vehicles and operations conducted by them need satellite signals. Robots could be totally autonomous with predetermined tasks, but controlled by satellites/computer instructions. Alternatively, farmers may alter course of automated vehicles in between, using computer-based decision support systems. Auto-guidance of farm vehicles and series of operations performed by them using satellite guidance is perhaps most crucial to 'Push Button Agriculture' that is proposed in this book. Auto-guidance using satellite signal is an evolving technology in farming zones (Adamchuk, 2008; Zhang and Pierce, 2013). There are several models of tractors and other farm vehicles that are either autonomous totally or they are semi-autonomous, but their movement and operations to a large extent are guided using satellite signals, maps and digital data. Automated guidance reduces driver's drudgery, fatigue and effort to a great extent. Farm machines connected with RTK-GPS systems are being used many agrarian regions. They allow accurate steering control; sometimes with an accuracy of few cm. Otherwise 1–2 m accuracy is possible routinely during farm operations in the field (Adamchuk, 2008; Li and Yi, 2013; Prakash, et al., 2012). Following are few examples of commercial satellite-guided auto-guidance systems.

Auto-Guidance System	Company
Accutrak AX5	Accutrak, Inc.
RowGuide	AgGuide, Inc.
AFS Accuguide	Case IH, Inc.
Autotrac (Greenstar, Firestar)	John Deere Company
Intellistar	New Holland, Inc.
Autosteer-Saturn	Rinex Technology, Inc.
AgGPS	Trimble Navigation Systems, Inc.

Source: Adamchuk, 2008; https://stellarsupport.deere.com/en_US/support/Auto Trac_Universal.html; http://crossroadsgpsinc.com/product/trimble-aggps-fmx-2/; http://www.hpj.com/archives/2005/jul05/jul18/AccutrakreleasestheAX5Autos.cfm

A step further, development of integrated satellite-based controls and guidance systems that include drones and ground robots that take dictate and signals seems highly pertinent. There are situations encountered frequently when satellite imagery is hazy, less than accurate and may show up interference due to cloud cover. For example, satellite imagery may not be sharp and accurate enough while judging soil moisture, insect attack or disease incidence, if they are sporadic and highly variable. Satellite images are more suited when dealing with large agrarian zones or fields and when disease/insect attack is wide spread. In such cases, accuracy at cm or a couple of meters at margins may not matter. In other instances, an inter-connected system that includes low flying UAVs capable of close-up imagery and accurate depiction of disease and insect attack or soil moisture variation in the field is most useful. In general, there is also need for inter-phasing of computers, UAVs and satellite connectivity. Satellite images usually cover large areas. Such data could be later used by drones to fly over marked zones only and provide accurate images and digitized data. Satellite imagery and drones could also be used to mark and steer farm vehicles accurately, taking shortest routes and avoiding non-working distances. Such integrated systems could also be applied to large fields with several tractors/planters (swarms) that are themselves interconnected through satellite signals. This leads us to a complicated system in the field. Highly sophisticated computer system that considers intricate signals among farm vehicles, such as swarms of tractors or combine harvesters that are regulated by satellite-guidance are required.

Remote sensing and satellite imagery has been applied to study various aspects of crop production such as land use monitoring, cropping pattern and sequential adoption of various soil and agronomic practices. Tillage is among the earliest of farm operations that has influence on different aspects of soil nutrient availability, soil environment and gaseous emissions. In fact, type of tillage adopted has immediate relevance to nutrient loss and emissions from soil horizon (Krishna, 2002; 2013). Makar et al. (2011) state that about 8% of greenhouse gas emissions are due to farming. The conventional deep tillage has generally induced greater loss of nutrient via erosion, runoff and gaseous emissions (N_2, N_2O, and NH_3). Hence, conservation or no-tillage practices are advised. In the present context, Makar et al. (2011) state that, it is possible to monitor and estimate the extent of different types of tillage practiced in a given agrarian belt using remote sensing methods. For example, Hyperion data, imagery from Landsat, Advanced Space-borne Thermal Emission and Reflection Radiometer (ASTER) and Advanced Land imager (ALI) sensors were found to be helpful in assessing tillage systems adopted on the ground. Satellite imagery can indeed help policy makers to regulate tillage systems adopted in a given agrarian zone. We should note that vast information on soil management practices prior to sowing could be observed on the computer, just by focusing the satellite imagery and/or analyzing the digital data stored. Compared with it, a manual operation covering an agrarian belt or a large field will be costly, tedious and time consuming. The reduction in requirement of human skilled labor and drudgery are added advantages.

Daughtry et al. (2011) have used different spectral techniques to categorize the different types of tillage and their intensity on soil. They have also used spectral properties of crop residue cover on the soil surface to judge the organic matter inputs. They noticed that analysis of crop residue cover using multispectral and hyperspectral methods via Landsat TM bands were weakly correlated. However, with Hyperion data, crop residue cover was linearly related to cellulose absorption index (CAI). The CAI is indicative of cellulose and lignin absorption features measured at 2100 nm. The correlation between CAI values and crop residue cover was significant ($r^2 = 0.85$), if spectral analysis was conducted using Hyperion imaging spectrometer mounted on Earth Observing Spacecraft-1 of NASA (see Figure 4.1). Interestingly, using Hyperion imagery, at least three different classes of soil tillage namely

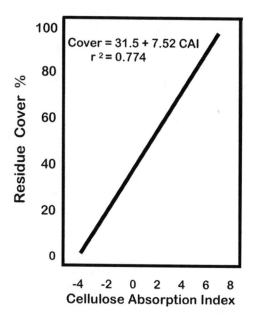

FIGURE 4.1 Relationship between Crop Residue Cover available after Tillage and Cellulose Absorption index (Note: Crop Residue cover left in the field is directly related to Tillage intensity and type. Conventional tillage allows low quantities of residue cover compared to restricted or conservation tillage. No-tillage system allows crop residue and stubbles to stay in the field. Cellulose Absorption Index measured using spectral sensors mounted on satellites is indicative of cellulose material still found on the soil surface. Therefore, higher the CAI greater is the crop residue and it is indicative of restricted tillage or low intensity tillage. Source: Drawn based on Daughtry et al., 2011).

intensive (<15% crop residue cover), reduced (15–30% crop residue cover) and conservation or restricted tillage (>30% crop residue cover) could be easily identified. The accuracy of detection of tillage class using satellite data was 75–82%. Further, Daughtry et al. (2011), state that combining previous year's crop/residue data generated using Hyperion and soil tillage intensities deciphered, we can organize management blocks that lead us to better results. Such demarcation of management zone using Hyperion imagery is said to help in soil conservation and soil C dynamics too could be managed appropriately with positive effects. Hyperion imagery also helps in keeping an inventory of crop residue and tillage operations conducted on a particular field.

Further, Daughtry et al. (2011) have stated that it is generally possible to map crop residue distribution on a field and even large agrarian regions using satellite imagery. Overall accuracy of crop residue distribution and tillage intensity ranged from 71–93%. Hence, they concluded that remote sensing could help in detecting variability in tillage type and intensity practiced by farmers in a given field or zone. In fact, a combination of estimates related to tillage intensity, crop residue left in the field and crop biomass productivity using satellite imagery could be effectively utilized to allocate biomass for biofuel production.

In addition to monitoring tillage operations and their impact on crop production in a given large field or an agrarian zone, satellite imagery could also be focused on a single field/patch to assess the tillage effects and prescribe agronomic procedures appropriately. For example, in Queensland, farmers have been provided the advantage of monitoring the progress of tillage in sugarcane field, on weekly and seasonal basis. Farmers could decide on tillage date, depth, intensity and extent based on satellite imagery. Tillage could also be varied based on topography and soil characteristics observed using satellite images on their computer screens. The GPS guided zonal tillage could actually avoid or at least reduce soil compaction and allow disruption of subsurface compactness by regulating tractor traffic in the sugarcane fields, prior to planting (Hughes et al., 2009).

4.2.4 SATELLITE-GUIDED SOIL SAMPLING

Soil sampling is a fundamental aspect of satellite-guided Precision farming. It has immediate impact on the accuracy of digital soil maps prepared.

Soil samples drawn have to be highly representative of actual situation. Any error during extraction and analysis of soil samples may have cumulative effects, and distort our inferences and maps prepared for use during precision farming. The rationale for locating sampling sites within a field has to be apt to the purpose. Historically, soil samples have been drawn mainly to assess physico-chemical properties such as texture, aggregation, bulk density, EC, CEC, SOM, available and total nutrients in different horizons of soil. Soil nutrient estimation and mapping their distribution has been the main purpose during past 5 decades and it continues to be so (Adamchuck et al., 2004; Ferguson and Hergert, 2009; Ferguson et al., 2007; DeGruijter et al., 2010). Within the realm of soil fertility management, there are at least four different types of soil sampling that could be adopted by farmers. They can all be tagged using GPS coordinates to enhance accuracy and digitized, so that they are amenable for vehicles with decision-support computers and variable-rate applicators. The four sampling systems are:

a) Bench mark sampling that involves collection of soil samples from different depths at locations that are unique. For example, a moist and highly productive zone, a location with low moisture holding capacity, a set of locations with low fertility and crop yield, region with a particular cropping system, etc.

b) Topographic sampling, as the name suggests, it involves soil sampling based on topography, say, on the hill, slope or depressions, etc.

c) Grid sampling requires formation of grid cells of definite size and identification of location of each soil sample, in a particular grid cell. Soil maps are constructed using data pertaining to each grid and satellite coordinates.

d) Management zone sampling involves sampling of different locations within a management zone (see Krishna, 2012b; Whelan and Taylor, 2013). There is no doubt that determining sampling approach is important. Prior to sampling we need to know the end use of data derived from these samples. For example, if fields are to be supplied fertilizer at uniform rates, then random sampling suffices. However, if variable-rate applicators are used, then dense sampling by adopting grid or management zones is necessary. Each soil sample is identified using GPS coordinates. Within this book, we are concerned more with details about Grid and Management sampling techniques for soils and crops.

4.2.4.1 Grid Sampling

Grid sampling is an apt method, if the intention is to adopt precision techniques to raise a crop. Grids are marked say at 0.2 ha, if the total area to be covered is 2–2.5 ha. Grids are marked in the entire field. So, grid system of sampling allows soil fertility variation to be documented in a digitized map without bias or any particular preference for locations within a field. Samples drawn using grids could be assessed for several physico-chemical properties and nutrients such as N, P, K and micronutrients too. It seems, grid systems are congenial, if new areas are to be brought under crop production. Each soil sample in a grid cell or entire grid cell could be GPS tagged, so that appropriate digitized maps are obtainable. Usually, smaller grid size and high-density grid sampling allows greater accuracy. Grid sampling techniques in North America have actually evolved from being a method that gives rough estimates or low-resolution nutrient distribution maps to accurate one. During 1990s, grids of 4.0 ha were formed. In due course, grid cell size decreased to 2 ha. At present, with the help of satellite service companies, grids of 0.2 ha are formed and relatively more accurate soil nutrient maps are prepared. Ferguson and Hergert (2009), for example, they say that during past decade, accuracy of satellite aided soil maps for soil-P distribution has improved 10 folds in resolution and accuracy. Shift from coarse grid sampling to densely drawn soil samples from smaller grids has induced the accuracy. Further, it has been opined that well placed grid system that is densely sampled provides detailed soil characteristics that could be mapped using satellite and GPS coordinates. Such maps are easily preserved for several years. Accurately prepared maps are advantageous while adopting variable-rate techniques. Sampling depth is an important aspect. Sometimes, soil nutrient availability to crop roots is depicted best, if both surface and subsurface horizons are sampled, analyzed and maps prepared accurately. For soil traits such as pH, CEC, SOM, available-P, K, S and Zn, obtaining surface samples and tagging them with GPS coordinates suffices. Surface samples means those drawn at a depth of 0–15 cm. Soil nutrient distribution along the depth of the profile is an important criterion while deciding on sampling. For example, as we reach deeper layers beyond 15 cm, SOM, soil-P and Zn may decrease. Surface samples are needed for all crops, but deeper samples are necessary, if the crop to be cultivated is deep rooted. While estimating nutrients that tend to percolate and accumulate in lower horizon, say for soil-N, both surface and subsurface sampling is required. For relatively shallow rooted crops such as wheat, beans, canola, millet and groundnut samples drawn at 60 cm depth suffices (see Jahanshiri, 2006; Ferguson et al., 2007; Krishna, 2012a, b).

Now let us consider a few examples from practical farming situations. During maize production in Minnesota, if soils were sampled at both surface and deeper horizons, then N requirement decreased. Actually, fertilizer-N needed for a maize crop at different stages of growth decreased perceptibly, if samples were drawn from 120 cm compared to 60 cm depth (Kasowski and Genereux, 1994). The decrease in cost of production due to reduced need for fertilizer-N ranged from 63 to 350 US$ ha. Mallarino (1998) studied utility of grid sampling for soil nutrients such as P and K. It seems in Iowa, soil P and K is optimum, but their spatial variability is high. So, crop response to P and K variability in soil is proportionate. Sampling the field densely and making grids of smaller size, then tagging each sample with GPS coordinates helped in removing the crop growth and yield variations within a field.

There is no doubt that denser sampling and small grids offer greater accuracy to soil nutrient distribution maps prepared. The variable-rate techniques that follow together have direct impact on crop productivity. Actually, nutrient dynamics in each small grid cell is affected. In some instances, cost of grid sampling could be high. Then, it is preferable to make larger grid cells and coarser sampling, but overlay it with satellite imagery. This allows farmers to arrive at appropriate fertilizer dosages but with a degree of approximation. This method is cost effective and profitable. An alternative is to form Management zones and restrict intensity of sampling based on known characteristics of management zone and crops to be grown. The grid sampling strategy suits to be adopted: (a) when measurement and mapping of non-mobile nutrients is intended; (b) when management practices are influenced by topography and other fixed factors in the field; and (c) when farmers have been using fertilizer and organic manures consistently for few years (Clemson University Cooperative Extension Services, 2014).

4.2.4.2 Management Zone Sampling

Management zone sampling is opted when prior knowledge about yield pattern or a map of the field is available. Several other characters of the field such as general soil fertility, moisture distribution, disease spread or insect attack are useful while deciding on management zone sampling. Sometimes soil traits such as compaction, depth, texture, subsurface nutrients, etc. are used to decide management zone sampling. Satellite imagery of crop growth or yield is over-layed and management zones are created. Soil sampling is confined to each management zone or definite blocks or strips and each sample is GPS tagged with coordinates. The density of soil sampling within a management block depends on facilities available for analysis of soil samples and purpose. Usually 8–10 core samples are picked from each management zone. Surface samples are drawn at 10–15 cm depth and subsurface samples are derived from 30–60 cm depth. The data is mapped and digitized, then utilized in variable-rate applicators. Management zone sampling could be denser when variation is high for soil traits such as texture, structure, soil fertility, nutrient distribution, moisture, EC and salinity. Farmers are often suggested to adopt management strips, if definite patterns of clayey or sandy textured soils are encountered. Quite often, management blocks or strips are formed based on grain/forage yield potential and economic benefits from different areas within a field (see Krishna, 2012b; Moshia et al., 2010). Management zone sampling could be adopted when the purpose is specific. For example, variable-rate application of N and P supply. Sampling could be denser so that digital maps of nutrient distribution are accurate. Management zone formation could also be focused to study soil physico-chemical properties. Usually, surface and subsurface samples are drawn to prepare soil maps for characters such as pH, texture, EC, CEC etc. Management strips are formed when adopting different cropping systems. Crops grown in each strip could be different. Soil sample timing, method, density and characteristics estimated, depends on each management strip. The digitized maps for each strip could be supplied to tractors and variable rate applicators, so that each and every agronomic procedure is aptly timed and accurate. We may also note that digital maps with soil variation known for each management block or strip could be stored for future use and drawn from GIS when needed.

We may note that several of the procedures involved in obtaining satellite imagery, formation of management blocks or grid cells, and excavating soil samples from surface and subsurface layers are mechanized, automated and works through electronic controls, at the touch of a button. During past decades, soil sampling was a procedure that involved really tedious drudgery. This was followed by complicated chemical analysis of each sample. Soil samples had to be marked and tagged manually using accurate numbering. However, during recent years, there are robotic soil samplers that are GPS tagged and operate based on pre-determined instructions held in computer. Autonomous soil samplers could also be guided using satellite guidance and I-tablets (see Plate 4.7). Over all, management zone sampling can be adopted on priority when cost of sampling and their

PLATE 4.5 'Precision Hawk'—a drone taking aerial survey of a cereal field in Indiana, USA (Source: Dr. Lia Reich, Sr. Communications Director, PrecisionHawk Inc., Noblesville, Indiana, USA; Note: Satellite imagery could be insufficient in terms of resolution and details while detecting crop production aspects such as crop stand and planting density, gaps in rows, water status of crop/soil, pest and disease. Drones that fly at low level, as shown above, need to be inter-connected or inter-phased and allowed to make a more detailed survey of crop field and even take necessary measures such as pesticide sprays, etc.).

chemical analysis is a major concern. Management zones are larger than grid cells; hence they reduce need for excessive sampling. Management zones are apt when measuring and mapping mobile nutrients in a farmer's field. Management zone sampling is also effective even if there is no history of fertilizer application, sampling, or knowledge of nutrient distribution (Clemson University Cooperative Extension Services, 2014).

4.2.5   SOIL SENSORS AND MAPPING

Sensors based on electronic circuitry and optics are used in agricultural crop production in several ways to accomplish variety of tasks (Thessler et al., 2011). Sensors ought to be most accurate and fail proof during operation, because any error may creep into series of other machine-based operations conducted in the field. Failures in sensors can create problems for series of other activities in the farm. Sensor's data is crucial to obtain accurate judgments using a range of computer software programs. At present, sensors are too common in farm equipment and they are used in networks. Therefore, optimum accuracy and functioning of each and every sensor is necessary, if not entire network may get affected. At present, most of the agricultural equipment used across farms in developed nations is attached with a few different types of sensors. Sensors are mounted on tractors that are deployed to plant seeds, spray pesticides and apply fertilizers. For example, liquid fertilizer sprayers are endowed with sensors that first verify, if the plant is green enough or it requires nutrition. Soil moisture sensors are among common attachments added to farm vehicles. Sensors with GPS attachment help in accurate mapping of variations of different properties of soil. Electronic sensors, no doubt, are becoming common in agricultural farms. They say, true precision farming procedures are dependent on accurate mapping and GPS-RTK elevation data. They determine accurate placement of inputs at each spot in the farm (Cropmetrics, 2014; Mask et al., 2011; Rovira-Mas et al., 2008).

Sensing soil properties and preparing digital maps depicting fertility factors is mandatory during precision agriculture. Satellite mediated soil-sampling procedures that tag each sample using GPS coordinates is becoming more common in the vast agrarian zones. The idea is to base variable-rate nutrient supply and other amendments using digitized maps (Gebbers and Adamchuk, 2010; Schirrmann and Domsch, 2011; Mask et al., 2011). Information on soil properties such as texture, SOM, nutrient status, particularly N and pH seems mandatory during satellite-guided precision agriculture (Wetterlind, 2009). There are at least four different principles on which most of the on-the-go soil sensors operate and collect GPS tagged information about soil and its variability. They have been classified by Bah et al. (2012) into:

a) Electrochemical sensors such as those used to measure soil pH, NO_3_N and K (Adamchuk et al., 2007, 2008; 2004);

b) Electrical and Electromagnetic sensors such as those used to measure soil texture, soil moisture content, CEC (Kim et al., 2009; King et al., 2005);

c) Optical and Radiometric Acoustic sensors are used to measure SOM, soil moisture, soil bulk density, formation of hard pans, etc. (Rossell et al., 2006); and

d) Mechanical sensors are often used to detect soil compaction and soil hard pans (Stafford and Werner, 2003; Hemmat and Adamchuk, 2008; Adamchuk and Jassa, 2014; Sudduth et al., 2008).

Casa et al. (2012) have explained that there is a clear need to study and obtain high-resolution digital maps of soil properties, prior to initiating precision techniques. Such maps should take into consideration variation of at least major factors such as soil fertility, water and vegetation growth, but it has to be accomplished at low cost and of course with greater ease. Adopting proximal methods for estimation of soil properties using a GPS vehicle, mounted with sensors for soil chemical analysis is a good idea (Rossell et al., 2011). However, Casa et al. (2012) and Ben-Dor et al. (2009) believe that adopting hyperspectral satellite imagery to ascertain soil physico-chemical variations is something that is quick, easy to pick the digitized maps, and feed the date to variable-rate applicators. Satellite imagery is low cost method compared to UAVs and ground based robotic vehicles. Satellite imagery is best suited, if the farms are large or the target area to study is a large expanse, say a couple of counties in Great Plains region, etc. The limitations are that soil fertility variations, if they occur at small distances, then resolution could be blurred or may even go undetected. Currently, there are several private agencies that utilize satellite data about soils and serve the farmers with well-demarcated maps and management blocks. They offer soil maps that depict 3-D versions of topography, soil texture, moisture holding capacity, macro-nutrient distribution and salinity (Trimble, 2014b). These soil maps are crucial to arrive at management decisions related to fertilizer inputs, irrigation and crop harvest. Digitized data about soil characteristics is almost mandatory while adopting precision farming methods such as variable-rate application of fertilizers.

4.2.5.1 Soil Electrical Conductivity (EC) Sensors and Mapping

Soil electrical conductivity is said to be a trait that statistically correlates and is indicative of several other physico-chemical properties. Soil EC itself is influenced by characteristics such as clay content, salinity, CEC, soil moisture content, depth of clay and abundance of certain ions. Soil EC is defined as ability of soil material to conduct (transmit) electrical current. It is measured as milliSiemens or deciSiemens per meter (mS m^{-1}) (Barbosa and Overstreet, 2010). One of the earliest proximal soil sensors to be developed and popularized is related to those measuring soil EC (Lund et al., 1998; Shaver et al., 2012). Kweon et al. (2012) state that of the many sensors developed and used by farmers, soil EC sensors are more popular. The soil EC signal is caused by soil texture changes and in locations where salinity is elevated it delineates such changes, in addition. Currently, farmers can avail two different types of soil EC measurement instruments. Many of these are GPS connected and offer well prepared soil EC maps for use during precision farming. For example, Veris Technologies of Kansas, in USA manufactures EC sensors that provide EC data at one or two soil depths at approximately 1 to 3 ft. (Plate 4.8; Schirrmann et al., 2011). These are contact sensors. The vehicle has 2 or 3 pairs of coulters. One pair is used to impinge electrical current into soil and the other is used to measure the voltage reduction between them, in order to measure soil EC. There are several models of non-contact soil EC sensors. Private companies such as Geonics, Aeroquest, Sensortech offer a range soil EC sensors that have satellite guidance and GPS connectivity for accurate location of spots measured on a map. The non-contact

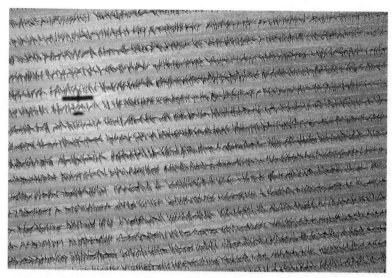

PLATE 4.6 A drone providing a close-up view of Planting density of Cereal in a field (Source: Dr. Lia Reich, Sr. Communications Director, PrecisionHawk, Noblesville, Indiana, USA; Note: Planting density and gaps if any could be monitored. A close-up view also helps in monitoring diseases, if any, using hyperspectral imaging of crops. Such drones could be inter-phased using computers that guide drones to areas of interest, identified using a satellite that provides coarser imagery of the crop fields).

PLATE 4.7 'AutoProbe' (left) and RapidProbe (right) are autonomous soil sampling robots (Source: Drs. Jim and Jeff Burton, AgRobotics, Little Rock, Arkansas, USA; Note: These soil samplers actually dig and extract cores at a rapid rate of 10–20 cores per hr and then prepare a composite a soil sample. They move autonomously based on signals from GPS connectivity. They help farmers in developing soil fertility map).

PLATE 4.8 A Soil Electrical Conductivity Sensor (Source: Mr. Eric Lund, Veris Technologies Inc., Salina, Kansas, USA http://www/veristech.com/products/soilec.aspx (August 1st, 2011); Note: Soil EC Mapper detects changes in soil EC and maps it 'on-the-go' using soil surface contact technique. Soil EC is supposedly indicative of several soil traits such as texture, SOM, salinity and soil moisture holding capacity. Indirectly it measures soil productivity).

soil EC sensors work on the principle of electromagnetic induction. As the name suggests, it does not have to be in contact with soil surface or particles. The method is actually based on mutual impedance between a pair of coils situated just above the surface of soil. Usually 2–3 pairs of coils are attached to sensors. A transmitter coil generates electromagnetic field at a specific frequency. This produces electrical current that flows in subsurface soil material and induces formation of eddy current. The eddy currents generate secondary magnetic field that is perceived by the receiver coil (see Sudduth et al., 2001).

There are several reasons to use soil EC maps prepared using GPS coordinates. Firstly, it is inexpensive and relatively rapid to move the sensor vehicle across the field. Soil EC correlates with several soil traits that provide useful insights to farmers about SOM, pH, salinity, moisture, etc. (Doerge et al., 2011; Upadhyaya and Teixeira, 2011). The soil EC sensors are utilized to delineate management zones. Soil EC is said to be correlated to productivity of soil, in other words, crop growth and grain yield to a certain extent. Hence, soil EC maps are used to decide on cropping systems and variable-rate seeding of fields. It is also useful in obtaining a rough estimate of soil fertility variation and hence fertilizer application too could be regulated using soil EC maps. Kerry and Oliver (2003) believe that while preparing soil EC maps, for use in precision agriculture, it is advisable to add few related information. For example, ancillary data and satellite imagery helps in judging soil EC maps better. Incidentally, Barbosa and Overstreet (2010) have stated that soil EC maps could also help farmers to direct variable-rate herbicide applicators. This is because, rates of herbicide supply is based on soil texture and organic matter. Both SOM and texture are related to soil EC. Therefore, soil EC maps could be easily calibrated and converted into herbicide supply maps and used appropriately. Often, pre-emergent soil application of herbicides is contemplated using soil EC maps. Satellite-guided variable-rate applicators serve excellently during herbicide application. Shaver et al. (2012) state that one of the most common variability detected in soil is caused by textural characteristics. Soils with different textures conduct electricity differently. Clay is more conducive to electrical conductance than sand. So, correlation between soil types, field variability and EC is possible. Soil maps depicting EC could be used to obtain a semblance of idea about soil type.

4.2.5.2   Sensors to Detect and Map Soil pH Variations

Soil pH influences several aspects of physico-chemical properties. Soil microbial activity is immensely influenced by pH. Nutrient absorption by crops is also is affected by soil pH. In fact, bio-availability of nutrients to crops is affected by soil pH and its fluctuations. Crop species flourish and produce luxuriantly at certain acclimatized soil pH range. Logdson et al. (2008) state that, aspects such as nutrient absorption by crops, legume nodulation and herbicide effectivity may all be influenced by soil pH. Crops are detrimentally affected if soil pH is uncongenial. Most importantly, we have to recognize that soil pH is a highly variable character that needs to be measured accurately in order to take appropriate remediation. Soil pH variability could occur at distances much shorter than typical grid sampling distances. Hence, on-the-go soil pH measurement that provides sharper maps that could be digitized and used in variable-rate lime applicators is needed. It is said that in some locations within Great Plains, a change of 2 units of pH has occurred in two samples drawn 12 m apart. In practical farming, grids of 1 ha are common and soil pH variation could be conspicuous and cannot be overlooked. Lime addition to match soil pH variation is needed (Brouder et al., 2005; Bianchini and Mallarino, 2002). Hence, GPS guided autonomous or semi-autonomous farm vehicles that sample soil at very closely situated locations or perform on-the-go soil pH analysis is required. The digitized data obtained from such soil pH sensors could be relayed to variable-rate applicators that apply lime based on computer-decision support systems.

A soil pH mapping vehicle with on-the-go pH sensor mounted is available, since past 5–8 years (Schirrman et al., 2011). The Veris Technology's soil pH Manager operates on any kind of soil surface and uses proximal analysis using ion selective antimony electrodes. It has GPS connectivity; hence, detailed soil pH map could be prepared. The Veris Technologies soil pH Manger has been tested for its suitability with different types of soil. Schirrmann et al. (2011) have reported that soil pH measured on-the-go using Veris Soil pH manager correlates with actual laboratory pH estimations significantly, ranging from $r^2 = 0.8$ to $r^2 = 0.84$. However, field specific calibrations are recommended to reduce systematic errors. At this juncture, we have to note that entire procedure of soil sampling at different fields or grids within them, actual soil pH analysis using proximal techniques, preparation of digitized soil pH maps and appropriate remediation by adopting matching variable-rate lime inputs could all be accomplished using GPS and electronic controls. Hence, this aspect clearly forms part of 'Push Button Agriculture' touted in this book. These procedures reduce requirement of human scouts, skilled technicians and almost eliminate human drudgery during soil excavation, sampling, soil sample preparation, laboratory chemical analysis and tedious mapping of data.

4.2.5.3 Soil Nitrate Sensing and Mapping

An on-the-go soil nitrate mapping system offers great advantage to farmers who consistently apply fertilizer-N as basal and split dosages. Basically, soil NO_3-N variability has to be sensed accurately and fertilizer-N supply has to match the NO_3-N variations in space and intensity. This concept though easy to suggest, it involves drudgery and tedious effort to collect large number of soil samples from small grid cells or management blocks. The chemical analysis performed in the general course requires skilled technician's time and effort. Then, the results need to be mapped accurately showing the NO_3-N variations in the field. Reports suggest that it is possible to assess NO_3-N variations using on-the-go 'Soil Nitrate Mapping Systems (SNMS).' The SNMS is essentially an autonomous electrochemical machine that collects soil samples from top soil, mixes it with water, and analyzes NO_3 concentration using a NO_3-ion electrode. It seems each sample takes 6 sec to analyze. The location of soil sample and NO_3 data could be easily GPS tagged with accurate coordinates (Sibley et al., 2008a, 2009; Adsett et al., 1999). In its simplest form, a SNMS consists of six sub-sections. They are soil sampler, soil metering and conveying, NO_3 extraction and measurement, auto-calibration, electronic control and GPS connectivity (Sibley, 2008). Field scale evaluation of SNMS by Sibley et al. (2008a) suggests that it could be used to draw soil samples, analyze for NO_3 N using electrodes and map them using GPS co-ordinates.

Regarding relevance of SNMS, it is said, it helps farmers to link soil-N variation to crop growth. It then allows them to understand the consequences of high NO_3-N accumulation in the soil profile and need for constant monitoring of soil NO_3-N. It allows farmers to know the soil-NO_3 variation and so it helps them in allocating fertilizer-N accurately using variable-rate applicators. Basically, soil-NO_3-N variation is caused by soil parent material, chemical and microbiological processes, crop production systems adopted and environment. During crop production, however, farmers have to ascertain soil-NO_3 variations accurately and at short intervals even within a crop season. Therefore, automatic and rapid soil NO_3-N monitoring becomes essential. Soil-N monitoring using tractors mounted with SNMS is also required, if farmers have to avoid undue accumulation of N in soil, loss via percolation or emissions and to avoid ground water contamination. SNMS could be used to assess soil-NO_3 N on-the-go and using a computer decision-support system, we could apply fertilizer-N even without recourse to GPS.

Further, a study by Sibley et al. (2008b) suggests that SNMS provides data regarding soil NO_3 N quickly, accurately and at low cost to farmers. The SNMS could be used effectively in assessing soil-NO_3 variations in small scale. The SNMS was utilized to judge NO_3 N variations in soil and relate it to performance of wheat and carrot, in terms of growth and yield. The tractor mounted sensor for NO_3 provided accurate data throughout the crop season. Sensor technology has been adopted to manage soil-N during cultivation of corn, cotton and wheat (Mask et al., 2011).

Sensors that assess macro-nutrients in soil are essential during farming. In addition to N, sensors have also been designed that measure P and K in soil. Kim et al. (2009) have made a critical review of our knowledge about sensors that are based on electrochemical and electromagnetic principles, particularly those that help us in assessing macronutrients in soil. There are also sensors developed for measuring soil Ca and pH. Field evaluations suggest that data derived from such sensors correlates significantly with those derived using regular laboratory chemical analysis using atomic absorption flame photometry (Lemos et al., 2007).

4.2.5.4 Satellite Aided Sensing to Measure and Map Soil Organic Matter

According to Shaver et al. (2012) satellite imagery can provide information about soil variability through colors. Dark areas in the field are usually related to heavier texture, higher clay and better water holding capacity. Lighter colors are related to sandy textures, low organic matter and low water holding capacity. Natural color aerial or satellite imagery could be used to judge organic matter distribution. There are also optical devices with GPS connectivity that help in accounting for soil organic matter distribution. The optical device is inserted into soil and reflected light is measured. The reflectance is supposedly affected by characteristics, primarily related to texture and organic matter content. There are also on-the-go soil organic matter mappers that operate based on optical sensors. Incidentally, there are several reports about *in situ* ground based data from sensors and laboratory evaluations using different wavelength bands. Reports suggest that reflectance from soil could be used to measure and predict soil organic matter (Hummel et al., 1996; 2001).

Regarding instrumentation for detecting SOM and mapping, Veris Technologies situated in Kansas, USA has developed an Optic mapper. The optic mapper is mounted on a vehicle, mostly a planter or one that has EC mapper. It assesses organic matter on the surface of soil and underneath the crop residue that has been spread in the field. The Veris Optic Mapper for soil organic matter is a very useful data collector. The data could be used along with those for soil texture, salinity, soil moisture, etc. (Veris Technologies, 2014; see Krishna, 2012b).

4.2.5.5   Satellite Mounted Sensors to Measure Soil Moisture

Knowledge about soil moisture is essential during crop production. Famers have to constantly monitor soil moisture status and match it with crops' need by adopting irrigation, if rainfall is insufficient. Soil moisture detection could be made using conventional gravimetric methods, ground based techniques such as gypsum blocks, etc. Ground based sensors mounted on vehicles or perches could also be used to assess crop water status.

Landsat satellites have been fitted with short wave infrared sensors (SWIR) and thermal infra-red sensors (TIRS) that are useful in estimating soil moisture. In addition, integrated sensor systems such as wireless, satellite and airborne imagery have been used to detect soil moisture distribution (Phillips et al., 2014).

In addition, there are reports that MODIS sensor placed on NASA's Terra satellite offers excellent data about soil moisture status in agrarian zones. Doraiswamy et al. (2004) have reported that MODIS on Terra Satellite provides imagery at 250 m resolution about soil moisture status of fields smaller than 25 ha. Most recent reports suggest that NASA's SMAP satellite is supposedly endowed with sensors that detect soil moisture at higher resolution. It provides information about soil moisture storage in the top 5 cm soil surface. Thermal infrared imagery helps in detecting soil moisture status. A new initiative by NASA-USDA has aimed at obtaining daily soil moisture 'snap shots' that helps farmers. They use Advanced Microwave Scanning Radiometer (AMSRE) for EOS sensor that is present on NASA's 'Agua' satellite. AMSR-e uses different frequencies to detect amount of emitted electromagnetic radiation from field surface (Dunbar, 2009). Within the microwave spectrum, radiation is closely related to amount of water that is present in the surface soil. This allows agricultural agencies and researchers to remotely sense soil water with a degree of accuracy, currently acceptable. Agricultural agencies in United States of America are now aiming at a network that verifies SMAP collected data with ground measurements and exchanges information rapidly about accuracy of sensors on the satellite (Yang et al., 2014).

4.2.6   SATELLITE-GUIDED SOIL FERTILITY AND FERTILIZER SUPPLY MANAGEMENT

Several aspects of soil fertility and their interactions with water may affect crop yield. These factors actually cause the crop yield variation within a field. Therefore, site-specific techniques should consider several of these innumerable soil fertility traits to achieve uniformly higher yield. Prior to it, farmers will have to resort to detailed grid sampling of soils using satellite-guided GPS coordinates. Then, map the variations for each of the many fertility characters. Some of these agronomic procedures are easier said than done. For instance, detailed soil sampling and analysis could be tedious and economically not feasible, if this procedure has to be conducted too often. For example, several soil nutrients such as P, K, NO_3-N, Zn, then soil pH, SOM are all highly variable even within a field. When these traits interact with other relevant aspects of soil profile such as profile depth, horizons, soil hard pan formation, aeration, CEC, EC etc. (Magri et al., 2005), it really ends up in a highly variable soil map. The digital maps prepared using these factors may constrain GPS connected variable-rate applicators to move very slowly. The accuracy of agronomic procedures too could get affected. It is basically cumbersome, if farmers have to tackle soil fertility variability for too many traits, during precision farming. Hence, most studies suggest use of management zones, based on single most important trait or clustering a few traits. Farmers are generally advised to collect yield maps of few years in succession, overlay them and study the variability in crop stand, growth traits and grain yield. Then, mark the management zones using a few other characteristics such as topography, gradient, general soil fertility expression, etc. Once, a rough sketch of management zones is done, farmers may adopt satellite-guided grid sampling for major nutrients or factors that affect the crop yield and prepare digital maps. The computer decision support could utilize these digitized data to accurately and rapidly guide the variable-rate applicators.

4.2.6.1 Management Zones and Fertilizer Prescription

Satellite technology is apt to be applied to large farms and agrarian expanses. Imagery derived could be used to identify variations in natural vegetation, crops and crop productivity trends. The imagery could be used further to analyze the constraints to crop production in greater detail. Crop productivity may vary widely across a field, a large farm or an expanse based on several factors related to soil, crop or environment. When several constraints operate at different intensities and vary enormously within a field, managing farm inputs and attaining uniformity across fields is not an easy task. Satellite imagery may actually highlight variations related to topography. Generally, topography may be undulated in certain area and flat in other region of a farm. Factors such as gradient of slope in a field and soil type may differ. Soil fertility variations that are too common may also affect agronomic procedures such as fertilizer supply, its quantity and methods adopted. Similarly, irrigation timing and quantity too may have to be varied, to match the variation encountered in soil moisture in the profile. The need for irrigation could be deciphered using satellite imagery derived using infrared thermal sensors. Weed infestation observed using satellite imagery too may vary enormously, both in terms of weed species and intensity. In order to match the situation, farmers may be required to adopt deep tillage, or application of specific herbicides at different concentrations. From the above description of causes, extent, intensity and intricate interactions of various factors that affect crop productivity, it should be clear that, methods that offer greater control and accuracy in adopting remedial measures is required. We can use satellite imagery and digital data pertaining to a particular field, farm or even an agrarian patch and demarcate the area into 'Management zones.' We may define 'management zone' in different ways depending on the context and factors that need emphasis. Management zone is actually a sub-region of a larger field that expresses a certain degree of homogeneity for a particular character. Such a management zone generally requires a fixed rate of input or remedial measure. Management zones should be carved out after studying the satellite imagery, digital data or if it relates to soil chemical constraints, then the area should be quantitatively and densely sampled and analyzed (see Doerge, 2010; Krishna, 2012b; Lakes, et al., 2007). Vrindts et al. (2005), again, emphasize on homogeneous combination of soil fertility factors and define management zones as a region that expresses similar yield limiting factor(s). Ferguson and Hergert (2009) suggest that it is easier to overlay yield maps of yester years, or soil fertility maps known for a few years and satellite imagery showing crop productivity trends and mark the 'management zones.' The number of management zones carved out may depend on size of the field, each zone and extent of variation for a particular soil fertility factor. Generally, satellite imagery should be consolidated and only a few management zones should be managed. We should note that a particular management zone, say that marked based on soil-N may reappear in discontinuous fashion and make it cumbersome while applying fertilizer at variable rates. Such confusing trends should be avoided using approximations on the satellite imagery. Management zones help in attaining greater accuracy during agronomic management of crops. Farmers could routinely observe satellite imagery, note the factors that vary and then mark the zones on a touch screen and preserve the demarcations. Such demarcated zones could then be used to supply various inputs in the field using variable-rate methods. Tractors with planting devices or fertilizer inoculators or irrigation equipment could be controlled using GPS connectivity and variable-rate applicators. These could be adopted to achieve uniformity.

During practical agriculture, farmers frequently encounter several different constraints and at different intensities for varying lengths of time. Management zones that they demarcate could be actually emphasizing any one or a few of them. Mostly, constraint that severely affects crop yield is considered first and others later. A satellite image may show-up, crop loss due to low inherent soil fertility, at the same time show-up effects of drought, disease or weed infestation, etc. However, there are suggestions that clustering of a few soil characteristics, topography, yield data, drought or flood effects, soil erosion trends could all be considered for preparing management zones (Rub et al., 2010). Let us consider some examples that depict use of soil analysis and satellite imagery to prepare and mark 'management zones' in greater detail.

Soil forming factors and topography are among important factors that vary and cause proportionate variations in crop productivity. Topography and soil surface characteristics could be easily studied using satellite imagery. Satellite images showing details of surface relief, soil color, texture and vegetation is obtainable from agencies. Fields showing slope are prone to erosion of surface soil and nutrients. The soil erosion could be highly variable based on gradient differences. Crop productivity is known to vary in response to slope of field and erosion. Nutrient and organic matter loss could affect the crop productivity. Management zones formed based on gradient of slopes in a field seem to match with crop productivity (Goddard and Grant, 2001).

Site-specific variation in crop productivity is the main reason for forming sub plots or management zones. Mzuku (2005) suggests that spatial variability in soil properties may actually be the prime reason for productivity differences within a field. Therefore, it is sensible to study site-specific variation of the factors that cause fluctuation in productivity. There are reports that soil characteristics such as EC, Ca, Mg, Na, and SiO_2 may show significant variation, even within a small topographic field unit. Soil bulk density and compaction too vary within small distances (Chung et al., 2001). Similarly, spatial distribution of soil properties may vary enormously (Yasrebi et al., 2008). For example, soil-P varied at short distances but Soil-Ca content varied after longer intervals of sampling distances. In other words, even within a small field, factors such as topography, slope, soil physical traits and chemical content are in complex interaction and this causes the fertility variability. Maps showing variations for such soil characteristics match with those for grain/forage productivity, if over-layed. In fact, Mzuku et al. (2005) report that for soils from several locations in Colorado, USA, soil physico-chemical properties such as bulk density, texture, compaction, moisture holding capacity, color, and organic matter content were used to form management zones. They found that maps depicting variations of crop productivity coincided with those of several of the above soil characters, if over-layed. Therefore, satellite-guided techniques could be focused to judge as many soil properties and productivity, while forming management zones for precision agriculture.

Soil Nitrogen, its physico-chemical transformations, availability to crop roots, its utility in plant tissue, accumulation in grains/forage, although complex, it has been among well researched topics in the realm of crop production. Nitrogen is the key element that is required in relatively larger quantities by crops. It's deficiencies are reported in most of agrarian belts, but there also regions experiencing excessive fertilizer-N application resulting in its accumulation. Soil-N fertility variation is also caused by factors such as erratic N losses from soil profile, leaching due to soil erosion; percolation to lower horizon or even ground water and volatilization as NH_3 or emissions as N_2O. Soil-N mineralization rates and soil-N fractions may also vary quantitatively in a field. Crops may remove soil-N at variable rates. Depletion rates for soil-N are not uniform even within a single small field. Somehow, blanket prescriptions of fertilizer-N do not seem to be apt. They may either over or underestimate N needs of crops. Hence, Franzen and Kitchen (2010), state that determining crop-N requirement accurately, despite such variations is a challenging task. More recently, soil-N corrections have been effected using precision techniques that involve formation of 'management zones.' Management zones that aim at achieving uniform soil-N levels can be demarcated using several different soil traits and methods. A few methods listed by Franzen and Kitchen (2010) are based on topography, aerial photography of crop growth/yield, satellite-guided sampling and detailed analysis of soil samples for total and different fractions of soil-N. Further, satellite imagery to estimate NDVI, chlorophyll content (crop-N), soil electrical conductivity, general crop yield maps are also adopted. Satellite imagery using multispectral cameras and GPS guided variable-rate supply of fertilizer-N based on digital data, computer-based decision-support systems and yield goals are important with reference to 'Push Button Agriculture' that is touted in this book. Tedious grid sampling and wet chemical analysis, then deciphering fertilizer-N dosages for grid cell seems out of place, since they do not provide rapidity or ease of operation in fields. Of course, on-the-go soil fertility analysis using indictors such as EC and variable-rate application of N avoids use of even satellite imagery and data.

Coming back to formation of Management zones using soil-N and yield goals, Franzen and Kitchen (2010) and Franzen et al. (2005) suggest that more than one layer of soil-N data may be needed. Therefore, if farmers have management zones already marked, then grid sampling could be restricted and focused properly. Sometimes, farmers may use soil-N data and compare soil-N fertility maps for a few years in a series, by overlaying them. Then, they mark the management zones. Cropping history and N depletion rates of each crop species grown in yester years need due attention, while deciding on the next crop in a particular management zone. Satellite imagery could be used to study the soil-N depletion rates. A lush green and highly productive crop would have literally exhausted fertilizer-N applied. Satellite imagery for topography, soil EC, and soil-N may coincide roughly. So, this information could be used to mark management zones. For example, Franzen and Kitchen (2010) have clearly shown that imagery from French satellite SPOT-5 showing topography, EC, soil NO_3 and plant-N data for wheat grown in South Dakota just coincide, if over-layed. A different report by Franzen et al. (2005) states that demarcation of management zones aimed at accurate fertilizer-N supply could be done using NDVI from Landsat-7 imagery at 30 m resolution and soil EC measurements done using Veris EC sensor connected to GPS. Generally, use of multiple data sets seems to correlate better with residual-$NO_{3-}N$ maps and yield maps. In the Hammocky terrain region of Manitoba, in Canada, Moulin et al. (2003) studied soil fertility variation, particularly soil $NO_{3-}N$ and its impact on crop yield in detail. They prepared management zones based on yield maps for several years. Yet, soil $NO_{3-}N$ and P were highly variable even within a field. Therefore, it required site-specific techniques such as satellite-guided variable-rate fertilizer applicators to achieve uniformity in soil fertility and crop yield.

We may note that management zones could be formed during satellite-guided precision farming using several other traits such as soil color, soil organic matter and soil moisture (Table 4.2). The satellite imagery could be obtained using multi-spectral sensors that includes IR-thermal sensors. The imagery could be over-layed with crop yield and management zones could be formed. However, we may also realize that some of the traits are transitory and keep changing in intensity and expanse. Hence, we have to draw imagery periodically from satellite companies and update the management zones (Moshia et al., 2010).

Soil electrical conductivity (EC) measured using sensors with rolling electrodes (e.g., Veris Technology) or electromagnetic induction (Geonics Ltd) could be used to demarcate management zones (Table 4.2). Soil EC maps however needs to be over-layed or data has to be correlated to other soil fertility traits or crop yield map *per se*. Franzen et al. (2005) and Franzen and Kitchen (2010) have shown that maps for soil EC, topography and wheat grain yield coincided. They have also proved that soil EC is related to corn yield and that grain yield decreased, if EC increased (Figure 4.2). However, soil EC maps correlated well with original management

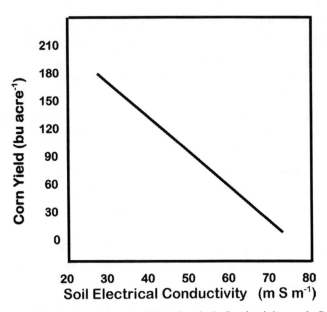

FIGURE 4.2 Relationship between Soil Electrical Conductivity and Grain yield variability of Maize grown on a Kastanozem in the Northern Great Plains of USA (Source: Redrawn based on Franzen and Kitchen, 2010; Note: Maize grain yield decreased as soil EC increased. This relationship has been used to form management zones during satellite-guided precision farming.

zones demarcated using topography. Li et al. (2008) have examined the relevance of soil electrical conductivity measurements, crop growth and yield data to prepare management zones. They have measured and mapped EC in soil, cotton growth parameters, yield and NDVI of 15 different fields utilizing over 200 soil samples, to erect management zones. They could create three management zones using clusters of parameters that included EC and cotton crop yield. Agricultural agencies prefer to use soil EC data (maps) because it is easy to measure it using on-the-go instruments. For example, soil electrical conductivity mapper produced by Veris Technologies, Kansas, USA could be used. Soil EC is related to other traits that affect crop growth and productivity, such as soil pH, salt concentration and water holding capacity. Doerge et al. (2011) have reported that soil EC maps coincide with those of crop yield. It can be verified easily by overlaying satellite imagery and ground data available. The EC maps may also show variations in productivity of natural vegetation. It is common to delineate large fields into management zones based on crop productivity, say as low, medium and high. In such a case, soil EC data could be used to demarcate management zones (Stafford et al., 1998, Blackmore 2000; Li et al., 2013). At this juncture, we may note that on-the-go EC measurement is accomplished using GPS controlled instrument hitched to tractors or other farm vehicles. Then, obtaining satellite imagery of crop growth and productivity is a matter of using computer connectivity with satellite agencies.

During wheat cultivation in Australia, farmers are forced to amend soils with several soil fertility factors such as nutrients, organic matter and lime to correct pH. Precision farming techniques adopted involve formation of management zones using characters such as crop yield maps, soil type and electrical conductivity. In fact, wheat researchers at Grain Research Development Centre (GRDC), Kingston, Australia state that using soil physico-chemical traits such as EC, texture, moisture and SOM, then over-laying them with yield maps offers best management blocks (GRDC, 2010). Dang et al. (2011) have assessed fields using EC sensors. Their aim was to ascertain the extent of constraints suffered by these fields and how far will EC measurements be indicative. On fields suffering due to salinity, sodicity, and high chloride ion concentrations, EC sensing of soils helped in marking out management blocks. Farmers could make decisions regarding yield goals and fertilizer inputs suitably for each management block. Using soil EC sensors and forming management blocks reduced cost on inputs. Profit due to formation of management zones ranged from 14–46 A$ ha^{-1}.

Oxisols that are sandy, low in fertility, acidic and show relatively higher levels of Al and Mn are common to many agrarian locations traced in Brazil, Columbia, Venezuela and Peru. The vegetation is constrained by several other soil related factors. Understanding physico-chemical variability of soils and their complexities is necessary, since they affect crop yield. Many of the production procedures for crops depend on extent of variation in soil physico-chemical properties. Soil sampling and analysis could be tedious. Therefore, Camacho-Tamayo et al. (2013) have suggested grid sampling with GPS coordinates and demarcation of fields into management zones. Soil physical properties such as predominant texture in surface and subsurface horizons of soil, saturated hydraulic conductivity, bulk density, particle density, porosity and water holding capacity were assessed to mark management zones (Table 4.2). Recognition of variability and demarcation based on groups of soil traits is necessary for higher productivity. Several of the soil physical properties estimated have direct impact on soil fertility expression and crop productivity (Pena et al., 2009; Amezquita et al., 2004; Cucuneba-Mello et al., 2011). According to them, management zones were actually required to make agronomic procedures easy to accomplish. Satellite imagery too could be used to study the crop production pattern and overlay the management zones marked using soil physical properties.

Management zone marked within a larger field has often been explained as a subfield region that is homogeneous with reference to different traits of landscape. Demarcation of such homogeneous management zones is said to enhance crop yield, when precision techniques are adopted (Li et al., 2007). As stated earlier, management zones could be marked using a single soil fertility trait or several of them that are directly related to crop productivity. For example, in case of coastal crop production zone in China, Li et al. (2007) have used traits such as NDVI image from SPOT-5, soil electrical conductivity, total nitrogen, organic matter and cation exchange capacity. Several other researchers have used characters such as topography, soil color and crop growth images obtained using aircrafts or satellites. One of the approaches described is statistical clustering of a bunch of traits relevant to soil fertility and crop management (Schepers et al., 2000; 2004). Li et al. (2007; 2013) have demarcated field into management zones using spatial variability of soil chemical properties, nutrient distribution, particularly, N and a few others. They have used geo-referenced soil samples to obtain high accuracy while mapping and management zone formation. Results indicate that aggregated data could be used to form management zones. Further, at a later period, when soil sampling and analysis has to be done, such management zones have helped in reducing sampling intensity and chemical analysis. A report by Zhong et al. (2009) suggests that typically, fertilizer-based nutrient supply to cash crops such as tobacco is done using uniform blanket recommendations for entire field. Again, this leads to over or under estimation of nutrient needs of the crop. Hence, they used a series of soil characteristics such as total-N, alkalytic-N, available-P, available-K, CEC and EC, then, they obtained

GPS coordinates for each data point and mapped it. They suggested that preparing management zones first, using yield maps or previous data or satellite imagery (NDVI), helps in reducing sampling intensity and cost of inputs. Therefore, using satellite images and other soil traits to mark management zone during precision farming seems necessary. It reduces drudgery, enhances crop productivity and economic benefits. Also, such efforts take us closer to achieving a degree of sophistication during farming that could be called "Push Button Agriculture."

Precision farming that requires use of management zones and satellite-guided soil sampling, remote sensing and crop monitoring is gaining in popularity in East European plains. For example, in Czech Republic, agricultural service companies that utilize satellite imagery and help farmers in soil/crop monitoring, formation of management zones and timely prescriptions about various agronomic procedures are proliferating (Gnip and Charvat, 2003). Further, it has been pointed out that subsidies offered by governments of EU countries plus conversion of almost all farm vehicles and equipment into GPS guided systems, use of digital sensors and variable-rate applicators has enthused farmers to use precision techniques, for example, it requires formation of management zones. Agricultural service companies perform accurate grid sampling, soil analysis for a range of characteristics, prepare vario-grams and mark the management zones. Such management zones simplify agricultural operations. Farm inputs are supplied more accurately. Management zones also provide economic advantages to farmers. According to Gnip and Charvat (2003), most commonly requested services from satellite agencies that relate to management zones are: (a) background satellite imagery, (b) soil type maps, (c) soil sample grids with GPS coordinates, (d) digitized soil test maps for nutrients such as P K, Ca, PH and EC with accurate GPS coordinates, (e) NDVI maps and digital data, (f) crop sample grids, yield prediction maps, and (g) recommendations regarding variable-rate application of nutrients, mainly N, P, K, lime.

Kitchen (2003) and Elstein (2003) state that, during precision farming, establishment of management zones is a very useful proposition. It allows farmers to reduce costs on inputs and makes it easy to focus soil and crop management techniques on a single or few factors. Such a management zone could be established using computer aided analysis of previous data and of course obtaining appropriate digital data from satellite companies.

Sometimes farmers may need multiple sets of imagery or demarcations of the same piece of land. For example, management zones marked based on weed infestation may look entirely different to those prepared using soil-N data or topography or grain yield map.

Now, let us consider an example pertaining to plantation fruit crop. Soil fertility varies enormously within citrus farms. Tree growth and productivity aspects actually respond to fertility and moisture variation that they encounter in the soil profile. Soil fertility corrections effected may not match the fertility variations. It may result in fertilizer supply at rates that under-estimate or over-estimate the nutrient dearth. Citrus trees are deep rooted and absorb nutrients and moisture from deeper horizons. Hence, soil analysis should include lower horizons of soil. The soil analysis data has to be GPS tagged in order to infuse accuracy to data about soil fertility variations. Soil sampling done using GPS-guided sampling could still be tedious and at times confusing to handle very large number of samples. Hence, it is useful to prepare soil fertility maps using leaf nutrient status and actual fruit productivity. Satellite imagery could be effectively used to study tree growth, canopy and productivity, and then utilized to map the citrus grove. Satellite derived digital maps could then be used to prepare management zones.

Citrus production, like any other crop, depends on assessing soil fertility variations, their apt management and evaluation of results. Satellite-guided Precision technique is one of the recently devised methods to manage such variations in citrus fruit productivity. Now, let us consider another example. In case of citrus grown in Central India, researchers have tried to integrate and super impose the productivity zones and ground data obtained regarding nutrient status of leaves (Srivastava et al., 2010). They have adopted DRIS (Diagnosis and Recommendation Integrated System) and GIS to develop a variogram for nutrients such as N, P, K, Ca and Mg. Citrus orchards were delineated into 3 zones based on nutrient constraints, including Zn content. Decision support systems that considered leaf nutrient optima and yield goal was used to supply correct dosages of nutrients to each management zones. It is believed that integrating ground data, satellite imagery and appropriate computer aided decision-support systems will reduce human drudgery in soil/plant sampling, chemical analysis and

TABLE 4.2   Field Characteristics Utilized to Demarcate Management Zones

Field Characters	Soil/Crop Characteristics
Quantitative Stable:	Topography, soil type and its texture, soil organic matter, pH or $CaCO_3$, CEC, WHC, electrical conductivity
Quantitative Dynamic:	Crop canopy, soil moisture, salinity soil-N, plant-N, weed density, yield data
Qualitative Stable:	Soil color, texture, soil survey records, immobile nutrients such as P and K. soil pathogen/insect attack soil aeration
Historical Records:	Yield patterns of yester years, soil tillage practices followed, cropping history, fertilizer supply trends, disease/pest incidence records.

Source: Doerge (2010); Krishna (2012b).

during fertilizer application. Demarcation of citrus groves into management zones and use of satellite imagery could also reduce costs on human labor.

Let us consider an example that describes how a 'Private Agricultural Company' offers services that helps farmers to adopt precision farming, using satellite imagery, management zones and variable-rate techniques. In South Dakota, a company named 'Precision Soil Management LLC' offers detailed satellite imagery, also conducts soil fertility analysis on actual soil samples. The agency maps the field using high-resolution satellite imagery to obtain yield maps and topography. Soil samples drawn from each 'management zone' is analyzed by laboratories. The productivity of management zones could then be classified into low, medium and high. The information could be transmitted to farmers through internet. Once, the agency gets an idea about the yield goals that farmer has envisaged, it works out the prescriptions regarding crop species that suits a particular management zone, variable-rate seeding, fertilizer inputs, harvest dates, etc. (Precision Soil Management LLC, 2014; Table 4.2). Some of the characteristics based on which management zones were created by this private agency are as follows:

a)   Level ground, thick soil, very high productivity;
b)   level ground, medium top soil, high productivity;
c)   gently sloping, medium thin soil, high productivity;
d)   strongly sloping, thin top soil, medium productivity;
e)   location in a depression, thick top soil, and high production;

f)   gently sloping, shallow gravely soil, and low productivity;

g)   strongly saline, low production;

h)   very sallow clay pan soil, medium production;

i)   deep soil with gravel, low productivity;

j)   sandy soil, low productivity;

k)   clayey soil, medium top soil and high productivity;

l)   poorly drained soil, shallow clayey and medium productivity;

m)   poorly drained, clayey soil, medium production; and

n)   gently sloping, thin top soil high salt and low production.

Some of the reactions of farmers about formation of management zones and their utility in hastening and making agronomic procedures easy, particularly variable-rate applications are favorable. For example, farmers in wheat belt in Dakotas and Minnesota state that use of satellite imagery, yield data of few years, yield maps and topography has helped them immensely in managing the large farm and still obtain uniformly high wheat productivity. Training farmers in reading and digitizing satellite imagery, operation of GPS guided variable-rate applicators and usage of software for decision support systems is necessary.

Doerge (2010) states that, there are at least three steps to evaluate performance of management zones. First, we have to start simple with available spatial data. Best quality information is usually derived from quantitative data based on dense sampling. Usually, field topography, bare soil photograph, soil survey maps, EC maps, etc. are used to demarcate

Crop Inputs	Management zone factors to asses and map
Immobile nutrients	Topography, grid sample soil nutrient maps, soil survey maps, soil EC maps
Nitrogen and manure	Soil texture, organic matter, yield zone, soil $NO_3.N$ map, crop canopy reflectance
Lime	Soil pH, CEC, texture
Gypsum	Soil EC map, yield pattern, pH and Na map
Seeding rate	Historical yield data, top soil depth and texture
Herbicides	Weed infestation maps crop species
Waters	Soil texture, topography, soil organic matter, yield zones

Source: Doerge, 2010

management zones. Secondly, management zones could be sharply defined using maps of a few more traits. Thirdly, evaluate the management zone strategies using crop yield. Some of the management strategies adopted are as follows:

Over all, during practical crop production in large farms, farmers tend to consider at least soil EC deciphered using on-the-go Electrical Conductivity measuring vehicle, satellite imagery of the field showing crop production trends and multi-year yield data to form management block and apply appropriate dosages of fertilizers (Clemson University Extension Service, 2014).

4.2.6.2   Sensing Soil Using Multiple Sensors

It seems adoption of remote sensing techniques in precision farming began with development of sensors for soil organic matter. However, during past 25 years, wide range of other sensors, for estimating different aspects soil and crops have been developed and used (Mulla, 2013). Soil sensing using multiple sensors mounted on a single vehicle and capable of estimating a few relevant soil traits in one stretch is perhaps most useful during precision farming. Satellite mediated location of soil samples and digital maps could be rapidly used to direct variable-rate applicators. Alternately, the data from multiple sensors could also be used on-the-go to correct soil fertility factors (See Adamchuk and Jassa, 2014; Krishna, 2012b; Taylor et al., 2010; Lee, 2010). Adamchuk et al. (2005) have reported that a multiple sensor built using ion-selective electrodes was tested for its efficiency, practically, for direct measurement of soil chemical properties on-the-go. The measurements using on-the-go system correlated with those derived from regular laboratory chemical analysis. The correlation for soil pH was $r^2 = 0.93$–0.96, for potassium (K) it was $r^2 = 0.63$–0.67, for NO_3N it was $r^2 = 0.51$, and for Na it was $r^2 = 0.31$. The multiple sensor performed best if utilized to measure soil pH, soil-K and NO_3N. Such multiple sensors are of utility during precision farming. The data from multiple sensors could be tagged with GPS and soil maps for traits such as soil pH could be prepared. It is pertinent to integrate sensing of major nutrients such as N, P and K (Sinfield et al., 2009). They have suggested using different sensing techniques based on physical, chemical and optical methods to detect all three macro-nutrients in one stretch. They emphasize use of minimal sample preparation and use of mid-infrared and infrared reflectance and electrochemical methods.

Multiple Sensing is also possible using satellite based techniques. Basic requirement is the transit of satellite above the field or zone, appropriate multi-spectral sensors that operate at visible, NIR, red-edge and Infrared/ thermal bandwidth. Multiple sensors have gained in acceptance in many of the European regions. Casa et al. (2012) have reported that CHRIS PROBA satellite imagery could be used for detecting a few aspects such as topography, vegetation, soil characteristics such as clay and sand content, soil water plus soil organic matter. Further analysis suggested that, soil texture and SOM measured using hyperspectral bandwidth from CHRIS PROBA, tallied excellently with that obtained using laboratory analysis of ground samples. Soil water measurements using satellite imagery could be utilized to supply water at variable rates using the model CropWat 8.0. The digitized data about soil water could also be used to study the soil water balance at various stages of the crop. Simultaneously, at the beginning of the crop season, satellite imagery from CHRIS PROBA could also be used form management zones using soil textural classes. This helps farmers to adopt apt agronomic procedures considering soil type, its texture and fertility.

4.3 SATELLITE-GUIDED CROP HUSBANDRY

In North American agrarian regions, particularly in the 'Corn Belt' and other cereal regions of Great Plains, large seeders with GPS guidance and variable-rate planting systems are becoming common (Plate 4.9). These semi-autonomous tractors are able to accomplish a range of tasks using GPS signals and computer-based decision systems, based on digital data supplied. For example, Barnhisel et al. (2014) state that semi-autonomous tractors fitted with variable-rate seeders dibble seeds at a range from 37 kg ha^{-1} to 85 kg ha^{-1} based on soil type, texture and soil-N fertility. Such variable-rate seeders could also be used to sow management strips. Interestingly, experiments on fertilizer-N rates by seeding rates could be initiated using such variable-rate fertilizer dispensers (Plate 4.10) and seeders. GPS guided variable-rate seeding of cotton has been in vogue for

PLATE 4.9 Variable-Rate Planters (Source: Dr. Daniel Danford, CEO, CaseIH Inc., Oakes Rd., Sturtevant, Wisconsin, USA).

PLATE 4.10 Strip cultivation and variable rate nitrogen fertilizer supply (Note: Strip cultivation is one of the Management Zones concepts adopted as part of Precision Farming. Management zones help in rapid and easier application of fertilizer-N (NH_3) to soil. Source: Mr. Dean, Twin Diamond Industries, Minden, Nebraska, USA).

over 10 years now in Central Great Plains and California (Cline, 2002). Initially, it was a tedious task conducted by human scouts. However, currently, variable-rate seeders are used to dibble cotton seeds based on management blocks, soil fertility and yield potential data. Currently, farmers adopting precision farming techniques use tractors/planters that possess GPS connectivity and variable-rate seeding facility. These planters consider soil fertility and yield goals stipulated by farmers. Such vehicles also possess ability to plant seeds based on sections. Seeds are delivered at variable-rates. Seeding density is monitored on-the-go and they can also provide maps showing different varieties and genotypes planted at various stretches/management zones (Gibbs Equipment Inc., 2014; CaseIH Inc., 2014). It is opined that variable-rate seeders, that are often semi-automatic, reduce requirements of labor. Seed costs are low and labor costs too get reduced. Lowenberg-DeBoer (1998) had forecasted this; way back a decade ago, that variable-rate seeding of cereals and oilseeds would reduce seed requirements and hence reduce costs.

In the European farming zones, large cereal farms are adopting planters that are connected with GPS guided variable-rate seeders. For example, 'Techneat intelli-rate controller' is one such system. It automatically adjusts seed rate based on yield potential depicted in the digital maps obtained using satellite sources. Usually such seeders are compatible with a range of vehicles that are semi-autonomous. Perhaps, in future, totally autonomous vehicles attached with seeders will rule the European plains regarding seeding of cereal crops. It then leads to 'Push Button planting of crops' using robotic seeders. Some of these variable-rate seeders are sensitive and adjustments allow them to plant seeds at low seed rates and identify increments of 0.1 kg ha^{-1}. A few models of seeders are also equipped with computer software and programs that allow them to mark management zones automatically using previous data on soil texture, depth, moisture and a number of other factors. Reports suggest that oilseed planters fitted with variable-rate planting devices are able to plant at 3.25 kg seeds ha^{-1} in low fertility lighter soils, but seed rate increases to 3.96 kg seed ha^{-1} in high fertility heavy soils, based on satellite signals and digital maps (Farming UK 2012).

Satellite-guided seeding of cereal fields is indeed an efficient proposition, particularly, if it involves autonomous vehicles connected with GPS and variable technology. It reduces labor, hastens seeding, takes care of soil fertility variability and judges apt seeding densities. This procedure of variable rate seeding is supposedly more efficient than erstwhile popular techniques. Report by UNCSAM-MARDI (2012) states that in rice producing regions of Malaysia and other Southeast Asian nations, crops are produced on soil with high degree of variability, with regard to fertility and nutrient availability. Factors such as topography, soil moisture and weeds are highly variable. Establishment of optimum plant density and uniform plant stand is a problem. It seems non-uniform seedling establishment is a major problem in rice fields. UNCSAM-MARDI (2012) suggests that variable-rate seeding techniques using digital maps of leveling index removes the effects of uneven land surface and plant stand. Practically, variable-rate seeding needs to be adopted using maps of factors that afflict the fields drastically. It could be topographical or soil type differences or soil fertility variations.

Farmers in Australia are currently adopting variable-rate seeding of crops such as corn, wheat, cotton and legumes. It seems high density planting of corn is accomplished better using GPS connected variable-rate seeders (Precision Seeding Solutions Inc., 2014).

4.3.1 SATELLITES IN MEASURING CROP REFLECTANCE, NDVI AND CROP GROWTH

Satellite sensors used during crop production could be categorized based on resolution and the specific purposes it serves. For example; low-resolution images of 1 km are suitable for depicting weather above cropping zones. A medium resolution imagery of 30–100 m is suitable to assess crop development in the field, particularly, providing in-season images. High-resolution 1–25 m imagery is useful in small scale crop analysis, developing NDVI maps and detecting crop productivity changes (GDA Corp, 2014d).

The NDVI data and maps prepared using different wavelength bands allow farmers to judge vegetation, its density and distribution, and most importantly the crop growth and canopy traits. Several of the crop management decisions could become easier and accurate, if NDVI data is available easily on the desk through internet. There are several satellite

service companies that offer periodic imagery of crops, general vegetation in the surroundings and detailed digitized data/maps depicting NDVI.
These could also be over-layed with soil maps obtained at different periods in the season and analyzed (Satellite Imaging Corporation (2014 d, e).
Such services reduce cost on detailed scouting using human labor. They
are rapid and accurate.

Let us consider another example, where in, satellite agencies provide
farmers with details about 'green vegetation index.' This suggests farmers
about the status of crop and biomass accumulation. A green vegetation
index map shows spatial distribution of vegetation (crop growth) in a field/
region. The green vegetation index is derivative of crop vigor and amount
of biomass accumulated in response to fertilizers, water and pesticides.
The satellite image is actually calibrated using algorithms that remove
noise from soil and surface water resources held in the field/region. Such
maps and digitized data could be transmitted to farmers for each field or
a zone specified. For greater accuracy, there are computer programs that
sharpen the image by removing interferences and making color codes
for different densities of vegetation. Chlorophyll content could also be
marked, using color codes on the maps. Green leaf area index, canopy
traits and biomass are usually supplied to farmers. Such green vegetation
index maps are also known to help in identification of damages to crop due
to soil erosion, insects or nutrient deficiencies. Farmers could reach the
spot using maps to navigate. In the general course, if satellite imagery is
absent, human scouts may be required to walk a lot of area per day. Their
accuracies of identification of spots with nutrient deficiency or other maladies are not high. An aerial image, a map on computer screen provides
excellent perspective of crop; its growth and maladies if any (see Satellite
Imaging Corporation, 2014 g, h, j).

4.3.2 SATELLITES TO MEASURE LEAF AREA, CANOPY CHLOROPHYLL AND CROP-N STATUS

It is not the intention of this chapter on satellite mediated crop monitoring to
go into greater details of hand held instruments. However, we may note a
few examples where in hand-held instrument and aerial surveys could be

utilized, in case satellite imagery gets obstructed due to clouds. In addition to satellite imagery, there are ground-based methods that measure NDVI, which is indicative of leaf chlorophyll content. A hand-held spectroscope based on optical sensing device could measure NDVI value of crop canopy and/or leaves accurately. The hand held instrument measures spectral reflectance at two bands 619 nm and 1220 nm to calculate NDVI accurately. It measures chlorophyll content using UV-VIS spectrophotometry (Cui et al., 2009). Now, let us consider satellite-mediated methods for crop monitoring.

Leaf area is a key determinant of growth and grain productivity of field crops. This is true with several horticultural plantations such as grapes, apple, etc. Johnson (2003) has shown that spectral imagery from IKONOS satellite could be used to assess leaf area and related traits of grape vineyards. Geo-referenced NDVI data were collected from grape orchards. Leaf area and planting density were then used to validate the satellite data with ground measurements ($R^2 = 0.72$). It is said that despite discontinuous nature of grape canopies and different leaf angles, leaf area could be mapped. Such data, it seems, serve farmers in deciding irrigation schedules and fertilizer inputs.

Reports from Chinese wheat producing zones indicate that satellite imagery could be adopted to trace the growth and monitor increase of leaf area index of a wheat crop. For example, Xiaoyu et al. (2014) have reported that, in case of winter wheat grown around Beijing area, satellite techniques were useful in showing LAI increase. Satellite images from a 3500 ha field showed that between April and May, LAI increased by less than 1 within 14% of field area. About 64% area showed LAI increase between 1 and 2, and 20% area showed LAI increase of about 2. It is suggested that LAI increase could be used as indicative of crop growth. Therefore, satellite-guided monitoring of wheat growth is possible. Further, Song et al. (2010) have shown that in the wheat belt, satellite imagery could also be adopted to monitor uniformity of wheat crop growth. Remote sensing allows farmers to screen a very large area of wheat belt for uniformity and other characteristics, in one stretch, using high-resolution imagery. For example, they used images from 'QuickBird' high-resolution sensors to study wheat crop at various stages of growth from seedling to booting and then until senescence of leaves. It seems wheat crop growth data at boot leaf stage provides important information about grain yield, since it is positively correlated with grain production. The correlation coefficient between OSAVI (Optimized Soil Adjusted Vegetative Index) with wheat grain yield was 0.651 (R^2). The GNDVI (Greenness-Normalized Vegetative Index) was negatively correlated with grain protein content. It is opined that satellite imagery procured from agencies connected with QuickBird or other satellites could be used by farmers to trace wheat growth and assess grain productivity and quality.

Satellite mediated imagery and digital data accrued has been used to assess soybean and corn production zones in Ohio. Several reports suggest that Landsat thematic mapper data that depicts regions with soybean and corn could be used during farming. Spectral responses of crops could be obtained at various stages of growth, to assess soil moisture component, soil organic matter, plant canopy characters, plant chlorophyll and N status, influence of fertilizers, etc. (Senay et al., 2000).

Remote sensing using high-resolution spectral imagery has been adopted in tropical rice growing regions. For example, in rice production zones of Malaysia, spectral reading such as canopy size, green vegetation index and chlorophyll content has been used and compared with SPAD reading obtained using hand-held chlorophyll meters. It seems they are highly correlated ($r^2 = 0.95$). Hence, remote sensed data could be mapped and used during variable-rate application of fertilizers (UNCSAM-MARDI, 2012). Nitrogen deficiency, sufficiency or excess could be easily recognized using green vegetation index and canopy traits.

Daughtry et al. (2000) state that, farmers may have to adopt tedious soil and plant sampling to decipher N dynamics and assess fertilizer-N requirements of crops. In case of large expanses of corn, wheat or soybean grown in North America, such processes will be time consuming and economically not so appealing to farmers. Since spectral reflectance is related to plant leaf chlorophyll content and that in turn is directly proportional to leaf-N status, it is possible to obtain an accurate estimate of plant-N status. Maize farmers could utilize such spectral data to assess crop fertilizer-N requirements using appropriate computer software (Daughtry et al., 2000).

Several experts dealing with maize crop nutrition and fertilizers have stated that side-dressing fertilizer-N in 2 or 3 splits is more efficient, than uniform application of entire quantity in one stretch to the field (Noh and

Zhang, 2012; Krishna 2012a). Farm scouting for N deficiency requires skilled human labor, time and cost. Each split application requires a map of the field showing spread of N deficiency and its intensity. Satellite aided monitoring and periodic imagery of maize crop is useful since it helps in detection of N deficiency. The crop canopy reflectance, NDVI, leaf chlorophyll and N status could be measured using hyperspectral imagery.

Cereal production strategies are mostly aimed at maximizing fertilizer-N use efficiency, matching N need at various stages of crop growth, avoiding soil-N accumulation that could become vulnerable to loss via emissions. Generally, uniform application of fertilizer-N has been discounted since it induces inefficiency. Synchrony between fertilizer-N supply and crop-N status could be achieved better through satellite monitoring of leaf chlorophyll, NDVI and canopy at short intervals. Shanahan et al. (2008) have suggested use of management zones marked using soil analysis data, satellite imagery and crop-N status data. It helps in reducing N supply to crop to reach the same yield goals.

Mapping soil fertility, crop growth and yield using remote sensing and reflectance is now common. Long et al. (2008, 2013) and Thylen et al. (2014) state that farmers are adopting yield monitors and yield mapping devices in the combines rather routinely. A step further, they are also being exposed to assess grain protein on-line while harvesting. Then, map actual protein harvested at different spots in the field. Protein yield maps of entire field are also possible. This data may help farmers to grade the grains of say wheat or barley based on protein concentrations in the grain lot derived from different locations in the field. Of course, we can also calculate nitrogen extracted into grains by the crop.

Reports suggest that site-specific techniques could be useful in attaining uniform grain yield and grain protein status. Wheat crop grown in the plains of Montana were harvested using combines equipped with grain yield sensors and GPS connectivity. Grain yield maps were also used to plot grain protein maps. The NDVI values derived at mid-season through aerial imagery correlated with wheat kernel protein status. The grain protein value ranged from 15–16% on dry grain basis. It is believed that farmers could take remedial measures using in-season adjustments in fertilizer-N supply to wheat crop, based on NDVI readings (Long, 2006).

Feng et al. (2014) have studied the relationship between NDVI, plant-N and grain-N in a wheat crop using satellite imagery. They aimed at forecasting grain protein content using satellite imagery at various stages of the wheat crop, grown under irrigated and rain fed conditions. They used a few different computer models to judge grain maturation and protein content, using satellite data. They suggest that we have reasonable scientific basis to connect satellite data about a crop, at its various stages of growth and maturity and grain protein that it may accumulate. Further investigations and standardizations are obviously needed before adopting it, to assess grain protein of winter wheat, routinely.

4.4   SATELLITE-GUIDED SOIL/CROP WATER STATUS AND IRRIGATION MANAGEMENT

Satellites have been used to assess water resources of several agrarian regions, on a wider scale. Satellites such as Landsat are efficiently used to collect digitized data and map water resources of the Earth. Landsat imagery about water resources, river flows, and cropping systems that demand water from various sources has been of great help to farmers across different continents, in addition to USA (USGS, 2014; Plate 4.3). Water held in forests, cropping belts and riparian zones have been consistently detected, mapped and relayed to farmers by Landsat and a few other satellites. In North America, satellite imagery has been used to monitor water resources for agricultural cropping, to estimate evapo-transpiration from riparian zones and crop belts, identification and estimation of area occupied by different cropping systems and their potential water needs (USGS, 2014). Landsat imagery showing water resources of a region has also been used to decide legal disputes and allocation of water to different states/regions. Monitoring drought, dust bowl, low storage of water in reservoirs and reduced riparian flows using Landsat satellite helps in adopting proper remedial measures. Landsat imagery using infrared and thermal bandwidths have been used to categorize and identify irrigated and non-irrigated dry arable cropping zones in many part of the world. There are now many private companies that hire transponders and estimate water needs of different crops. For example, farmers with grapevine yards in

France, wheat crop in USA or Europe, citrus plantations in Florida could purchase satellite imagery periodically and estimate water requirements. They can revise irrigation intensity based on yield expectations during the season.

Reports from Stanford University explain that satellites are efficient in tracing and monitoring ground water resources. Ground water is of immense utility to agricultural policy makers and of course famers. Other than data collected by satellites, ground water regulators have very little authentic information or forecasts about ground water resources that could be allocated to crop production. Ground water levels fluctuate, it raises and shrinks seasonally based on replenishments from rain, percolation, surface flows, etc. In some regions, snowmelts causes an increase in ground water level. Ground water forecasts are of great value to farmers intending to grow summer crop or those affected by droughts. There are specialized satellites that trace fluctuations in ground water levels and movements of ground water (Stanford University, 2014). It amounts to seeing for water under the surface of ground. Satellites are also focused to study the ground above the aquifers, so that water could be allocated appropriately for crop production.

Buis and Murphy (2014) state that, currently, worthwhile ground or satellite based networks that monitor soil moisture in the agrarian zones is lacking. This is despite knowing that information on soil moisture is critical to farming. The European Space Agency's Soil moisture and Ocean Salinity mission does track soil moisture, but its resolution is pretty low at 50 km. Soil moisture, as stated above is highly variable parameter even within a small field. However, NASA's SMAP (Soil Moisture Active Passive) is supposed to be relatively highly accurate due to microwave sensors that monitor the top 5 cm of crop field and map the moisture content. In future, droughts of both mild and intense versions could be detected and farmers warned. In the general course, knowing soil moisture storage and its availability in the top horizon is crucial. In future, with the launch of SMAP, satellite imagery helps us in this activity with better accuracy.

Soil moisture is a highly variable parameter that affects global agriculture at different intensities. As such, maintenance of optimum soil moisture all thorough the crop period, right up to maturity of grains and senescence is not an easy task. Knowledge about soil moisture status, quantum that could become available to crop production and authentic forecasts about water receipts are almost essential during crop production. Forecasts about drought and floods are also useful to farmers. In addition to above facts, we may note that farmers may often require authentic data of soil moisture fluctuations, at timely and frequent intervals. Traditional method of sampling and fixing electrodes in soil, noting the readings and converting the data into maps requires technicians, skilled labor and extended period of field scouting. It is a labor intensive exercise indeed. A recent report by National Aeronautic and Space Agency (2014) states that soon a satellite that utilizes microwave and radiometer is set to assess soil surface moisture and freeze and thaw levels of the entire land area of earth, ones every three days. It is called 'Soil Moisture Active Passive Satellite (SMAP). The SMAP can forecast droughts. Suggestions about alternative cropping systems and water management practices using imagery and computer decision support are possible. Agricultural droughts occur when precipitation is drastically low compared to crop's demand and when other sources of water from lakes, rivers, dams, etc. are dried out. Some of these aspects are also considered by SMAP, in addition to regular monitoring of soil surface moisture. Researchers at NASA/USDA believe this satellite's data has immense value for farmers' worldwide who experience water shortages. Plus, it helps any farmer with almost regular data about soil moisture in the upper layers of soil horizon. Incidentally, moisture in the surface horizon is perhaps most important while channeling irrigation and setting yield goals. This soil moisture detecting satellite is said to collect data about soil surface moisture and disseminate it through internet to help farmers. They say, since it measures soil moisture, at frequent intervals and offers imagery, it will be of use during water supply and scheduling irrigation at short notice. This service is something most farmers would like to utilize, since it offers better maneuvering and aids in revising irrigation quantity and timing. At this juncture, we have to note that all this is done using computers software, monitors and push button systems. It reduces requirement for human labor at frequent intervals. Currently, in the field, farmers do adopt sensor networks that determine timing and quantity of irrigation. It seems integrating such systems with satellite guidance could be helpful (Vellidis et al., 2008).

4.4.1   VARIABLE-RATE CENTRE-PIVOT IRRIGATION

Traditional irrigation procedures involve monitoring of soil moisture using neutron moisture meters. Major limitations are need for skilled labor, tedious placements of probes and instrumentation. Farmers have indeed adopted a wider range of simple to complex methods, in order to estimate soil moisture and decide on the quantum of irrigation. First step is to feel the soil clump and recognize its moisture content. Next, there are hand–push probes, also called Paul Brown Probe, that determine soil wetness based on the depth to which the probe can be pushed into soil profile. Gravimetric methods are tedious and time consuming. The soil sampling is also tedious and GPS coordinates have to be marked on each sample. Most commonly used soil moisture sensors are the 'soil moisture blocks.' They could be electrical resistance gypsum blocks or granular matrix sensors. The data could be noted or recorded on a hand held data logger. There are 'tensiometers' with porous ceramic tip that could be placed in the soil and soil moisture tension could be recorded. Soils are saturated with water at 0–10 centibar (cb) soil moisture tension. At 40–60 centibars is the typical range for medium irrigation. At water tension of 70 cb irrigation is necessary and at 100 cb and above crops experience drought stress. All the above listed methods are tedious, time consuming and need detailed field/soil sampling procedures (see Morris, 2006). However, during recent years there are reports that suggest use of Electromagnetic Induction (EMI) techniques to measure soil moisture. The EMI provides a measure of bulk EC of the soil profile. It has correlation with variations in soil moisture distribution. Huth et al. (2012) have reported that EMI techniques are rapid in estimating soil moisture content. Such information could be utilized to decide on irrigation schedules in crop fields. During satellite-guided precision farming, these data points obtained could be tagged with GPS coordinates and utilized appropriately in variable-rate irrigation systems. The EC technique is also amenable for on-the-go variable-rate irrigation.

During recent years, farmers in North America, particularly those in semi-arid regions of Great Plains, have been insisting on release of greater portion water resources. Water resources, no doubt are limited by several factors. Hence, water use efficiency measures are applied at various stages during crop production. In fact, in most locations, farmers have been applying uniform rates of water to a crop that is cultivated on fields with inherently variable soil moisture distribution. The variable nature of soil moisture availability is caused by several factors related to topography, soil texture, aggregation, depth, percolation and seepage rates, weather and seasonal changes, crop species, its root spread and moisture depletion ability. It is generally believed that for a variable soil moisture distribution, apt method is to apply water at variable rates, to correct moisture dearth, if any, and to attain uniform moisture availability within a field or management zones. This procedure is generally termed Variable Rate Irrigation (VRI). Nationwide, center-pivot irrigation systems are fairly common and there are currently over 150,000 such systems operative in USA alone (Clemson University Cooperative Extension Services, 2014). Variable-rate supply of water using center-pivot works on the principle that specific management zone should be provided water in quantity based on requirements of crops, at a particular stage. It seems by optimizing VRI, we can save several million gallons of water, that otherwise would have gone as excess into channels or got percolated or evaporated. One of the earliest VRI systems operated using GPS connectivity and wireless guidance systems were operative in University Experimental farms (Clemson University Cooperative Extension Services, 2014). Currently, VRI is adopted in several states of North America. Some of the reasons for installation of satellite-guided Centre Pivot systems are soil moisture conservation, improved productivity, economic benefits, and environmental concerns. There are several companies offering satellite controlled VRI. For example, Reinke Inc., Zimmatic Inc., Gilford-hill Inc., and Valley Inc., provide a range of options for regulating soil moisture supply to match variations encountered in fields. Variable-rate irrigation has been installed to prop up the erstwhile center-pivot system. It has generally involved alterations such as attaching variable rate nozzles and GPS connectivity, etc. Within the farm, forming management zones based on soil moisture distribution is yet another procedure that is common. In Australia and New Zealand, farmers adopting such variable-rate irrigation systems have opined that they reap certain clear advantages. For example, variable-rate irrigation offers better utilization of water. Better utilization of water lessens water logging in tracks and gateways. Variable rate water supply protects soil structure, particularly if soil is heavy. The cost on water supply and maintenance of equipment is lessened. Farmers say they have expanded area under irrigation since the VRI techniques require less quantities of water compared to blanket applications.

There are VRI applicators of water that suits lateral irrigation system. This is used during site-specific water management in few locations of South Carolina. This system involves supply of irrigation water based on actual soil moisture recorded at allocation. Generally, soil moisture is monitored using Clemson VR lateral irrigation systems. Information provided by soil moisture sensors and thermal sensors are pooled and soil moisture map is prepared. The center-pivot system is operated using wireless and signals from computer decision-support systems for variable-rate nozzles (Clemson University Cooperative Extension Services, 2014). Farmers in North America are able to enlist services of private satellite and UAV related companies to survey crops for water scarcity, drought, measure soil moisture and then prepare digitized maps that could be used as signals during precision irrigation. Centre –pivot irrigation systems are controlled using wireless and satellite data, so that irrigation is accurate and commensurate with soil fertility and yield goals envisaged by farming companies (Trimble, 2014c). Fertigation is also accomplished using satellite data and variable-rate applicators. Ground based remote sensing systems are available for irrigation scheduling. For example, using sensors (Infrared Radiometers) mounted on linear moving irrigation systems provide ET data (Asher et al., 2013).

Water needs of horticultural crops have also been estimated using satellite techniques. They have been used to determine growth stage of trees, size, canopy traits and irrigation needs. Reports by USDA/NASA state that remotely sensed NDVI values and CC (canopy cover) are correlated ($R^2 = 0.95 < 0.01$) to water needs of crops. Therefore, monitoring growth and NDVI potentially suggest about water and irrigation demand by horticultural crops. Field specific and regional estimates of CC have been used to obtain an estimate of water needs of trees (American Society for Horticultural Science, 2008; Trout et al., 2008).

4.5　SATELLITE-GUIDED VARIABLE-RATE TECHNIQUES AND PRECISION SOIL FERTILITY MANAGEMENT

'Matching variable-rate inputs for variable fields is an important theme to follow during satellite-guided Precision Practices.' Historically, farmer's fields were relatively small, so it was proportionately easy to manage and accomplish various soil and crop management procedures. Fertilizer supply at specific rates based on field location was practiced. In other words, variable-rate application of fertilizer and other inputs is not new. Researchers at Clemson University, Clemson, USA state that, currently, fertilizer supply is done with great accuracy using computer-based decision support systems, digitized soil fertility maps, satellite guidance of tractors and variable-rate equipment. In addition, farmers are being exposed to use of EC meters and soil EC mapping to alter nutrient supply and attain uniform grain/forage yield (Clemson University Cooperative Extension Services, 2014). Butzen and Gunzenhauser (2009) aptly state that prior to adoption of variable-rate technology, it is necessary to study the yield map and pattern for previous years and delineate management zones or regions of distinct yield potential. Formation of management zones reduces complexity of agronomic operations.

Fertilizer applicators with facility to apply at variable rates are becoming increasingly popular in North America, particularly with farmers/companies having large areas under crop production (Plate 4.11A-D). It has been sought with enthusiasm by farmers specializing in both field and horticultural crops (see Schuman, 2010). Basic principle is of course matching fertilizer inputs with nutrient needs of the crop, at various stages of growth and grain/fruit formation. Its adoption reduces fertilizer inputs considerably when compared with blanket applications. Schuman (2010), explains further that variable-rate is at the core of precision agriculture. It adds accuracy to fertilizer supply by considering the within-field variations and avoids accumulation/dearth of essential soil nutrients, particularly N, P and K. Variable rate technique involves use of high-speed computers with software for rapid decision making. Computers use of previous data and appropriate programs, based on which they calculate fertilizer supply rates for each spot. Computer takes signals from GPS connectivity and maintains accuracy using GPS coordinates for each spot in the field. Computers consult GIS extensively, particularly yield and soil maps of yester years, prior to offering decisions. It uses electronic sensors to forecast grain yield. Most importantly, variable-rate techniques give weightage to each and every spot/management zone location, regarding crop yield goals and fertilizer needed at that spot. Definitely, it is not an easy task to perform in the absence of satellite imagery, computer-based rapid decision makers and electronic controls on nozzles.

In summary, introduction of GPS, global navigation system technologies and well-tuned variable system that operates based on satellite imagery/digital data, that together forms the precision agriculture has potential, to revolutionize crop production. This system allows collection of large amounts of useful data in a short time through drones/satellites. Most importantly, it allows farmers to move away from sampling a large number of representative plants, say 1 per thousand, followed by chemical analysis for nutrients. It is replaced by a more authentic method. Satellite imagery examines a large patch of crop, at one stretch and provides data about NDVI, leaf chlorophyll and N status, which can be effectively used to apply fertilizer-N (see Murguia, 2014). Such satellite-guided methods are known to reduce fertilizer-N need compared to blanket applications, to achieve similar grain productivity.

Variable-rate aerial application systems that are essential to adopt site-specific management are in vogue since a decade (Lan et al., 2010; Plate 4.11A-D). Variable rate application allows farmers to channel fertilizers, pesticide and herbicides in quantities that are required exactly by the location. Variable rate method matches plant's ability and need for nutrients exactly. Extra inputs are totally avoided. Accumulation of nutrients/pesticides in the profile or canopy is avoided all together. *Therefore, VRT offers a great opportunity to regulate soil nutrient dynamics in the agroecosystem.* Blanket application of fertilizers that were adopted hitherto does not offer this opportunity. Nutrient loss via leaching, percolation, emissions and diversion to weeds could be regulated effectively using VRT. Many of the consequences of climate change effects such as emission of NH_3, N_2O, CH_4 could be effectively checked. Satellite-guided precision farming is actually at the threshold of regulating agriculture related climate change effects. We have to adopt this technique as efficiently and in large scale if the intention is to reduce on fertilizer inputs by improving its use efficiency. It also avoids soil deterioration and restricts greenhouse gas emission. There are in fact several reports that suggest that VRT is environmentally friendly, and that it lessens untoward effects of chemicals (Robinson, 2007; Seelan et al., 2003; Krishna, 2012b).

Generally, satellite-guided variable rate application of fertilizer brings about a certain degree of uniformity to soil fertility of the fields. Incessant adoption of VRT should lead us to fields with uniform crop stand and productivity. Adoption of VRT is said to reduce fertilizer requirements, sometimes marginally compared to blanket recommendations. However, reports suggest that reductions in fertilizer need due to adoption of precision farming could be about 8–43 kg N ha^{-1}, depending on crop and geographic location. Reduction in fertilizer-P and K has also been noticed (see Krishna, 2012b). Satellite-guided precision farming improves N-use efficiency from 13 to 28% compared to blanket applications. Therefore, fertilizer-N required to achieve the same yield level reduces markedly. When extrapolated to large farms of over 10,000 ha, reduction of inputs and cost is perceptible (Krishna, 2012b).

Over all, we should realize that adoption of satellite-guided Precision farming techniques, just do not minimize supply of fertilizer-based nutrients such as N, lime or other chemicals such as pesticides and herbicides (Robinson, 2007; Krishna, 2012b). In addition, satellite-aided techniques allow farmers to attain a semblance of uniformity for soil fertility related traits that directly affect crop yield within a field. Formation of management zones, based on soil nutrient/moisture dearth or excess, allows farmers to judge factors that may affect soil environment. Farmers could regulate soil nutrient supply so that they are held within the profile, in the upper layers. Since nutrient inputs are matched exactly chances for excess nutrients to accumulate is least, if not nil. Nutrient loss through seepage, leaching or even emission is minimized enormously. Ground water contamination with fertilizer-N is almost avoided. Hence, satellite-guided tractors, planters, fertilizer inoculators with variable technology really provide us with a great opportunity to reduce fertilizer and chemical usage, reduce N-emissions and ground water contamination. Since, many of the agronomic methods are amenable to steer-less automation, satellite guidance and economically advantageous, they should lead us to better grain/forage yield, economic gains and still maintain soil environment.

PLATE 4.11A–D Continued

PLATE 4.11A–D   A variable rate applicator set for fertilizer supply based on digital maps and satellite guidance (Plate 4.11A: Tractor in front is GPS connected and is hitched to a Variable-rate attachment. Plate 4.11B: The Variable rate fertilizer Nitrogen supplier. The tracks and chains offer better grip and movement. Plate 4.11C: A close-up view of Variable-Rate Applicator. Plate 4.11D: Lateral view of Variable-rate fertilizer N applicator. Note that storage drum is connected to nozzles that are regulated by computer decision support systems. Source: Dr. Rebecca Montag, Vice President, Montag Manufacturing Inc., Emmetsburg, Iowa, USA).

Farmers worldwide, initially utilized plant leaf color, its health, leaf-N analysis obtained either by chemical analysis or SPAD meter readings. The SPAD meter provides an idea about green color, in other words, chlorophyll content of leaves. It is directly related to leaf-N status. Digitized data and maps prepared using SPAD readings could be used, but it is tedious and needs regular and labor-intensive scouting. Currently, in case of rice fields of South Asia, farming agencies are providing farmers with digitized maps depicting plant canopy size, leaf area index, leaf greenness index and chlorophyll. These could be utilized to calculate fertilizer dosages at each spot. NDVI and chlorophyll maps could be used by GPS connected tractors and fertilizer inoculators. This way, it reduces chances of fertilizer excess in the rice fields. In some cases, satellite imagery has been utilized to mark out management zones based on soil fertility and crop growth potential. This step allows easier adoption of variable-rate technology in rice blocks (UNCSAM-MARDI, 2012).

In addition to above satellite mediated imagery for estimation of crop canopy, its reflectance, chlorophyll and N status, there are several brands of hand-held or vehicle mounted instruments that perform similar tasks. Hand–held spectroscopes that are based on VIS/NIR wavelengths are used on field crops and tree plantations. For example, Rotbart et al. (2012) have reported about a VIS/NIR spectroscope. It estimates N content of leaves of Olive orchard. Luminar-5030, a spectrometer could detect leaf-N accurately with a correlation of $r^2=0.91$, when compared with usual chemical analysis of leaves.

Ishola et al. (2013) state that GPS-guided variable-rate input does help us in reducing chemical spray and fertilizer supply to soil. It avoids accumulation of chemicals in soil and pollution of ground water. However, there are situations when GPS connectivity is not available, due to weather conditions (clouds) or other causes. In such situations, Radio Frequency Identification (RFID) Reader could be used effectively to detect each tree in the plantation and apply fertilizers/pesticides accurately. The variable rate applicator could be mounted with RFID, so that it detects the RFID tag ID on the tree. For example, in Malaysian Palm plantations, Ishola et al. (2013) found that RFID mounted variable applicators could detect the trees accurately using ID. The accuracy of detection and variable-rate

application could be influenced by the transit speed of the VRT applicators. They have used RFID mounted VRT to band fertilizers in the palm
plantations with acceptable accuracy. The vehicle could cover about 6 km
h^{-1}. The efficiency of the RFID-VRT could be improved further by including automatic fertilizer fillers and appropriate computer programs. Plus,
it could be adapted several other tree plantations such as rubber, cocoa,
coconuts, citrus, apple, etc.

4.6   SATELLITE-GUIDED CROP DISEASE, INSECTS AND WEED MANAGEMENT

4.6.1   SATELLITES TO DETECT PESTILENCE

Satellite mediated mapping of pest-attacked zones is a good addition to set
of techniques that reduce human drudgery and make farm operations more
efficient. In a large agrarian belt or even a large farm scouting for pests is
not easy. It is time consuming, needs skilled human labor and costs could
increase if scouting has to occur a few times during the crop period. Push
button techniques employing satellite imagery are perhaps best bets for
the future years when insect control is achieved predominantly using precision techniques and variable- rate pesticide application. Recent reports
suggest that a '*pestMapper*' which is amenable to be operated using internet connectivity allows farmer to regularly monitor their fields or even a
large zone for insect pest attacks on crops (Xia et al., 2009). The program
allows farmers to concentrate and focus on the area of interest, for example,
region, state, county/district or field and identify insects prevalent and pest
damage, if any, at any time during the crops season. This is an excellent
example for the evolving 'Push Button Agriculture,' since it allows farmers to check the field conditions for pestilence using computers. It may
find rapid acceptance since this push button satellite-guided technique is
economically efficient. It needs low human labor hours. Pesticide sprays
could be guided accurately to locations affected. Digital maps can also
help farmers to apply pesticides at variable rates. This procedure reduces
on cost of pesticides and its application. There are satellite based devices
such as 'Crop Advisors' that are excellent data collectors about pest occurrence and spread, disease and weed incidence. They also help farmers to

map the pest-affected zones in their farms using satellite-guided imagery (GPS, 2014). Such instrumentation allows farmers to develop predetermined flight path for drones or robots and spray chemicals accurately.

As stated earlier, satellite techniques could be used effectively to detect and map insect attacked zones in crop fields. However, digital imagery of pest-affected zones may not be of high-resolution compared with those derived from UAVs. Yet, it could be used during variable-rate application of pesticides. Generally, it is said that farmers derive better assessment of insect attack and digital maps depicting it when they used UAVs (Faical et al., 2014). Ground based remote sensing and spectral imagery is also used to detect insect attack and extent of pest damage (Prabhakar et al., 2011). Airborne techniques using low flying planes with hyperspectral imaging sensors are also possible. As an alternative, we have to interconnect satellite imagery with UAV operation. Deploy UAVs when satellite imagery is insufficient regarding details. UAVs provide close up data about a crop area.

4.6.2   SATELLITE-GUIDED CROP DISEASE MONITORING

Crops are afflicted with a range of diseases during the period from seedlings to maturity. Field crop disease management has been tackled using several procedures, some singly or several others are applied in an integrated system. A single crop may become vulnerable to a few different types of disease, caused more commonly by bacteria, fungi and viruses. Crop disease control is achieved partly by using disease resistant cultivars. However, extent of resistance may vary. Also, most importantly, a single genotype may not harbor genes for resistance to several diseases. Hence, farmers periodically scout the field, note the locations/patches afflicted with diseases, if any. They also take note of extent and intensity of disease. In a large field, crop scouting for disease is tedious, cumbersome, time consuming and production costs increase. Several procedures that are rapid and offer maps showing disease spread have been explored and used with certain advantage. Air borne surveys and drones have been used to judge and survey crop fields for various diseases. For example, airborne surveys have been used to map the blight disease progress on wheat, rice (rice sheath blight) (Qin and Zhang, 2005); tomato (late blight) (Zhang et al., 2003); sugarcane yellow stripe (Grisham et al., 2010); powdery mildew of grapes (Oberti et al., 2014); and citrus greening disease (Pereira et al., 2011; Sankaran et al., 2011) etc. However, in the context of this chapter, we are concerned with disease identification and mapping using satellite imagery. At present, we know that quite a few crop diseases, particularly those affecting leaf have been investigated using satellite imagery. Periodic imagery has helped in tracing disease in a field.

Crop production during winter is prone to suffer due to several weather mediated deleterious factors. Cropping systems adopted in fact depends on frost-free period and ability of crop genotype to withstand cold conditions. Genetic traits that impart cold hardiness are essential for winter wheat grown in typically cold regions of North America, Europe and Fareast. Wheat grown in Shanxi province of China, for example, has been regularly monitored for diseases using satellite imagery. Freeze injury and disease onset could be detected using high-resolution satellite images. Freeze injury afflicted zones could be mapped in large area of wheat belt (over 10,000 ha) in a matter of few minutes using digitized data. The winter wheat freeze injury in some locations could be detected and mapped using multi-temporal moderate resolution imaging spectro-radiometer (MODIS) data. Satellite data could be easily authenticated using ground sampling and analysis. Observation of NDVI, prior to and after freeze injury is a good indicator of weather mediated frost injury and related disastrous grain yield decrease (Feng et al., 2009). Further, it has been shown that satellite images can provide information on severity of frost injury and its distribution across a winter wheat field. The wheat crop growth recovery rate monitored using satellite images were correlated with final growth and grain productivity ($R^2 = 0.665$).

Wheat grown in China is vulnerable to affliction by powdery mildew. Traditional methods of scouting for disease affected zones and marking management zones prior to spray of chemicals is relatively costly. During recent years, satellite imagery with multispectral sensors in the range of IR, NIR, and Red is in vogue. For example, employing CCD sensor on Huanjing Satellite has offered digital maps of disease affliction, progress and extent. The digital data could be used for variable-rate inputs of disease control chemicals. Field trials showed that satellite imagery was correlated (78%) with ground data about powdery mildew. Zhang et al.

(2014) further suggest that multi-spectral data about wheat disease, particularly those affecting canopy and leaf could be monitored and mapped using multispectral imagery.

Satellite imagery using high-resolution sensors could be used regularly to assess, if fruit trees are afflicted with diseases. In case of Citrus grown in Florida or in other regions, trees are vulnerable to attack by citrus canker. It seems hyperspectral images of trees could help in tracing variety of citrus peel affecting maladies, such as cankerous tissue on ruby red grape fruits, greasy spots, insect attacks, melanose scab, wind scar, etc. Qin et al. (2009) have reported that citrus canker disease could be identified rapidly using satellite imagery. The hyperspectral images need to be compared with canker reference images. Canker intensity and spread could be quantified by comparing imagery with standard reference images. They have obtained 95.2% accuracy in detecting canker-affected regions within citrus orchards. At this juncture, we may have to note that, if scouting and mapping were done manually at different stages of tree growth and fruit maturation, it would cost farmers exorbitantly. A satellite image shows up and covers large areas and trees at one stretch. It costs relatively marginally to procure and to overlay satellite image it with growth/yield data. Satellite imagery of field and plantation crops to monitor crop health is now rather routine and is available from private agencies (Galileao Geo Inc., 2014). There are reports that suggest that images from low flying drones show greater details about disease/pest attack on trees. A coordinated effort, first to obtain satellite imagery covering large areas, then to focus on disease-afflicted spots using drones is perhaps shrewder. Such coordinated effort could provide accurate control of disease/insect attack.

4.6.3 SATELLITE-GUIDED WEED DETECTION AND ERADICATION

Weeds are wide spread all over the agrarian zones of the world. They compete with main crop(s), deplete and divert soil nutrients and water that occurs inherently in the soil profile and that otherwise is basically meant for crops to absorb. Weeds cause economic loss to farmers and if extrapolated to large expanses they may actually account for perceptible loss to crop yield. Weed species that affect crops may depend on several factors related to geographic location, topography, cropping systems, weather pattern, agronomic inputs, etc. Weed detection and eradication has to be timely, so that it does not out-compete the crop species and suppress crop's canopy and growth parameters. In a large farm, weed detection by human scouts and herbicide application is a time consuming and costly process. There are currently, ground robots being developed that are efficient in detecting weeds and spraying herbicides. Robots also take satellite guidance to locate weed-infested zones and apply herbicides. Satellite imagery with high-resolution and ability to distinguish the weeds from crops based on spectral signatures are in vogue. Hunt et al. (2012) state that weed infestation zones could be mapped faster using satellite imagery. However, remote sensing for weed detection may not be the best option in all locations and situations. We have to understand the limitations and advantages of using satellite imagery for weed control (Table 4.3). Some of the points to be considered are: what weed species are eradicated best using satellite imagery and herbicides? What is the threshold density of weeds that get identified by sensors on the satellites? What resolution and spectral bandwidths are required? What weeds are easily identified using their spectral signature? Rapid identification of weeds requires computer programs that quickly detect specific weeds using spectral signature data. We may also use GIS to compare and detect weeds by their spectral signatures.

Techniques used to detect and discriminate the weeds as different from crops and ability to relay this information in digital form is the crux of the satellite-guided weed control. Studies using detached leaves and other parts of vegetation have shown that multi-spectral and hyperspectral remote sensing could be utilized to detect several weed species and two or three crops simultaneously in the field. Stepwise discrimination analysis indicates that for a reliable identification of weeds and crops adoption of spectral bandwidths at visible and red-edge regions are necessary. Using spectral signatures, weeds such as *Amaranthus retrofluxes, Avena fatua, Brassica kaber, Chenopodium album, Setaria viridis* and crops such as wheat and canola could be easily discriminated (Anderson et al., 1993; Backes and Jacobi, 2006; Smith and Blackshaw, 2003; Decastro et al., 2010; Lopez-Granados et al., 2006). However, they caution that during detailed analysis of individual weed species, errors

TABLE 4.3   A Comparison of Weed Detection Using Aircraft Mounted Sensors and Satellite Imagery

Platform	Sensor	Resolution		Advantages	Disadvantages
		Spatial	Spectral		
Aircraft					
	Digital camera	Very high	Low	Digital, quick turn-around	Needs rectification, No color
	Multi-spectral scanner	<5 m	15 bands	Digital, Good Spectral quality	Needs rectification
	Hyperspectral imagery	<5 m	100 bands	Accurate, maps several weed species	Very expensive, Difficult to process
Satellite Imagery					
	High-resolution imagery	<5 m	4 bands	Large area coverage	Covers small area compared to other types of satellite imagery; Cloud cover affects imagery
	Moderate resolution imagery	<20 m	4–7 bands	Inexpensive, Large area coverage, Continuous imagery	Low spatial resolution, Cloud cover interferes

Source: Hunt et al., 2012; GPS, 2014; Note: Weed identification and mapping their infestation in a field requires computerized digital data pertaining to spectral signatures of several weed species, particularly those common to the area in focus. The computer programs have to pick different species of weeds accurately, using their spectral reflectance. Basically, sensors/computer programs have to firstly distinguish crop species and weeds through their spectral signatures, only then herbicides could be decided.

may occur. Perhaps, crop specific and weed specific spectral signatures have to be carefully evaluated using established libraries. Crop specific spectral library for each cultivar is available in some cases (Rama Rao, 2008). In fact, in addition to discrimination of crops and weeds, detailed spectral analysis is used to identify cultivars and species ofcrops. Everitt and Yang (2007a,b) have reported that QuickBird imagery and spectral data could be used to map weeds such as Spiny Aster and

Snakeweed. Yang (2012) has stated that it is possible to detect and map weeds such as giant reed, hydrilla, giant salvinia and water hyacinth using multi-spectral imagery. Several aquatic weeds occurring in the Rio Grande valley of Texas, USA could be monitored using satellite imagery.

In the Cordoba region of Spain, multi-spectral high-resolution satellite imagery obtained using 'Quickbird satellite system' has shown good potential for satellite-guided weed management. Field scale evaluation of a large agrarian belt was examined taking at least 263 winter wheat fields as representative samples. The ability of QuickBird's multispectral sensors to detect commonly occurring weeds of *Brassica* species was evaluated. The cruciferous weed patches in the wheat field were mapped and digital images were transmitted to computers along with GPS coordinates. The classification of weed infested areas and weed species was accurate to the extent of 83–91% (Ana Isabel de Castro, 2013). The digital imagery from QuickBird could also be used to operate variable-rate herbicide applicators. Incidentally, reduction of herbicide application due to use of satellite imagery and guidance was reported to be about 61% compared with blanket applications. No doubt, in addition to satellite imagery, a well-tuned computer decision support system with broad scale GIS about spectral reflectance traits of crops and various weed species is essential. For example, SEMAGI expert system helps detection of weeds and crops (Castro-Tendero and Garcia-Torres, 1995). Spectral imagery from QuickBird satellite has also been used to detect and discriminate cereal weeds, even within a crop field, such as barley. For example, patches of cereal weeds such as *Avena sterilis* in winter barley fields could be discriminated with great accuracy using satellite imagery (Martin et al., 2011). Similarly, Lopez-Granados et al. (2006) have reported that in wheat fields of Spain, satellite maps for weed patches with species such as *Avena sterilis*, *Physalis paradoxa*, *Physalis minor* and *Lollium rigidum* could be digitized and used during precision weeding. Satellite imagery has also been utilized to detect and map weeds such as *Ridolfia segetum* patches in sunflower crop using Remote sensing (Pena-Barragan et al., 2007).

4.7 SATELLITE-GUIDED HARVESTING AND YIELD MAPPING

About a decade ago, agricultural experts had opined that use of GPS-guided combine harvesters with yield monitor attached to them are still rudimentary in many parts of the world, including Europe (Lowenberg De-Boer, 2004). There were harvesters that recorded yield and mapped it but without co-ordinates. Final yield could only be known. However, at present most combines, if not all sold to farmers are provided with GPS connectivity and grain yield monitors. Grain productivity maps with accurate coordinates for each spot or patch is possible. In fact, Sylvester-Bradley et al. (2006) have stated that yield maps and forecasts about potential grain productivity have been effectively used in deciding inputs, precise application of chemicals and water. Most strikingly, yield maps are being used by agricultural markets, merchants, funding agencies and policy makers to decide on quantum of investment, credits and in channeling facilities to farms in Northern Europe. The above situation is also true with North American farming belt.

Yield monitoring and mapping are perhaps the first steps during adoption of precision farming using satellite guidance. By definition, Yield mapping refers to the process of collecting geo-referenced yield data while the crop is being harvested using a combine harvester or similar farm equipment, that has GPS connected yield monitor. The vehicle records quantity of grain/forage harvested at each location and tags it with exact location in the farm using GPS coordinates (Adamchuk et al., 2004; Adamchuk et al., 2008; Gunzenhauser and Shanahan, 2014; University of Nebraska Crop Watch, 2014; EGNOS, 2014; GPS, 2014; New Holland Agriculture, 2014). Yield monitoring and mapping devices were first installed into combines around 1990s. Currently, combines with grain yield monitors and mapping facility are used mandatorily in most of the agrarian regions of developed world. Some of the most common yield monitors are those produced by companies such as New Holland Agriculture (New Holland Agriculture, 2014), AgLeader (e.g., Agleader Insight, AgLeader-2000), John Deere, Massey Ferguson, etc. Such combines are also in vogue in many Asian nations. For example, combines fitted with automatic yield monitoring systems that consist of yield sensor, global positioning system, field computer and related software to read maps are being used in the wheat belts of Northwest India (Singh et al., 2012). Generally, components that make up a 'grain yield mapping unit' are:

a) grain flow sensor that determines the volume of grain harvested;
b) grain moisture sensor that is used to remove errors due to variations in grain moisture content;
c) grain elevator speed sensor;
d) GPS antenna that receives satellite signals;
e) yield monitor display system with geo-reference and grain record data;
f) combine Harvester position sensor;
g) combine harvester speed sensor and GPS receiver system that supplies coordinates.

(See University of Nebraska Crop Watch, 2014).

Forage harvesting combines that are autonomous and connected to satellite are useful. However combines having facility for noting forage harvest on-the-go with GPS coordinates began sometime in mid-1990s. A forage harvesting combine with yield monitoring ability basically has electronics for measuring mass-flow, harvested width, speed and location derived from global navigation satellite system. The forage harvest is recorded using mass that enters with feed rolls. Essentially feedroll displacement is measured to judge yield (Digman and Shinners, 2014).

How do we process yield maps or satellite imagery derived about a crop field? Actually, grain yield harvested at each location can be plotted on the map and shown on the computer screen using GIS software package. There are of course several other mapping packages. Usually, errors related to raw data such as few points at the beginning and at the end of the pass are removed. Start-up and end pass delays are also removed. This step is called data filtering and enhancing accuracy of grain yield data (see Schearer et al., 1997; Sudduth, and Drummond, 2007). Yield monitor accuracy that depends on grain flow and Combine harvester's speed is also important (Colvin ad Arslan, 2002). Often historical data sets about grain yield maps are considered. Temporal variations (year-to-year) is compared with current year's data/map. Temporal maps can be compared by calculating normalized yield. Normalized yield is the ratio of actual yield to the field grain yield average (University of Nebraska Crop Watch, 2014). This data can then be over-layed while analyzing and arriving at decisions during precision farming. Inputs can be channeled accordingly. Interpretation of yield maps based on long historical data is usually preferred. Typically, at least five-year data/maps are collected about a field and analyzed. The final map/data could then be used to mark management zones and fix grain yield goals.

Yield maps derived from combines connected to GPS usually form the basis for the next season. The cropping pattern, planting plans and allocation of inputs such as fertilizers, water, pesticides and herbicides are discussed, analyzed and decided, based on previous season/year grain/forage yield maps. A grain yield map clearly depicts the areas with higher crop productivity and those with depression. It amounts to identifying areas in field, where crop has extracted greater quantities of nutrients and water in order to attain higher productivity. By studying the yield maps we can also make appropriate inferences about the causes for yield variability. Most often, farmers focus on amending the factors such as topography, soil compaction, fertilizers, crop cultivar/hybrids, plant protection procedures, etc. Presently, agronomists try to overlay maps of grain yield with those of different factors to judge and decide on variable supply of nutrients, water and pesticide sprays. Yield of previous season could also help farmers to decide on the crop and yield goals possible in the next season or even a complete crop rotation or rotation.

Shaver et al. (2012) suggest that yield maps of fields can show us the areas with potentially higher grain/forage productivity and where we can manage fertilizer inputs with better productivity. However, yield maps could be transitory within and between two seasons. Therefore, crop growth and/or yield maps need to be measured often may be on seasonal or yearly basis. Crop growth maps derived from satellite imagery that depict within season variations is also used. Satellite connected leaf color meters could also be used to measure in-season growth/biomass accumulation and map them. There are ultrasonic plant height mappers. They map plant height, based on ultrasonic sound wave that is reflected and captured by sensors. They are also used to measure canopy height in general (Shaver et al., 2012). Consequently, satellite connectivity can be used to produce a geo-referenced map of plant height in the field.

At present, satellite companies develop crop area maps with clear identification of location, crop species and specific traits possible using imagery. They also compare historical data about cropping pattern and grain yield. The grain yield data for each year or season could be obtained from GIS. Then, base line grain yield forecasts are added to the information requested by customers. At times, depending on the agency, farmers are also supplied maps depicting cost per metric ton. Grain yield maps depicting costs vs. profits obtained in yester years or that forecasted is also possible. Farmers can obtain the satellite images plus crop cost vs. possible profit maps on-line based on their requirements, in-season and as many times (GDA Corp, 2014a). Yield maps of crop species such as tubers and vegetables are also being utilized. For example, in case of potato, it is widening interplant space, reducing planting density and improving fertilizer/irrigation inputs that enhances yield quality. In addition, such measures remove within field productivity variations (Clarke, 2014).

In addition, we may realize that well equipped software that converts raw imagery to grain productivity distribution can consider series of data sets from previous years for a particular administrative unit, for example, a county in state or a state in a country. Grain yield forecast and actual grain productivity of previous years could be arranged in sequence and arranged, showing per season grain productivity changes in each county/ state or even a region. It helps during analysis of pattern of grain yield in a particular field/region. There are established satellite imaging companies that conserve crop related data about a particular field/farm for several years. They could use it to forecast and prepare detailed maps about crop density, leaf area index, plant chlorophyll content and water content. Fortnightly forecasts about crop growth progress and forecasts for future are possible. Satellite maps depicting growth rates at start of the crop season (seedling growth), mid-season changes in canopy and chlorophyll content, late season senescence pattern and ripening of grains are clearly depicted (GDA Corp, 2014e). Satellite imagery also helps in developing detailed digital data for a field, depicting variations in crop productivity. This can be used directly in site-specific techniques and variable-rate applicators.

4.7.1 MULTI-YEAR YIELD ANALYSIS AND MAPPING

Crop yield mapping using satellite imagery or combine harvesters fitted with GPS connectivity and yield mapping devices are now a common procedure in many of the high input farms of developed nations. Yield maps of immediate yester years are prerequisites to demarcate management zones during precision farming. Multi-year yield mapping is a process by which grain/forge yield of few years, for a well-defined field is combined into a single composite, in order to view spatial yield patterns across many years. This spatial data, which is often digitized could be utilized to create management zones (Gunzenhauser and Shanahan, 2014).

Now, how is a multi-year cumulative data analyzed, so that it is feasible to be adopted during management zone formation and variable-rate seeding, fertilizer supply, irrigation, etc.? Firstly, yield data for each year is normalized to 100% of the field average. This creates relative yield values. The relative values will be higher if grain/forage productivity is more than the average, but low if it is less than average. The relative yield values are then allocated appropriately into each grid cell of correct location. This procedure of locating grid cell and placing relative yield is done for each year data. Then, a composite multi-year layer is prepared by combining all the years' yield values from the same grid location. The average relative yield value is calculated and coefficient of variation is also shown, if available. Based on composite relative yield value, a grid cell is classified as low, medium or high yielding zone. Grid cells of similar yield level are grouped, then mapped and utilized during variable –rate application of inputs such as seeds, fertilizers or irrigation.

Researchers at Pioneer Hybrid Inc., state that one of the advantages of multi-year yield maps is that it is based on yield history of a field. Such multi-year maps supposedly reflect actual productivity better than that based on previous one year data. However, there are disadvantages too. The yield for previous several years, all of them may not be from normal years. Drought affected or low rainfall years will pull the average yield levels to lower levels, even if the previous data is good. Such multi-year maps may also be used for allocating management zones. A high quality multi-year map could lead to well-judged management zone formation.

Usually yield data for 3 to 4 previous consecutive years are included to arrive at multi-year yield maps. We should note that yield data utilized should be complete and authentic. Any raw data should be filtered and removed from the data set. Errors in data sets should be carefully avoided. The below- average and above average cut off for normalized value should be appropriately labeled as low or high. For example, if the average yield goal is 100 bushels grains and yield recorded is 120 bushels then it should be correctly allocated into cells that occur in the management zone marked as 'high.' The grid cell size should be apt. A larger grid cell may show varied effects and could be unwieldy. Grid cell sizes of 30 sq. ft. are preferred in some locations, but it depends on resolution available with satellite agencies. Over all, multi-year yield maps provide good flexibility, but careful planning of management zones is necessary, to make variable-rate applications less cumbersome. At the same time, we may also note that this entire procedure of yield data collection, formation of grid cell and management zones, followed by variable-rate application of inputs as deciphered by computer decision-support systems is accomplished, using computers. They form important portion of 'Push Button Agriculture' as suggested in this book.

Now, there are indeed several government supported and private agencies that specialize in acquiring crop yield maps developed using satellite imagery and drones. Generally, for most farmers, obtaining crop yield maps and analyzing the reasons for variations for grain/forage productivity is the first step in satellite-guided precision farming. Agronomists offer suggestions that often relate to uses of appropriate crop species, hybrids, fertilizer combinations and dosages, herbicide and pesticide sprays, etc. Yield maps are effectively used by satellite companies to prepare management zones for farmers. They often use multi-year grain yield data. Most importantly, multi-layered data is super-imposed with final grain yield to demarcate management zones and adopt variable-rate techniques. For example, a satellite based private company uses imagery depicting soil type and fertility variations, crop species and hybrids grown in the field, maps of Veris- EC, Veris-pH, soil color and organic matter content. Then, such maps are over-layed with grain yield to construct an authentic multi-layered map. Such multi-layered maps are highly useful during precision farming. It helps in judging which factor is affecting the crop productivity most in a given location and to what extent it needs to be corrected (CropQuest Inc., 2014c; GPS, 2014).

4.8   SATELLITE GUIDED FARMING: SOME EXAMPLES

At this juncture, we should note that several aspects of satellite-guided farming currently adopted in commercial farms are actually, only portion of a larger concept that aims at complete sophistication of farm operation that is termed as "Push Button Agriculture" in this book. Satellite guidance, indeed forms a major share of such electronically regulated farming. Other techniques based on concepts such as autonomous vehicles, robotics, drones and computer decisions form the rest of push button approach. We may note that both governmental agencies and private ventures have been striving to rapidly convert as many of the farm operations into electronically controlled autonomous systems (robots). Such systems reduce farm labor requirement, remove human drudgery, make farm operations easier, and reduce cost of crop production. There are currently several private farm advisory companies that serve farmers and cooperatively operated large farms. They supply satellite imagery, conduct surveillance of farms, crop monitoring, supplying digital maps for variable-rate application, disease/pest monitoring and final grain yield maps (see Table 4.4; GPS, 2014). With the advent of drones and satellite-mediated crop husbandry, our dependence on large-scale soil sampling that includes surface and subsoil layers, wet soil chemical analysis on well processed soil samples, etc. could be diminished. Agricultural consultancies and farmers now prefer to conduct on-the-go soil analysis. They adopt easier techniques based on soil EC, pH, SOM and leaf-N status using tractors connected to GPS and remote sensing satellites. These techniques could be used to supply digitized data to variable-rate applicators.

4.8.1   FIELD CROPS

Wheat crop in different regions of the world experiences wide range of weather related and biotic factors as it grows to maturity. Grain productivity is dependent on the growth pattern, particularly leaf area index (LAI), canopy traits, nutritional status, grain formation and maturation. At present, large wheat belts are exposed to satellite mediated growth monitoring. It helps agricultural consultants and policy makers to decide upon a range of options regarding cropping systems, genotype, timing of planting, in-season agronomic procedures, particularly, split-N inputs, irrigation and of course timing of grain harvest. Wheat crop is also monitored for canopy and LAI that correlates with final yield. Hence, farmers could use it as an indicator to forecast or fix yield goals. In addition, wheat that is exposed to vagaries of weather such as drought, frost bite, and periodic floods has also been monitored using satellite imagery (Xiaoyu et al., 2014; Feng et al., 2009; Seelan et al., 2003).

The South Asian rice belt thrives on diverse topography and soil types with variable fertility and moisture distribution. Fields support rice crop that is often uneven in grain productivity. Seeding, plant stand, canopy size, fertilizer supply and irrigation are all affected due to field variability. For example, blanket supply of fertilizer to a field may generate spots with high/low nutrient availability. Therefore, farmers in Malaysian rice farming zones are being exposed to use of satellite-guided variable-rate techniques. Preparation of digitized maps of field topography, soil type, fertility status, previous crop yield maps, etc. are essential. Satellite agencies do supply such information through internet to farmers (see Aisha et al., 2010). Farmers are supplied with digitized maps of green vegetation index, canopy size, chlorophyll content and leaf-N status. This information is used by computer-based decision support systems available on vehicle and fertilizer/irrigation is applied at variable rates to match the inherent levels and yield goals stipulated (UNCSAM_MARDI, 2012; Chuang et al., 2014). Satellite-guided farm operations during rice production is also in vogue in China (Qin and Zhang, 2005). Reports suggest that mapping of rice production zones and estimates of productivity has been possible using remote sensing. A few studies have attempted to examine canopy, leaf area index and plant growth using spectral imagery, so that rice grain yield could be forecasted using computer models. Leaf chlorophyll estimates have been used to know plant –N status and to obtain an estimate of plant-N requirement. Remote sensing has also been used to detect disease affliction, if any. The spectral differences of healthy and disease affected patches of rice is used to estimate disease intensity and to warn farmers about possible spread of a particular disease.

TABLE 4.4 Satellite Companies and Services offered to Farmers Involved in Crop Production-Examples

---

Ag Business and Crop Inc., Ontario, Canada

Soil sampling and field maps with GPS co-ordinates to help in the demarcation of management zone. Crop input planning and production advice based on satellite imagery. Field monitoring throughout the season from seeding till maturity. Precision management processes involving supply of digital data to variable-rate applicators, depicting soil fertility and crop growth. Supply of aerial images that depict incidence of disease and pest. Green seeker is a set of data about crop growth, chlorophyll content and crop-N status. Planning application of split-N to crops. Supplying high-resolution crop images for specific reasons is another task. Historical data about a field depicting soil and cropping systems followed, yield data, etc.

Rx-VRT Providers, LLC Aberdeen, South Dakota, USA

Rx-VRT provides soil maps from extensive satellite imagery data. This company helps in management block formation, determining fertilizer needs and provides digitized maps for use in variable-rate applicators. Company offers suggestions on variable-rate seeding and determination of cropping systems. The company also offers satellite derived yield maps with GPS coordinates.

CropQuest Inc., Dodge city, Kansas, USA

This company offers a range of agricultural services pertaining to Central Great Plains region. They are: Digital field reports derived using satellite imagery (topography, vegetation, soil type, etc.); GPS locations for grid sampling, formation of management zones for variable-rate application of fertilizers; rapid delivery of satellite imagery for instantaneous adoption of farm operations; satellite aided monitoring of irrigation systems, storm damage, soil erosion, etc.; satellite aided surveillance of disease and insect prevalence in the farms and supply of digital data or maps for variable-rate application of pesticides. Yield mapping and analysis is an important satellite aided service offered.

Trimble Agriculture Inc., Sunnyvale, California, USA

They offer a range of satellite-guided and drone services to farmers. Some of their products are satellite data management, guidance and GPS tagging of soil related procedures. Variable-rate application using GPS coordinates. Soil mapping (including 3D images of topography) using advanced sensors is an important service offered. Soil characteristics such as texture, compaction, moisture content, salinity could be over-layed with yield maps to form management blocks. Irrigation control using soil moisture maps and variable-rate applicators is another important service offered. Trimble Inc., also provide touch screen display systems (Trimble CFX-750) that improve guidance, steering and maximize efficiency during variable strip tilling, planting, spraying and fertilizer application. They also help in production of specific crops such as grapevines and sugarcane employing satellite-guided vehicles and precision techniques.

TABLE 4.4 Continued

Satellite Imaging Corporation, Houston, Texas, USA

Satellite imaging corporation (SIC) offers aerial imagery for different purposes such as defense, industry, agriculture, etc. In case of agriculture, services include supply of spectral data related to crop assessment, crop stand and health, environmental data, irrigated landscapes, yield forecasting and determination, soil maps, etc. They also supply spectral data about NDVI, soil maps that depict color, SOM and water content of a field or a large area, say a county. Tree grading using satellite images is another service useful for foresters and fruit plantations. There are also 'Agrowatch' services that help farmers to continuously monitor fields. Satellite imagery of natural resources and climate change effects such as soil erosion, floods and droughts could also be obtained from this satellite company.

Land IQ, Sacramento, California, USA

This company offers a range of satellite imagery and guidance related services to agricultural agencies and farmers. Primarily, it offers Multi and Hyper-spectral imagery that could be decoded using appropriate software to obtain a clear image of happenings in the landscape and farms. Thermal imagery and water resources are also mapped and offered. General topographic images, landscape changes and greenhouse gas emission patterns are mapped and provided to agricultural agencies. This company also provides models and advices to offset C and N emissions from crop fields. Maps depicting land use and cropping pattern, invasive species, weeds, and forest vegetation are supplied. Maps of counties showing crop production trends are provided regularly.

Geospatial Corporation, Science Park Road, Pennsylvania, USA

The GDA Corp Inc., is a women-owned and maintained concern specializing in satellite imagery and related business. It provides services such as satellite images of agricultural crops, forests and natural resources. It helps in obtaining images, processing them using different software and supplying to agriculturists, based on their needs. It is directly involved in providing data about agricultural business intelligence and obtaining data about crop acreage. Crop condition monitoring, assessment of health and forecasting are other functions. It helps policy makers with global agricultural cropping trends and forecasts. To individual farming companies, this agency also provides crop and yield maps each season, as many times, based on requirements.

BDAEL for Agriculture Co Ltd, Dubai

This company located in Middle-East offers a range of agricultural services that are based on ground data, Unmanned Aerial Vehicles (drones) and extensive satellite imagery. They serve farmers and farm companies in the African continent. Some of their services concerned with satellite imagery and GPS connectivity relate to soil mapping, nutrient mapping, formation of management blocks, variable-rate application of nutrients, yield monitoring and maps. BDAEL also offers maps showing disease or pest incidence.

Precision Seeding Solutions, Premer, New South Wales, Australia

This company offers high precision variable seed planters for corn and wheat grown in New South Wales. They specialize in high density planting with accuracy. They also supply monitoring systems for seedling emergence, weeds in the field. Planters tailored specifically for cotton seed planting are also offered.

TABLE 4.4   Continued

RoboFlight Systems Inc., Denver, Colorado, USA

RoboFlight Inc., supplies satellite imagery that has geo-referenced aerial data, about the fields and crops. Multispectral data covers both crop and livestock aspects. Farm data is collected using Visible, Near Infra-Red images and Thermal bandwidth. They also provide appropriate computer software such as AgPixel to decode and utilize satellite images.

Vega-Pro, Space Research Institute, Russian Academy of Sciences, Moscow, Russia

This is a satellite services agency that supplies farmers in Russian Plains and Steppes with imagery. It is used by policy makers to obtain information on soil, cropping systems and water resources. Satellite imagery is also processed using appropriate computer software. Vega-Pro service is mainly aimed at monitoring crops at various stages. It offers suggestions to crop insurance companies about the stage of the crop and yield forecast. It also helps farmers by providing timely warning about various agronomic operations required. (Vega-Pro, 2014).

Scanex, Moscow, Russia

Scanex is a Russian satellite agency that offers services to agricultural and related professionals. It specializes in the area of crop production, forest vegetation, cartography, weather and natural disaster warming. It provides maps and digitized information about land use pattern and changes. It monitors agricultural crop production, acreage and fluctuations in cropping zones. This company offers maps and information about crop growth, spectral data during crop season, NDVI, disease/pest attack, and yield forecasts. It also provides useful management decisions to farmers using satellite imagery (Scanex, 2014).

Source: Ag Business & Crop Inc. (2014); Rx-VRT Providers LLC (2014); BDAEL for Agriculture Co Ltd (2014); CropQuest Inc. (2014a,b); GDA Corp (2014a,b,c,d), Satellite Imaging Corporation, (2014d,e); Trimble (2014a,b,c); Precision Seeding Solutions Inc. (2014); RoboFlight Inc. (2014); Land IQ (2014); Galileo Geo Inc. (2014); Scanex (2014); Vega-Pro (2014).

   Maize production zones have been under constant surveillance and monitoring by a range of satellites belonging to different agencies. Maize production is world-wide, and each zone may experience different kinds of constraints. Factors that induce high crop yield may also vary. Maize crop could be monitored to assess its phenology and growth traits using remote sensing (Vina et al., 2004). Maize production zones in North America utilize satellite services to monitor crop growth, decide on inputs such as fertilizer, irrigation and pesticides. Satellite-guided operations are common in the Corn Belt of USA. Multi-spectral imagery is used to detect nitrogen deficiency and leaf-N content. The data collected is used to judge fertilizer-N requirements of the crop. Satellite data regarding chlorophyll and leaf-N is useful in deciding on the quantum of in-season split-N dosages for corn. Evaluations have shown that, data from satellite images about crop-N status and SPAD readings recorded by human scouts do correlate to a great extent (Noh and Zhang, 2012).

In the 'Corn Belt of USA,' farmers have opted for satellite-guided precision farming techniques in large numbers. They have deployed farm machines/vehicles with GPS and digital mapping facility. Variable rate seeders are used to achieve optimum planting density. Double seeding is effectively avoided. Fertilizer prescriptions are tailored to suit each field. Fertilizers are placed using variable-rate methods. Several of the procedures involving tractors are done using GPS controlled steer-less vehicles (Meersman, 2014). Such accuracy helps in reducing nutrient accumulation, greenhouse gas emission and loss of nutrients.

In Mexico, satellite-guided farming procedures and preparation of grain/yield maps are in vogue. Maize production techniques differ based on regions within Mexico. The grain yield varies from 1.0 t ha^{-1} to 12 t ha^{-1} but the average yield for entire country is 3.2 t ha^{-1}. Satellite-guided techniques are therefore adopted based on farmer's economic disposition, farm size and economic benefits. For example, Soria-Ruiz and Fernandez-Ordonez (2011) have used satellite imagery and geo-referencing for maize crop monitoring and preparation of grain yield maps, using SPOT satellite system. They have reported that yield maps derived from satellites are accurate to an extent of 90%, if checked with ground realities (see Soria-Ruiz et al., 2007; Soria-Ruiz et al., 2009). In many farms specializing in satellite-guided corn production, yield maps are over-layed with those of terrain and particular topographical features, soil fertility traits, soil moisture distribution, pestilence/disease incidence, etc. Such an analysis helps farm personnel to judge factors that affect grain yield at various spots. Hence, inputs and remedial measures could be adopted accordingly. We can obtain 3D maps of terrain showing the elevation of crop field in great detail (Rovira-Mas et al., 2008; Galileo Geo Inc., 2014). Perhaps, overlaying 3D maps of field with various other characteristics of soil could be helpful.

Let us consider an example depicting satellite-guided mapping of soil fertility variation, mainly nutrients such as N, P, K and soil pH. Papadopoulos et al. (2014) have reported that in Greece, techniques for site-specific soil fertility management is possible using GPS techniques, soil mapping and application of fertilizers based on inherent soil nutrients and yield goal. They have used fertilizer advisory software to provide decision support to vehicles used for inoculating fertilizers-N, P. and K. Further, it is opined that such soil fertility maps could also help policy makers to organize soil fertility programs to larger area, with greater accuracy.

Satellite mediated techniques are still being standardized. However, in some places, it is already used with a certain degree of advantage, in terms of crop management and economic benefits. For example, potato crop productivity shows significant within-field variations. Both, quality and size of the tubers tend to vary in patches. It is said that such variations may often account for yield reductions ranging from 30 to 40%, if compared to a uniformly moderate or large tubers. Plant spacing and fertilizer are factors that affect a good potato harvest. Hence, satellite-guided yield mapping and recording variations in tuber size has been adopted using GPS techniques.

Let us consider a few examples of vegetable production zones. Field trial covering large patches of vegetable production that includes cabbage has shown that, it is possible to monitor this crop using spectral imagery from satellites such as QuickBird, Landsat, etc. Oki et al. (2011) have reported that accuracy of satellite imagery is optimum to monitor the cabbage crop, at various stages of growth and formation of bulb. They further state that, satellite imagery could be used to channel various inputs based on crop stand and conditions prevailing in the soil/field. Such information could be obtained periodically using QuickBird satellites endowed with multi-spectral imagery. In case of other common vegetable crops such as Bell Pepper, there are ground based techniques that utilize multi-spectral imaging. Hand-held spectrophotometers and those mounted on vehicles are also used. Imagery is usually done at visual range, Near Infra-Red and Shortwave Infra-Red bandwidth. It is believed that satellite imagery too could be adopted, especially during crop growth and maturation, and then compared with ground data. Fusion of satellite and ground based spectral imaging may lead us to better judgment of the crop growth and matured fruits that could be harvested mechanically (Ignat et al., 2014).

Sugarcane is a long duration field crop that requires several agronomic procedures beginning with deep soil tilth, ridging, plantings setts, inter-culture, periodic inputs of fertilizer, irrigation, pesticides, etc. Farm labor is required at relatively higher rate per unit area. Fertilizer and water requirements are again higher; therefore, their efficiency is to be enhanced. Perhaps ground robotics and drones are most welcome during sugarcane production. Further, we should note that satellite-guided sugarcane farming is already in vogue in southern USA, Brazil, Caribbean, China and India. Several procedures adopted during sugarcane production could be accomplished with greater ease and higher efficiency using satellite and drone technology, along with precision methods. Satellite agencies offer services related to soil mapping, ridges and furrow formation and steer-less GPS guidance of tillage equipment, planting setts using GPS co-ordinates and/or tractor mounted cameras. Monitoring sugarcane crop growth until its maturity is an important aspect offered by satellite agencies (Trimble 2014 d, e). Variable rate application of fertilizers and disease/pest control chemicals also utilize satellite guidance (Trimble 2014 e, f). Satellite agencies also offer sugarcane yield maps and preserve data for several years, so that, it could be retrieved and used during management block formation (see Trimble 2014 d, e, f).

4.8.2 PLANTATIONS AND FORESTS

Grapevine management requires constant surveillance, scouting for growth parameters, disease/insects, drought effects if any, natural effects such as soil erosion, cold injury, etc. During past decades, planting till harvest, all the activities were monitored and accomplished using human scouts and skilled farm labor. This has been a tedious task and at times costly. However, with the advent of satellite aided GPS technology and robotics, a portion of agronomic procedures has been guided and handled efficiently. There are several satellite agencies and private companies that provided services such as satellite (GPS-RTK) guided planting with 2 cm accuracy (Trimble 2014 g). They help in planting grapevines and maintaining row spacing for easy transit of steer-less vehicles such as autonomous weeders, canopy monitors (green seeker) etc.

Soil information is also crucial. Satellite companies help grape farmers with soil maps and well-marked management blocks. Soil maps showing compactness, pH, salinity, pH and nutrient distribution could be used to decide, grapevine varieties, setting yield goals and inputs. Soil moisture data and maps are used to regulate variable-rate irrigation (Trimble 2014 h, i).

Santangelo et al. (2013) have attempted to use satellite derived NDVI data at various stages of grapevine growth during a year, to relate it to various characteristics relevant to crop growth, fruit quality and productivity. In fact, they have aimed at scheduling various agronomic procedures in the vineyard, based on satellite imagery. They have estimated NDVI values at berry set, pre-veraison and ripening. They have reported that NDVI data from 'RapidEye' satellite could be statistically related to fruit parameters such as sugar and anthocyanin. However, they suggest that a few more seasons or years data are required to standardize the satellite-based technique. Such a technique, no doubt reduces on human scouting for grape bunches in a large farm to judge quality and ripeness. It reduces drudgery by skilled farm labor. Grapevines have also been mapped for NDVI, LAI and planting density using satellites such as IKONOS. Spectral data has helped in developing crop growth models and developing decision support systems for irrigation and canopy management (Johnson et al., 2003). Several aspects of grapevine management and their automation require mediation through satellite imagery and GPS signals. Farm vehicles and robots used in grape orchards are mostly connected to satellite guidance. Robotics relevant to fertilizer and pesticide spraying, irrigation and harvesting need satellite guidance at various stages (see Zhang and Pierce, 2013).

Citrus plantations are important economic enterprises in Florida, USA. Satellite-guided monitoring of citrus plantations for tree health, nutritional status, moisture availability, pest and disease spread if any, are in vogue (Shrivastava and Gebelein, 2007). Remotely sensed data about citrus and other vegetation is utilized to regulate tree planting programs and replanting procedures. Satellite mediated detection of nutrient and moisture status of citrus orchards are also practiced. The data obtained are used in precision techniques, particularly during variable-rate fertilizer and water distribution to plantations. Satellites are also used to obtain information about each individual tree in an orchard. It helps growers to manage the orchard more efficiently using images and digitized data about the entire plantation. Trees missing could be easily located and measures for replanting could be adopted. Tree growth rate, canopy size, leaf biomass and fruiting too could be forecasted by adopting suitable computer software (Satellite Imaging Corporation, 2014 f). For example, in the Citrus belt of Florida, it seems satellite-guided variable-rate supply of fertilizers and pesticides are important tasks. Trees with impaired growth not in tune with others and blanks without tree in the spot are frequent. Satellite guidance absolutely avoids fertilizing spots without trees. Treeless spaces are often created by Citrus greening disease and Canker that force farmers to remove the tree and destroy. Satellite-guided precision techniques, it is said, improves uniformity across plantations with every season. Satellite imagery and spectral analysis (Landsat Enhanced Thematic mapper) of entire orchard that takes into account trees with different foliage, canopy size and fruit bearing potential is effectively used to forecast citrus fruit yield (Shrivastava and Gebelein, 2007). Regarding precision techniques and need for variable-rate inputs, it could be guessed that ultimately precision techniques may not find use, if the field or plantation is already uniform with regard to soil fertility and tree growth characteristics. We may also forecast, that it is a matter of time, before as many of the farm operations in citrus/apple fruit plantations are taken over by satellite-guided 'Push Button Systems.'

Managing natural stands and private forests is a tedious task in the general course, if top view from a satellite is absent. Many of the forestry applications need accurate data about fluctuations in land cover. It allows foresters to plan appropriately while initiating new planting. Many of the forest maintenance operations could be conducted with greater accuracy, using top view. Forest diversity, especially flora, could be managed better using satellite imagery. Mainly, forest management that involves regular mapping of plantations and ecosystems is conducted via satellites. Forest resources such as soil, its fertility, organic matter, water and biomass could be regulated much better using digitized images of forest flour and canopy. Satellites can alert the growers about fire, its spread and rate of loss of forest cover, rather rapidly at an early juncture. Forest logging and transport of forest wood could also be monitored using satellite imagery and monitoring. Deforestation rates could be measured using imagery and rectified appropriately. Many of the natural process that occur at the confluence of forest fringes, cropping zones and natural waste land could be monitored and understood more accurately using satellite images (Satellite Imaging Corporation, 2014g). There is a great interest in satellite imagery of various growth and biomass accumulation parameters of trees that form the commercial forest plantations. For example, LAI measured using ground or space borne sensors is used to estimate forest biomass and forecast yield. Multispectral imagery that includes infra-red bandwidth is known to provide useful estimates of LAI. Further, such spectral signatures of forest trees could also be used to classify and mark species diversity (Rautianen, 2005). There are reports that satellite imagery could be used to detect water stress in coniferous stands. Spectral reflectance of water stressed and normal stands of *Pinus* species differ and such data could be used to detect drought effects, if any (Riggs, 2001). High-resolution multispectral data of forest zones provides some excellent data about biomass and species diversity. It helps with maps showing fluctuations in forest vegetation in relation to season. Forest vegetation mapping is an important task performed using satellite imagery. It is cost effective, accurate and needs only few skilled human laborers as scouts, if any. Satellite mediated and airborne hyperspectral imagery has been consistently used to monitor tropical forests (Mugisha and Huising, 2008).

4.8.3 PASTURES AND RANGELAND

Remote sensing has been applied to pasture management to accomplish several aspects of its development. Precision techniques envisaged during pasture management do involve satellite imagery. The digital images obtained using satellites helps farmers to assess spatial coverage, NDVI, and productivity of pastures (Schellberg et al., 2008). Satellite imagery offers great advantage to farmers situated in remote locations, mountainous terrain and deserts. Farmers could obtain information regarding topography, natural vegetation, crops and pastures, influence of torrential rains, erosion, dust storms, soil erosion, river flow and water resources. Let us consider an example from a remote location in the mountainous districts of Tibet. Pastures found in the Tibetan Plateau are not easy to monitor and maintain. Lack of skilled scouts and navigation are hindrances. Such pastures may suffer due to over-grazing by livestock. In view of the above facts, Lehnert et al. (2013) have made an attempt to study the feasibility of hyperspectral imagery from satellites to manage forage pastures in Western Tibet. They estimated photo-synthetically active vegetation, leaf chlorophyll, proportion of grass and non-grass, etc. Previous data from 48 different locations in the region was used to predict biomass, chlorophyll and N content of pastures. The overall accuracy of estimates of pasture growth and chlorophyll from satellite imagery was high (r^2 = 0.85). Accuracy for separation of grass and non-grass was much higher at r^2 = 0.98. Farmers may obtain satellite images periodically from either governmental agencies or private agencies at definite intervals using internet.

In Northeast of USA, satellite-guided monitoring techniques of pastures and cover crops grown in the offseason is in vogue. The productivity of cover crops is evaluated against nutrients still available as residuals in soil. The site-specific techniques are employed to assess cover crop growth. In particular, biomass accumulation using NDVI values derived from satellite imagery seems useful (Hively et al., 2013). Trials have clearly shown that NDVI values from satellites are highly correlated with biomass accumulation pattern noted on ground.

Forage biomass in a grass-legume mixed pasture could be assessed using ultrasonic sensors. Fricke et al. (2011) have suggested that forage crops such as *Trifolium repens, T. pretense, Medicago sativa* grown in combination with perennial rye grass (*Lolium perenne*) could be assessed using GPS and on-the-go biomass recording gadgets. They suggest that a combination of ultrasonic sensors and other spectral reflectance could be effective in accurately estimating forage biomass.

4.8.4 SATELLITE-GUIDED TECHNIQUES TO MONITOR CLIMATE CHANGE EFFECTS

Satellite imagery offers opportunity to monitor climate change effects and develop appropriate measures using computer-based techniques. It helps to study, analyze, then re-analyze climate change and its effects on crops/forests using different soft wares and models and forecast forest/ crop growth and biomass productivity. Agricultural expanses world-wide are exposed to climate change and its effects such as global warming, N-emissions, loss of organic matter from soil profile, change in land use pattern and cropping systems, etc. Gaseous emissions such as N_2O, NO_2, NH_4, CO_2 and CH_4 are prevalent at high rates in most of the high intensity farming belts. They are caused by natural soil physico-chemical and biological processes. Fertilizer supply in excess has induced nutrient accumulation and contamination of ground water and aquifers. Precipitation pattern and amounts received by cropping zones also alter as a consequence of climate change. Satellite imagery is among most important methods employed currently to monitor climatic factors, decipher climate change and its consequences on several aspects of agricultural crop production (USGS, 2013). Satellite aided techniques are sophisticated and data could be accrued using computer programs, decoded using appropriate software and used during farming. These techniques easily become part of an elaborate concept envisaged, for example, 'Push Button Farming.' In general, satellite imagery is used to study:

a) Ice and Glacier movement in crop belts in North America, Europe, and wherever cropping zones experience snow melts, ice and cold water flow, frost damage and floods;

b) Atmospheric dynamics;

c) Tracing carbon emissions and mapping extent of loss of CO_2 from crop fields;

d) Soil and canopy temperature monitoring;

e) Erosion and surface soil loss monitoring (Kefi et al., 2009; Kefi and Yoshino, 2010; Yoshino and Ishioka, 2005; Le Roux et al., 2007);

f) Drought, dust bowl, desertification and water stress and their effect on cropping expanses and their productivity;

g) Forests, cropping systems and their productivity in relation to climate change (Devendra Kumar, 2011; Eniolorunda, 2014; Lele and Joshi, 2009).

Report by UNEP suggests that smallholder farmers need to be aware of climate change effects on crops and their productivity (IFAD, 2013). They are generally not fiscally endowed to undertake procedures that mitigate climate change effects. Satellite imagery covering large patches of small land holdings should be helpful in informing farmers to undertake certain procedures such as contouring, mulching, splitting fertilizers, etc. that reduce greenhouse gas emission. Satellite data about each farm too could be conveyed to small farmers.

Satellite imagery is used extensively in monitoring vegetation, fluctuations in water resources, river flow and erosion of its banks and crop fields in the delta zones of Bangladesh and India. The cyclonic storms and floods result in loss of paddy crop. Tidal waves and winds that lash the sandy plains and mangroves affect soil fertility and crop productivity. Satellite mediated monitoring of climate and its vagary is a common feature. It helps in warning the farmers to move to locations at higher altitude as river banks erode and fields are flooded (ICID, 2012; Bhattacharyya and Werz, 2012; Krishna, 2015). Satellite imagery and periodic monitoring has been adopted to study land use pattern, vegetation intensity, plant diversity, cropping pattern and soil deterioration, particularly gully erosion in the agrarian zones of Sahelian West Africa (see Krishna, 2015). Fluctuations in vegetation pattern, particularly shrinkage and extension of crops at the confluence of Sahel, Sudanian and Guinean regions of West Africa are best studied using satellite imagery. In Thailand, soil erosion is rampant because of high rainfall, floods and undulated terrain. Farmers are prone to loose soil fertility each year due to erosion. Landsat-TM imagery helps in assessing erosion of agricultural fields and help farmers in adopting remedial measures (Hazarika and Honda, 2011).

Satellite imagery is currently routinely utilized to monitor soil erosion risks in the farm land of South Australia. Farmers are supplied with decoded images of their fields with topography and deterioration of land surface if any. Elevation maps and erosion information is used to develop appropriate remedial measures (Lewis, 2009). Such push button techniques reduce need for tedious scouting and costs on preparing maps and action plans.

Agricultural cropland and forests are interspersed in many regions of the world. Forest cover is an important factor that has impact on global nutrient cycles and water recharge. Land use change from forests to cropland and vice versa is a common phenomenon. At present, policy makers do utilize a wide array of satellite data and maps to judge forests, fluctuations in their expanses and productivity. Computer softwares help in deciphering climate change impacts on specific aspects of forests/crops and land use pattern. For example, 'IDRISI' is a computer software devised recently. It helps in mapping land cover, forest carbon, deforestation and land use change scenarios. It is used to conduct REDD (Reducing Emissions from Deforestation and Degradation) mapping, prepare models for ecosystem services and their impacts on forest/crops, study forest/cropland biodiversity, invasive species, etc. (Clark Labs, 2012). Similarly, 'CLASlite' is another computer-based software system that helps in automatic mapping of tropical forest and land use change pattern (Asner et al., 2009). These computer based techniques offer information on forest decline, degradation, and land use change pattern.

Satellite-guided techniques, particularly precision farming and variable-rate inputs to meet the crop's demand accurately and exactly has some hidden advantages of great value. They reduce greenhouse gas emission, also, avoid deterioration of land and water resources. We ought to realize that satellite techniques make farmers add fertilizers, herbicides and pesticides in quantities exactly needed by crops. Chemicals are applied to match crops' need at various stages, hence, accumulation due to inaccuracies are absent. Basically, satellite techniques avoid undue accumulation of chemicals and ground water contamination. So, satellite techniques really offer a great opportunity to thwart climate change effects. It offers good control over various factors that may induce climate change. Accuracy that satellite mediated soil fertility maps, digital data and fertilizer supply via variable rates offer, literally avoids gaseous emissions, nutrient accumulation and water contamination. Satellite-guided techniques may also induce appropriate nutrient recycling by providing farmers with information on soil organic matter distribution. Satellite imagery showing topography and elevation (slopes, gradients) helps farmers in reducing soil erosion by adopting suitable procedures such as contouring, mulching and sprinkle irrigation, etc. Clearly, satellite-guided techniques should reduce greenhouse gas emission, land degradation and preserve water resources much better than what farmers/experts did during yester years.

KEYWORDS

- **LANDSAT**
- **Multi-Year Yield Analysis**
- **Natural Resource Mapping**
- **QuickBird**
- **Satellite Guided Variable Rate Applicators**
- **Satellites**
- **Spectral Reflectance**
- **Yield Mapping**

REFERENCES

1. Adamchuk, V. I. (2008). Satellite-based auto-guidance. Precision Agriculture. University of Nebraska-Extension services, Lincoln, Nebraska, USA, pp. 1–6.
2. Adamchuk, V., Dobermann, A., Ping, J. (2008). Listening to the story told by yield maps. University of Nebraska Cooperative Extension Publication EC704 http://digitalcommons.unl.edu/cgi/veiwcontent.cgi?article=1707 &context=extemsionhist, pp. 15 (October 3rd, 2014).
3. Adamchuk, V. I., Hummel, J. W., Morgan, M. T., Upadhyaya, S. K. (2004). On-the-Go soil sensors for Precision agriculture Computers and electronics in Agriculture 44: 71–91 DOI: 10113/16856/1/IND43638848 (October 23rd, 2014).
4. Adamchuk, V. I., Jassa, P. J. (2014). Future of Soil Sensing Technology. Precision Farming Dealer. http://www.precisionfarmingdealer.com/content/future-soil-sensing-technology, pp. 1–8 (November 27th, 2014).
5. Adamchuk, V. I., Lund, E. D., Reed, T. M., Ferguson, R. B. (2007). Evaluation of on-the-go technology for Soil pH mapping. Precision Agriculture 8: 139–149.
6. Adamcuk, V. I., Lund, E. D., Sethuramsamyraja, M. T., Dobermann, A., Marx, D. B. (2005). Direct measurement of Soil chemical properties on-the-go using ion-selective electrodes. Computers and Electrodes in Agriculture 48: 272–294.
7. Adsett, J. F., Thornton, J. A., Sibley, K. J. (1999). Development of an automated on-the-go soil nitrate monitoring system. Applied Agricultural Engineering 15: 351–356.
8. Ag Business & Crop Inc., 2014 Ag Services. http://www.agbusiness.ca/, pp. 1–4 (October 26th, 2014).
9. Aisha, A. W., Zauyah, S., Anour, A. R., Fauziah, C. I. (2010). Spatial variability of selected chemical characteristics of paddy soils in Sawah Sempadan, Selangor, Malaysia. Malaysian Journal of soil Science 14:27–39.

10. American Society of Horticultural Science, 2008 digital Cameras, Remote Satellites measure Crop Water demand. http://www.sciencedaily.com/releases/2008/07/080717140409.htm, pp. 1–5 (November 15th, 2014).

11. Amezquita, E., Thomas, R. J., Rao, I. M., Molina, D. L., Hoyos, P. 2004. Use of deep-rooted tropical pastures to build-up an arable layer through improved soil properties of an Oxisol in the Eastern Plains (Llanos Orientales) of Colombia. Agriculture, Ecosystem and Environment. 103: 269–277.

12. Ana Isabel de Castro, F. López-Granados, M. Jurado-Expósito. 2013. "Broad-scale cruciferous weed patch classification in winter wheat using QuickBird imagery for in-season site-specific control." Precision Agriculture, DOI 10.1007/s11119–013–9304-y. Factor de Impacto: 1.549, pp. 1–8 (October 16th, 2014).

13. Anderson, G. L., Everitt, J. H., Richardson, A. J., Escobar, D. E. (1993). Using satellite data to map False Broomweed (*Ericameria austrotexana*) infestations on South Texas rangeland. Weed Technology 7: 865–871.

14. Aron, J. (2013). How Nigeria has been using satellites. New Scientist, http://www.newscientist.com/article/dn24025-how-nigeria-has-been-using-its-satellites.html#.VGjFu_mUd0w, pp. 1–4 (November 16th, 2014).

15. Asher, J. B., Yousef, B. B., Volinskly, R. (2013). Ground-based Remote sensing system for Irrigation Scheduling. Biosystems Engineering 114: 444–453.

16. Asner, G. P., Knapp, D. E., Balaji, A., Pacz-Acosta, A. (2009). Automated mapping of Tropical deforestation and forest degradation: CLASlite. Journal of Applied Remote Sensing 3:1–9.

17. Backes, M., Jacobi, J. (2006). Classification of weed patches in Quickbird images: Verification of ground truth data. EARSel Proceedings 5: 173–179.

18. Bah, A., Balasundaram, S. K., Husni, M. H. A. (2012). Sensor Technologies for Precision Soil Nutrient Management and Monitoring. American Journal of Agricultural and Biological Sciences 7: 43–49.

19. Barbosa, R. N., Overstreet, C. (2010). What is Soil Electrical Conductivity. Louisiana Agricultural Centre. http://www.LSUAgCenter.com, pp. 1–7 (October 23rd, 2014).

20. Barnhisel, R., Bitzer, M., Shearer, S., Murdock, L., Howe, P. (2014). Variable rate seeding and Nitrogen Application. https://www.bae.uky.edu/precag/PrecisionAg/Development_and_Assessment/vrseeding.htm, pp. 1–4 (November 17th, 2014).

21. BDAEL for Agriculture Co Ltd, 2014 Introducing BDAEL. http://www.rxvrt.com/media/RxVRT-Dealer-Locations.pdf, pp. 10–8 (October, 2014).

22. Ben-Dor, E., Malthus, T., Plaza, A., Schlaper, D. (2009). Hyperspectral Remote Sensing. http://www.geo.uzh.ch/microsite/rsl-documents/research/publications/book-chapters/2012.EUFAR.Hyperspectral-3741993216/2012.EUFAR. Hyperspectral.pdf, pp. 239 (December 20th, 2014).

23. Bhan, S. K., Saha, S. K., Pande, L. M., Prasad, J. (2012). Use of Remote sensing and GIS technology in sustainable agricultural Management-Indian Experience. wgbis.ces.iisc.ernet.in/energy/HC270799/LM/SUSLUP/.../617/617.pdf, pp. 1024 (November 28th, 2014).

24. Bhattacharyya, A., Werz, M. (2012). Climate change, migration and conflict in South Asia. Centre for American progress. http://www.knowledgebank.irri.org/ckb/PDFs/croppingsystem/environmentalcarbon%20sequestration%20in%20%20soils%20of%20the%20indo-gangetic%20plains.pdf, pp. 1–98 (December 20th, 2014).

25. Bianchini, A. A., Mallarino, A. P. (2002). Soil sampling alternatives and variable-rate liming for a soybean-corn rotation. Agronomy Journal 94: 1355–1366.

26. Blackmore, S., 2000 The interpretation of trends from multiple yield maps. Computer and Electronics in Agriculture 26(1): 37–51.

27. Brouder, S. M., Hofmann, B. S., Morris, D. K. (2005). Mapping soil pH: Accuracy of common soil sampling strategies and estimation techniques. Soil Science Society of America Journal 69: 427–441.

28. Buis, A., Murphy, R. (2014). New Satellite data will help farmers facing drought. http://www.nasa.gov/jpl/smap/satellite-data-help-farmers-facing-drought-2014081#VJZHoF4AA, pp. 1–6 (December 22nd, 2014).

29. Butzen, S, and Gunzenhauser, B. (2009). Putting variable rate seeding to work on your farm. Crop Insights 19: 12–19.

30. Camacho-Tamayo, J. H., Rubiano Sanabrio, Y., Santana, L. M. (2013). Management zones based on the Physical properties of an Oxisol. Journal of Soil Science and Plant Nutrition 13: 1–8 http://dx.doi.org/10.4067/S0178–95162013005000061 (October 8th, 2014).

31. Campbell, J. B. (2002). Introduction to Remote Sensing. Guilford press, New York, USA, pp. 287.

32. Casa, R., Castaldi, F., Pascussi, S., Pignatti, S. (2012). Potential of hyperspectral remote sensing for field scale soil sampling and Precision Agriculture applications. Italian Journal of Agronomy 7: 331–336.

33. CaseIH Inc., 2014 CaseIH advanced Farming Systems http://www.caseih.com/en_us/afs/Pages/Home.aspx, pp. 1 (November 17th, 2014).

34. Castro-Tendero, A. J., Garcia-Torres, l. 1995 SEMAGI-an expert system for weed control decision making in sunflower. Crop Protection 14: 543–548.

35. Chenghai, Y., Everitt, J., Reginald, F. (2008). Using high-resolution SOT 5 Multi-spectral imagery for Crop identification. Proceedings of ASABE Annual International meeting, ASABE paper No 08–3598 http://www.ars.usda.gov/ research/publications/publications.htm?SEQ_NO_115=226511, pp. 1–4 (October 1st, 2014).

36. Chuang, T. C., Wan, C. C., Daud, A. H. (2014). Application of Remote Sensing technique for Rice Precision Farming. http://www.aprsaf19_data/eo/D2_07_Malaysia_MARDI.pdf, pp. 1–14 (April 3rd, 2014).

37. Chung, S. O., Sung, J. H., Sudduth, K. A. Drummond, S. T., Hyun, B. K. (2001). Spatial variability of yield, chlorophyll content, and Soil properties in a Korean rice field. In: Proceedings of the 5th International Conference on Precision Agriculture Robert, P. C. (Ed.). American Society of Agronomy, Madison, Wisconsin, USA, pp. 234–242.

38. Clark, S. (2009). Soyuz rocket sends new Russian weather satellites. http://www.space.com/738-soyuz-rocket-sends-russian-weather-stellite-orbit.htm., pp. 1–3, 9 December 3rd, 2–14).

39. Clarke, A. (2014). Precision Farming aims to Break Potato yield Barrier. Precision Farm Dealer, http://www.fwi.co.uk/arable/precision-farming-aims-to-break-potato-yield-barrier.htm, pp. 1–3 (November 22, 2014).

40. Clark Labs, 2012 Forest and land change Mapping. Clark University, Massachusetts, USA, http://www.clark;abs.org/ applications/forest-and-land-change-mapping.cfm, pp. 1–3 (November 30th, 2014).

41. Clemson University Cooperative Extension Service 2014. Introduction to Precision Agriculture Clemson University http://www.clemson.edu/extension/rowcrops/precision_agriculture.html, pp. 1–2 (October, 13th, 2014).

42. Cline, H 2002. Cotton seeding latest Variable-Rate goal. http://westernfarmpress.com/cotton-seeding-latest-variable-rate-goal., pp. 1–4 (November 17th, 2014).

43. Colvin, T. S., Arslan, S. (2002). Yield Monitor accuracy. The Site-Specific Management guidelines. Potash and Phosphate Institute, Norcross, Georgia, USA, pp. 1–7.

44. Cropmetrics, 2014 Up and coming Precision Agriculture Technology: drones. http://cropmetrics.com2014/01/up-and-coming-precision-agriculture-technology-farm-drones/, pp. 1–5 (October 23rd, 2014).

45. CropQuest Inc., 2014a Actionable Imagery. http://www.cropquest.com/wp-content/uploads/2014/04/Imagery-v2.pdf, pp. 1–2 (October 27th, 2014).

46. CropQuest Inc., 2014b Precision Ag Services. http://www.cropquest.com/precision-ag/, pp. 108 (October 27th, 2014).

47. CropQuest Inc., 2014c Yield Analysis. http://www.cropquest.com/precision-ag/yield-analysis/, pp. 1–4 (October 27th, 2014).

48. Cucunebá-Melo, J. L., Álvarez-Herrera, J. G., Camacho-Tamayo, J. H. 2011. Identification of Agronomic management units based on Physical attributes of soil. Journal of Soil Science and Plant Nutrition. 11, 87–99.

49. Cui, D. Li, M., Zhang, Q. (2013). Development of an Optical Sensor for Crop leaf chlorophyll content detection. Computers and Electronic in Agriculture 69: 171–176.

50. Dang, R. C. Dalal, M. J. Pringle, A. J. W. Biggs, S. Darr, B. Sauer, J. Moss, J. Payne and D. Orange 2011 Electromagnetic induction sensing of soil identifies constraints to the crop yields of north-eastern Australia. Soil Research 49: 559–571.

51. Davis, G. (2011). History of NOAA Satellite Programs. http://www.ossd.noaa.gov/download/JRS012504-GP.pdf (October 2, 2014).

52. Daughtry, C., Beeson, P., Hunt, R., Sadeghi, A. (2011). Remote sensing for Assessing Crop Residue Cover and tillage intensity. USDA-ARS, Beltsville, Maryland USA, http://www.slideshare.net/CIMMYT/remote-sensing-for-assessing-crop-residue-cover-and-soil-tillage-intensity, pp. 1–24 (October 27th, 2014).

53. Daughtry, C. B., Walthall, C. L., Kim, M. S., Brown de Coulston, E., McMurtrey, J. E. (2000). Estimating corn leaf chlorophyll from leaf or canopy reflectance. Remote sensing of Environment 74: 229–239.

54. DeCastro, A. I., Jurado-Exposito, M., GomezCasero, M. T., Gomez-Condon, d., Caballero-Novella, J. J., Lopez-Granados, F. (2010). Discrimination of Cruciferous weeds in wheat using Qucikbird satellite image. In: Proceedings of the 3rd International Symposium in Recent Advances in Quantitative Remote Sensing, Sobrino, A. (Ed.). Publicacions de la Universitat de Valencia, Italy, pp. 133–137.

55. DeGruijter, J. J., McBratney, A. B., Taylor, J. A. (2010). Sampling for high-resolution Soil mapping. In: Proximal Soil sensing. Rosell, V. (Ed.). Springer Science Business Media B. V., pp. 77–88.

56. Devendra Kumar, 2011 Monitoring Forest Cover changes using Remote sensing and GIS: A global Perspective. Science alert. http://scialert.net/fulltext/?doi=rjes.2011.105.123, pp. 1–15 (November 28th, 2014).

57. Dickson, P. (2009). Sputnik: The shock of the century. Bloomsbury Publishing Inc., Pennsylvania, USA, pp. 298.

58. Digmann, M., Shinners, K. (2014). Calibrating your Forage harvester's yield Monitor. Integrated Pest and Crop Management. http://www.ipcm.wisc.edu/blog/2011/08/calibrating-your-forage-harvsters-yield-monitor.htm., pp. 1–3 (April 25th, 2014).

59. Doerge, T. A. (2010). Site Specific Management guidelines: SSMG-2: Management zone concepts. Potash and Phosphate Institute, Norcross, Georgia, USA, pp. 1–4.

60. Doerge, T. A., Kitchen, N. R., Lund, E. D. (2011). Soil electrical conductivity mapping. International Plant Nutrition Institute, Norcross, Georgia, USA. http://www.ipni.net/ppiweb/ppibse.nsf/$webindex/article=BDICF45C852569 D700636EDAC9ADC4DC, pp. 1–8 (June 27th, 2011).

61. Doraiswamy, P. C., Hatfield, J. L., Jackson, T. J., Akhmedov, B., Prueger, J., Stern, A. (2014). Crop condition and yield simulations using Landsat and Modis. Remote sensing of Environment 92: 548–559.

62. Dunbar, B. (2009). NASA uses satellite to unearth innovation in Crop forecasting. National Aeronautic and Space Agency, USA, http://www.nasa.gov./topics/earth/features/crop_forecast.html, pp. 1–5 (December 22nd 2014).

63. EGNOS 2014 EGNOS for Yield Mapping: The power of knowledge. http://www.egnos-portal.gsa.europa.eu/sites/default/files/content/document/egnos-for-yield-mapping-eu.pdf, pp. 1–3 (April 25th, 2014).

64. Elstein, D. (2003). Using Management zones to help in Precision Agriculture. United States Agricultural Department, Agricultural Research Service, Beltsville, Maryland, USA, http://ars.usd.gov/is/pr/2003/030814.htm, pp. 1–3 (October 7th, 2014).

65. Eniolorunda, N. B. (2014). Climate change analysis and Adaptation: The role of Remote sensing (RS) and Geographical information system (GIS). International Journal of Computational Engineering Research 4:41–51.

66. Everitt, J. H., Yang, C. B. 2007a .Mapping Spiny Aster infestations with QuickBird imagery. Geocarto International 22:273–283.

67. Everitt, J. H., Yang, C. B. 2007b. Mapping Broom Snakeweed though image analysis of colored-infrared photography and digital imagery. Environmental Monitoring and Assessment 134: 287–292.

68. Faical, B. S., Costa, F. G., Pessin, G., Freitos, H., Torstein, B. (2014). The use of Unmanned Aerial vehicles and Wireless sensor networks for spraying Pesticides. Journal of Systems Architecture http://dx.doi.org/10.1016/ j.sysarc.2014.01.004, pp. 1–8 (November 12th, 2014).

69. FAO 2007 Procedures for Land Resource Inventory. Food and Agricultural Organization of the United Nations, Rome, http://www.fao.org/docrep/x5648e07.htm, pp. 1–17 (December 22nd, 2014).

70. Farming UK 2012 Oilseed seeding system to launch at Cereals. http://www.farminguk.com/News/Oilseed-seeding-system-to-launch-at-Cereals_23584.html, pp. 1–3 (November 17th, 2014).

71. Feng, M. C., Xiao, L., Zheng, M. J., Yang, W., Deng, G. W. (2014). integrating Remote sensing and GIS for prediction of Winter wheat (*Triticum aestivum*) protein contents in Linfen (Shanxi), China. PlosOne DOI: 10.1371/ journal.pone.0080989, pp. 1–8 (November 23rd, 2014).

72. Feng, M. C., Yang, W., Cao, L., Ding, G. (2009). Monitoring Winter wheat freeze injury using Multi-Temporal MODIS data. Agricultural Science in China DOI: 10.1016/s1671–2927(08)60313–2, pp. 1–9 (October 25th, 2014).

73. Ferguson, R. W., Hergert, G. W. (2009). Soil sampling for Precision Agriculture. Journal of Animal and Plant Sciences 5: 494–506.

74. Ferguson, R. B., Hergert, G. W., Shapiro, C. A., Wortmann, S. S. (2007). Guidelines for soil sampling. NebGuide. University of Nebraska-Lincoln, Extension Service. Publication No G 1740, pp. 1–4.

75. Franzen, D. W., Kitchen, N. R. (2010). Developing Management zones to target Nitrogen applications. Site-Specific management Guidelines http://www.ipni.net/publication/ssmg.nsf/0/A5CD47480DAF17D1852579E500765D7B/ $FILE/SSMG-05.pdf SSMG-5: 1–4 (October 9th, 2014).

76. Franzen, D. W., Nana, T., Casey, D., Ralsoton, J., Starckia, J., Halvorson, M., Hofman, V. Lamb, J., Sims, A. (2005). Zone delineation for Nitrogen Management. North Central Extension Industry Soil Fertility Conference 21: 25–32.

77. Fricke, T., Ritcher, F., Wachendorf, M. (2011). Assessment of Forage mass from grassland swards by height measurement using an Ultrasonic sensor. Computers and Electronics in Agriculture 79: 142–152.

78. Galileo Geo Inc., 2014 Agriscience. http://galileo-gp.com/applications/agriscience/?gclid=CPnficHgocICFVgnjgodurwARQ., pp. 1–8 (November 29th, 2014).

79. GDA Corp 2014a Global Agricultural Crop Monitoring. http://www.gdacorp.com/crop-mapping., pp. 1–5 (November 8th, 2014).

80. GDA Corp 2014b Crop condition Monitoring, Assessment and Forecast. http://www.gdacorp.com/crop-condition-and-health/, pp. 1–4 (November 8th, 2014).

81. GDA Corp 2014c Agricultural Business Intelligence. http://www.gdacorp.com/agricultural-intelligence., pp. 1 (November 8th, 2014).

82. GDA Corp 2014d Satellite Image Overview. http://www.gdacorp.com/satellite-image-prcessing, pp. 1 (November 8th, 2014).

83. GDA Corp 2014e Crop Yield Forecasts. http://www.gdacorp.com/crop-yield-forecasts., pp. 1–4 (November 8th, 2014).

84. Gebbers, R., Adamchuk, V. I. (2010). Precision Agriculture and Food Security. Science 327:828–831.

85. GeoEye Inc., 2011 All about GeoEye-1. http://wwwlaunch.geoeye.com/LaunchSite/about/, pp. 1–8 (October 5th, 2014).

86. Gibbs Equipment Inc., 2014 Precision Farming. http://www.gibbsequipment.com/precision-farming/default.aspx, pp. 1–2 (November 17th, 2014).

87. Gnip, P., Charvat, K. (2003). Management zones in Precision Farming. Zemedelska Ekonomika Praha 49:416–418.

88. Goddard, T., Grant, R. (2001). Landscape Management of Agronomic processes for Site-Specific Farming. http://www.ppi.ppic.org/far/farguide.nsf/$webindex/article=860FCI7685256A1000521799506AEE1C!opendocument (October 25th, 2014).

89. Goldfinch Inc., 2014 AgPixel Services. http://www.agpixel.com/company/, pp. 1–4 (December 19th, 2014).

90. GPS 2014 Global Positioning Systems and related applications. Http://www.gps/applications/agriculture., pp. 1–3, 9 October 2nd, 2014).

91. GRDC, 2010 Precision Agriculture fact sheet. How to put Precision Agriculture into practice. Grain Research and Development Corporation, Kingston, Australia. http://www.grdc.com.au, pp. 1–6 (October 3rd, 2014).

92. Grisham, M. P., Johnson, R. M., Zimba, P. V. (2010). Detecting Sugarcane yellow leaf virus infection in asymptomatic leaves with hyperspectral remote sensing and associated leaf pigment changes. Journal of Virological Methods 167: 140–145.

93. Group of Earth Observations 2014. Global Agricultural Monitoring. http://www. earthobservations.org/Cop-ag-gans/., pp. 1–24 (December 3rd, 2014).

94. Growing Nebraska 2014 New Constellation of Satellites to be launched for Agriculture. http://growingnebraska.com/news/2014/05/new-constellation-satellites-be-launched-agriculture/, pp. 1–8 (December 3rd, 2014).

95. Gunzenhauser, B., Shanahan, J. (2014). Crop insights: Using multi-year yield analysis to create management zones for variable rate seeding. Pioneer Hybrid Inc., https:// www.pioneer.com/CMRoot/Pioneer/US/Non_Searchable/agronomy/ars_pdfs/2011_ Corn_MYYA_for_VRS_HQ.pdf, pp. 107 (October 12th, 2014).

96. Harford, J. 1997. *Korolev how one man masterminded the Soviet drive to beat America to the Moon*, New York. John Wiley and Sons, Inc., pp. 121.

97. Hazarika, M. K., Honda, K. (2010). Estimation of soil erosion using remote sensing and GIS, its valuation and economic implications on agricultural production. In: Sustaining Global Farming. Stott, D. E., Mohtar, R. H., Steinhard, G. C. (Ed.) National Soil Erosion Research Laboratory, Purdue University, Indiana, USA, pp. 1090–1093.

98. Hemmat, A., Adamchuk, V. I. (2008). Sensor systems for measuring soil compaction: Review and Analysis. Computers and electronics in Agriculture 63: 89–103.

99. Hest, D. (2012). New driverless tractor, grain cart systems. TractorLife Systems Inc., http://www.tractorlife.com/new-driverless-tractor-grain-cart-systems-coming-this-year/.

100. Hively, W. D., Lang, M., McCarty, G, Sadeghi, A., Keppler, J., McConnell, L. 2013. "Remote Sensing of Cover Crop Nutrient Uptake on Maryland's Eastern Shore" Proceedings of Annual meeting of the Soil and Water conservation Society, Saddle brook Resort, Tampa, Florida, http://citation.allacademic.com/meta/ p174067_index.html, pp. 1–7 (April 12th, 2015).

101. Hughes, J. R., Coventry, R. J., DiBlla, L. (2009). GPS-guided: an option for managing soil compaction and yield variability in the wet tropics. Proceedings Australian Society of Sugar Technology, Poster papers 31: 1.

102. Hummel, J. W., Gaultney, L. D., Sudduth, K. A. (1996). Soil property sensing for Site-specific crop management. Computers and Electronics in Agriculture 14: 121–136.

103. Hummel, J. W., Sudduth, K. A., Hollinger, S. E. (2001). Soil moisture and organic matter prediction of surface and subsurface soils using an NIR soil sensor. Computers and Electronics in Agriculture 32: 149–165.

104. Hunt, R., Hamilton, R., Everitt, J. (2012). Mapping weed infestations using remote sensing a weed manager's guide to Remote Sensing and GIS-Mapping and Monitoring. http://www.fs.fed.us/eng/rsac/invasivespecies/documents/mapping.pdf, pp. 1–8 (October 15th, 2014).

105. Huth, N., Boulton, G., Dalgliesh, N., Cocks, B., Poulton, P. (2012). Electromagnetic Induction methods for monitoring soil water in irrigated copping systems. Proceedings of 16th Australian Agronomy Conference, pp. 1–8 http://www.regional.org.au/ au/asa/2012/precision-agriculture/8133_huthni.htm (November 16th, 2014).

106. ICID, 2012 Bangladesh. International Commission on Irrigation and Development. New Delhi. India. http://www.icid.org/i_dbangladesh.dpf., pp. 1–8 (December 19th, 2014).

107. IFAD, 2013 Small holders, Food Security and the Environment. International Fund for agricultural Development, Rome, Italy, pp. 146.

108. Ignat, T., Alchanatis, V., Schmilovitch, Z. (2014). Maturity prediction of intact bell peppers by sensor fusion. Computers and Electronics in Agriculture 104: 9–17.

109. Irons, J. R., Taylor, M. P., Rocchio, L. 2014a. Landsat 8 National Aeronautics and Space Agency, USA, www.landsat.gsfc.nasa.gov/?p=318, pp. 1–4 (October, 2014).

110. Irons, J. R., Taylor, M. P., Rocchio, L. 2014b. US uses Landsat Satellite Data to fight hunger, poverty. National Aeronautics and Space Agency http://www.landsat.gsfc. nasa.gov/?p=832, pp. 1–5 (October 3rd, 2014).

111. Ishola, T. A., Yahya, A., Shariff, A. M., Aziz, S. A. (2013). An RFID-based Variable rate Technology Fertilizer Applicator for tree crops. Journal of Applied Sciences 13: 409–415.

112. Jahanshiri, E. (2006). GIS-based Soil sampling for Precision Farming of Rice. Master's Thesis, University Putra Malaysia, Selangor, Serdang, Malaysia, 72 http://psair. upm.edu.my/612, pp. 1–9 (April 23rd, 2014).

113. Johnson, L. F., Roczen, D. E., Youkhana, S. K., Nemani, R. R., Bosch, D. F. (2003). Mapping vineyard leaf area with multispectral satellite imagery. Computers and Electronics in Agriculture 38:33–44.

114. Kasowski, M., Genereaux, D. (1994). Farming by the foot in Red River valley or Minnesota. Agricultural Finance 124:20–21.

115. Kefi, M., Yoshinao, K. (2010). Evaluation of the economic effects of soil erosion risk on agricultural productivity using remote sensing—case of watershed in Tunisia. International archives of the Photogrammetry, Remote sensing and Spatial information Science, 38: 930–935.

116. Kefi, M., Yoshino, K., Zayani, K., Isoda, H. (2009). Estimation of soil loss by suing combination of erosion Model and GIS: Case study of Water sheds in Tunisia. Journal of Arid Land Studies 19:287–290.

117. Kerry, R., Oliver, M. A. (2003). Variograms of ancillary data to aid sampling for soil surveys. Precision Agriculture 4:261–278.

118. Kim, H. J., Sudduth, K. A., Hummel, J. W. (2009). Soil macronutrient sensing for Precision Agriculture. Journal of Environmental Monitoring 11: 1810–1824.

119. King, J. A., Dampney, P. M. R., Lark, R. M., Wheeler, H. C., Bradley, R. L. (2005). Mapping potential crop management zones within fields: use of yield-map series and patters of Soil physical properties identified by electromagnetic induction sensing. Precision Agriculture 6: 167–181.

120. Kinze Inc., 2014. Autonomous tractor and harvest systems. http://www.kinze.com/ article.aspx?id=341&Kinze+Adds+New+Features+to+Its+Autonomous+Harvest+ System, pp. 1–5 (December 9th, 2014).

121. Kitchen, N. R. (2003). In: Management zones help in Precision Agriculture. United States Department of Agriculture. Agricultural Research Magazine. http://www. highbeam.com/doc/1G1–106650576.html, pp. 1–3 (October 7th, 2014).

122. Krishna, K. R. (2002). Agrosphere: Nutrient Dynamics, Ecology and Productivity. Science Publishers Inc., New Hampshire, USA, pp. 346.

123. Krishna, K. R. (2012a). Maize Agroecosystem. Nutrient Dynamics and Productivity. Apple Academic Press Inc., New Jersey, USA, pp. 348.

124. Krishna, K. R. (2012b). Precision Farming: Soil Fertility and Productivity aspects. Apple Academic Press Inc., Waretown, New Jersey, USA, pp. 137–142.

125. Krishna K. R. (2013). Agroecosystems: Soils, Climate, Crops, and Nutrient Dynamics. Apple Academic Press Inc., New Jersey, USA, pp. 558.

126. Krishna, K. R. (2015). Agricultural Prairies: Natural Resources and Crop Productivity. Apple Academic Press Inc., New Jersey, USA., pp. 500.

127. Kweon, G., Lund, E., Maxton, C. (2012). The ultimate soil survey in one pass soil texture, organic matter, pH, elevation, slope and curvature. Proceedings of the 11th International Conference on Precision Agriculture, Indianapolis, Indiana, USA. http://www.veristech.com/pdf-files/11th ICPA-Veris%20SP3%20Paper.pdf., pp. 1–6 (April 22nd, 2014).

128. Lakes, J. V., Bock, G. R., Goode, J. A., Mulla, D. J. (2007). Geostatistics, Remote Sensing and precision Farming. Proceedings of CIBA foundation symposium on Precision Agriculture http://onlinelibrary.wiley.com/doi/10.1002/ 9780470515419.ch7/ summary:js.htm, pp. 1–2 (July 12th, 2012).

129. Lan, Y., Thomson, S. J., Huang, Y., Clint Hoffman, W., Zhang, H. (2010). Current status and future directions of Precision aerial application for site-specific crop management in the USA. Computers and Electronics in Agriculture 74:34–38.

130. Land IQ 2014 Remote Sensing http://www.landiq.com/remote-sensing, pp. 1–18 (November, 29th 2014).

131. Landsat Program (2012). The Landsat program Ames Research Centre. National Aeronautic and Space Agency, Washington D. C. http://geo.arc.nasa.gov/age/landsat/daccess.html, pp. 422.

132. Landsat Program (2014). Technical Details. National Aeronautical Space Agency, USA. http://landsat.gsfc.nasa.gov/about/L7_td.html., pp. 238 (October 4th, 2014).

133. Landsat 7 Gateway (2012). The Landsat 7 program. Goddard Space Flight Centre. http://landsat.gsfc.nasa.gov., pp. 287.

134. Launius 2005 Sputnik and the Origins of the Space Age http://www.hq.nasa.gov/ office/pao/History/sputnik/sputorig.html.

135. Lee, W. S., Alchanatis, V., Yang, C., Hirafuji, M., Moshou, D., Li, C. (2010). Sensing technologies for Precision Specialty crop production. Computers and Electronics in Agriculture 7:2–33.

136. Lehnert, L. W., Meyer, H., Meyer, N., Reudenbach, C., Bendix, J. (2013). Assessing pasture quality and degradation status using hyperspectral imaging: A case study from Western Tibet. Proceedings of SPIE 8887 Remote Sensing for Agriculture, Ecosystems and Hydrology 15:888701. http://www.doi.10.1117/12.2028348, pp. 1–3 (October 25th, 2014).

137. Lele, N., Joshi, P. K. (2009). analyzing deforestation rates, spatial forest cover changes and identifying critical areas of forest cover changes in North-East India during 1972–1999. Environmental Monitoring and Assessment 156:159–70.

138. Lemos, S. G., Noguiera, A. R., Torre-Neto, A., Parra, A., Alonso, J. (2007). Soil Calcium and pH monitoring Sensor System. Journal of Agricultural Food Chemistry 55:4658–4663.

139. Le Roux, J. J., Newby, T. S., Sumner, P. D. (2007). Monitoring soil erosion in South Africa a regional scale: Review and Recommendations. South African journal of Science 103:1–11.

140. Lewis, M. (2009). Towards image-based monitoring of soil erosion risk in South Australia. Landscape Futures Program, University of Adelaide, South Australia, http://www.slideshare.net/Environment/towards-imagebased-monitoring-of-soil-erosion-risk-in-southern-australia, pp. 1–15 (November 29th, 2014).

141. Li, Y., Shi, Z., Wu, C., Li, F. (2008). Determination of potential Management zones from Soil electrical conductivity, Yield and Crop data. Journal of Zhejiang University Science B. 9: 68–78.

142. Li, Y., Shi, Z., Wu, H. X., Li, F and Li, H. Y. (2013). Definition of Management zones for enhancing cultivated land conservation using spatial data. Environmental Management 52: 792–806.

143. Li, Q., Wu, B., Jia, K., Dong, Q., Eerens, H and Zhang, M. (2011). Maize acreage using ENVISAT MERIS and CBERS-02B CCD data in the North China plain. Computers and Electronics in Agriculture 78: 208–214.

144. Li, Y. L., Yi, S. P. (2013). Improving the efficiency of spatially selective operations for agricultural robotics in cropping field. Spanish Journal of Agricultural Research 11: 56–64.

145. Li, Y., Zhou, S., Feng, L., Hong-Li, L. (2007). Delineation of Site-Specific Management zones using fuzzy clustering analysis in a Coastal saline land. Computers and Electronic in Agriculture 56: 174–186.

146. Lobell, D. B., Hicke, J. A., Asner, G. P., Field, C. B., Tucker, C. J., Los, S. O. (2002). Satellite estimates of productivity and light use efficiency in United States Agriculture 1982–1998. Global Change Biology, 8:722–736.

147. Logdson, S., Clay, D., Moore, D., Tsegaye, T. (2008). Soil Science step-by-step field analysis. Soil Science Society of America, Publication No 147, pp. 287.

148. Long, D. S. (2006). Innovating Site-Specific Management for Farming and Ranching. http://www.reeis.usda.gov/web/ crisprojectpages/0171168-innovating-site-specific-mangment-for-farming-and ranching.html, pp. 7 (October 4th, 2014).

149. Long, D. S., Engel, R. E., Siemens, M. C. (2008). Measuring grain protein concentration in-line using near infrared reflectance spectroscopy. Agronomy Journal 100:247–252.

150. Long, D. S., McCullum, J. D., Scharf, P. A. (2013). Optical-Mechanical system for on combines segregation of wheat by grain protein concentration. Agronomy Journal 105:1529–1535.

151. Lopez-Granados, F., Jurado-Exposito, M., Pena-Barragan, J.M and Garcia-Torres, l. 2006. Using Remote sensing for identification of late-season grass-weed patches in wheat. Weed Science 54:346–353.

152. Lowenberg-DeBoer, J. (1998). Economics of variable rate planting for corn. Department of Agricultural Economics, Purdue University, Indiana, Staff paper No 98–2. 1–14.

153. Lowenberg De-Boer, J. (2004). Precision Farming in Europe. http://www.agriculture. purdue.edu/rsmc/news/june03_ precisionAgEurope.htm, pp. 1–3 (April 3rd, 2014).

154. Lund, E. D., Christy C. D., Drummond, P. E. (1998). Applying soil electrical conductivity technology to Precision Agriculture, St Paul, Mn, USA, pp. 78–89.

155. Magri, A., Van Es, H. M., Gloss, M. A., Cox, W. (2005). Soil test, Aerial image and yield data as inputs for Site-Specific Fertility and hybrid Management under maize. Precision Agriculture 6: 87–110.

156. Makar, N., Rooks, K., Archer, S., Sparks, K., Trigg, C., Lourie, J., Wilkins, K. (2011). Monitoring Agricultural Tillage Practices with NASA Hyperspectral Satellite imagery. Earthzine, pp. 1–2 http://www.earthzine.org/2011/08/10/ monitoring-agricultural-tillage-practices-with-nasa-hyperspectral-satellite-imagery/ (October, 21st, 2014).

157. Mallarino, A. P. (1998). Using Precision Agriculture to improve Soil fertility management and On-farm Research. Iowa State University Extension Services. Integrated Crop Management 480: 12–14.

158. Martin, M. P., Barreto, L., Fernandez-Quintanilla, C. (2011). Discrimination of sterile oat (*Avena steriles*) in winter barley (*Hordeum vulgare*) using Quickbird satellite images. Crop protection 30: 1363–1369.

159. Mask, P. L.; Howe, J. A.; Fulton, J. P.; Shaw, J. N.; McDonald, T. P.; Adrian, A. M.; Taylor, S. E. (2011). Precision agriculture technologies to increase production efficiency in Alabama. United States Department of Agriculture-Agriculture Research Service, Beltsville, Maryland, USA http://www.aaes.auburn.edu/research/documents/precision-ag.pdf.

160. Maxwell, S. K., Nukolls, J. R., Ward, M. H., Hoffer, R. M. (2004). An automated approach to mapping corn from Landsat imagery. Computers and electronics in Agriculture 43:43–54.

161. Meersman, T. (2014). Precision Agriculture: GPS, Robots, Drones are Minnesota farmhands. http://www.startribune.com/ business/259320921.html, pp. 1–8 (November 24th, 2014).

162. Morris, M. (2006). Soil moisture monitoring: Low cost tools and methods. National Sustainable Agriculture Information Service. ATTA, http://www.attra.ncat.org/attra-pub/soil_moisture.html, pp. 1–12 (November 16th, 2014).

163. Moshia, M. S., Khosla, R. Davis, J. G., Westfall, D., Reich, R. (2010). Precision Manure Management across Site-Specific Management zones. International Annual Meetings of American Society of Agronomy, Long Beach, California, USA, pp. 1–8.

164. Moulin, A., Derson, D., McLauren, D., Grant, C. (2003). Spatial variability of Soil Fertility and identification of Management Zones on Hummocky Terrain. Brandon Research Centre, Agriculture Canada, Manitoba, Canada, http://www.umanitoba.ca/faculties/afs/MAC_proceedings/2002/pdf/P4.pdf, pp. 1–7 (October 11th, 2014).

165. Mugisha, S., Huising, J. (2008). Optimal resolution for large-scale vegetation mapping using air-borne multi-spectral data. http://www.isprs.org/proceedings/XXXIV/6-W6/papers/mugisha_huis_opt.pdf, pp. 1–10 (November 27th, 2014).

166. Mulla, D. J. (2013). Twenty five years of Remote Sensing in Precision Agriculture: Key advances and remaining knowledge. Biosystems Engineering 114: 358–371.

167. Murguia, J. (2014). UAVS: The key for Precision Agriculture. Precision Farming Dealer, http://www.precisionfarmingdealer.com/content/uavs-key-precision-agriculture, pp. 1–3 (December 5th, 2014).

168. Mzuku, M., Khosla, R., Reich, R., Inman, D., Smith, F., Macdonald (2005). Spatial variability of measured Soil properties across Site-Specific Management zones. Soil Science Society of America Journal 69: 1572–1579.

169. NASA (2013). Soyuz/Resurs launch. National Aeronautics and Space Agency, USA, http://www.nasaspaceflight.com/2013/06/soyz-2–1b-resurs-P-launch.html, pp. 1–4 (December 2014).

170. National Aeronautic and Space Agency 2014 NASA to launch soil moisture active satellite. http://www.agprofessional.com/news/NASA-to-launch-soil-moisture-active-satellite-274361731.html, pp. 1–8 (November 16th, 2014).

171. Natural Resources Canada (2014). Land observation satellites and sensors. Government of Canada. Canada, http://www.nrean.gc/earth-sciences/geomatics/satellites-imagery-air-photos/satellites-imagery-products/educational-resources-9375, pp. 1–7 (October 4th, 2014).

172. New Economist 2009 Artificial satellites are helping farmers boost crop yields. Science Technology, New Economist, http:// www.economist.com/node/14793411, pp. 1–4 (October 1st, 2014).

173. New Holland Agriculture 2014 PLM Precision Farming- Plm maps software. http://agriculture.newholland.com/uk/en/PLM/data/software/Pages/Mapping.aspx, pp. 1–3 (November 9th, 2014).

174. Noh, H., Zhang, Q. (2012). Shadow effect on Multi-spectral image for detection of Nitrogen deficiency in Corn. Computers in Agriculture 83: 52–57.

175. Oberti, R., Marchi, M., Tirelli, P., Calcante, A., Inti, M., Borghese, A. N. (2014). Automatic detection of Powdery mildew on grapevine leaves by image analysis: Optimal view-angle range to increase the sensitivity. Computers and Electronics in Agriculture 104: 1–8.

176. Oki, K., Mizoguchi, M., Noborio, K., Yoshida, K., Osawa, K., Soiochi, O. (2011). Accuracy comparison of cabbage estimated from Remotely sensed imagery using an unmixing method. Computers and Electronics in Agriculture 79:30–35.

177. Okwu-Delunzu, V. U., Enete, C., Abubaker, A. S., Laundi, A. S. (2013). Monitoring gully of Enugu State South-eastern Nigeria, using Remote sensing. Proceedings SPIE 8887. Remote Sensing for Agriculture, Ecosystems and Hydrology 15: 1–13 doi: 10.1117/12.2035967.

178. Papadopoulos, A. Papadopoulos, F., Tziachris, P., Metaxa, I., Iatrou, M. (2014). Site Specific agricultural Soil Management with the use of new technologies. Global NEST Journal 16: 59–67.

179. Pena, R., Rubiano, Y., Peña, A., Chaves, B. (2009). Variabilidad especial de los atributos de la capa arable de un Inceptisol del piedemonte de la Cordillera Oriental (Casanare, Colombia). Agronomy de Colombia 27, 111–120.

180. Pena-Barragan, J. M., Lopez-Granodos, F., Jurado-Exposito, M., Garcia-Torres, I. (2007). Mapping Radolfia segetum patches in Sunflower crop using Remote sensing. Weed Research 47: 164–172.

181. Periera, F. M. B., Milon, D. M. B. P., Filho, E. R. P., Venancio, A. L., Russo, M. T., Cardinelli, M. C., Martins, P. K., Fetias-Ashia, F. (2011). Laser-induced fluorescence imaging method to monitor citrus greening disease. Computers and Electronics in Agriculture 79: 90–93.

182. Phillips, A. J., Newlands, N. A., Liang, S. H., Ellert, B. H. (2014). Integrated sensing of Soil moisture at the field scale: Measuring, modeling and sharing for improved Agricultural decision support. Computers and Electronics in agriculture 107: 73–88.

183. Pixel Mapping Inc., (2014). Photogrammetry and Geospatial mapping Consulting. Http://www.pixel-mapping.com, pp. 1–3 (December 19th, 2014).

184. Prabhakar, M., Prasad, Y. G., Thirupathi, G., Sreedevi, G., Dharajoti, B., Venkateswarulu, B. (2011). Use of ground based hyperspectral remote sensing for detection of stress in cotton caused by leaf hopper (Hemiptera: Cicadellidae). Computers and Electronics in Agriculture 79:189–196.

185. Prakash, N. R., Kumar, D., Nandan, K. (2012). An Autonomous Vehicle for Farming using GPS. International Journal of Electronics and Computer Science Engineering 7: 1695–1697.

186. Precision Seeding Solutions Inc., (2014). Precision Planting in Australia. Precision Seeding Solutions Inc., Premer, New South Wales, Australia, http://www.precision-seedingsolutions.com.au/contact/, pp. 1–7 (November 23rd, 2014).

187. Precision Soil Management LLC (2014). The prescription for more profitable yields. http://www.precisionsoil.com/ ourprocess/php, pp. 1–4 (October 3rd, 2014).

188. Qin, J., Burks, T. F. Ritenour, M. A., Bonn, W. G. (2009). Detection of Citrus Canker using Hyperspectral reflectance imaging with Special information divergence. Journal of Food Engineering 93:183–191.

189. Qin, Z., Zhang, M. (2005). Detection of Rice sheath blight for in-season disease management using multispectral remote sensing. International journal of Applied Earth observation using Multispectral Remote sensing. 7: 115–126.

190. Rama Rao, N. (2008). Development of a crop-specific spectral library and discrimination of various agricultural crop varieties using hyperspectral imagery. International Journal of Remote Sensing 29: 133–144.

191. Rautianen, M. (2005). The spectral signature of Coniferous forests: The role of stand structure and Leaf Area index. Faculty of Agriculture and Forestry, University of Helsinki, Helsinki, Finland, Dissertation, pp. 48.

192. Riggs, G. (2001). Measuring Near infra-Red spectral reflectance changes from water stressed conifer stands with AIS-2.

193. Robinson (2007). GPS, GIS, V R and Remote Sensing technologies continuing to evolve. South-western Farm Press 34:12–14.

194. RoboFlight Inc., 2014 Aerial Precision Ag acquired by RoboFlight Systems. Precision Farming Dealer, http://www.precisionfarmingdealer.com/content/aerial-precision-ag-acquired-roboflight-systems., pp. 1–5 (November 28th, 2014).

195. Rossell, R. A. V. Adamchuk, V. I. Sudduth, K. A. McKenzie, N. J., Lobsey, C. (2011). Proximal soil sensing. An effective approach for soil measurements in Space and time. Advances in Agronomy 113:237–282.

196. Rossell, R. A. V., Walvoort, D. J. J., McBrattney, L. J., Janik, L. J., Skemjstad, J. O. (2006). Visible, infrared, mid infrared or combined diffuse reflectance spectroscopy for simultaneous assessment of various soil properties. Geoderma 131:59–75.

197. Rotbart, N., Schmilovitch, Z., Cohe, Y., Alchanatis, V., Erel, R., Ignat, T., Schenderey, C, Dag, A., Yermiyahu, U. (2012). Estimating olive leaf nitrogen concentration using visible and near-infrared spectral reflectance. Biosystems Engineering 114: 426–434.

198. Rovira-Mas, F., Zhang, Q., Reid, J. F. (2008). Stereo vision three-dimensional terrain maps for precision agriculture. Computers and Electronics in Agriculture 60: 133–143.

199. Rub, G., Kruse, R., Schnieider, M. (2010). A clustering approach for management delineation in Precision Agriculture. fuzzy.cs.uni-magdeburg.de/aigaion/index.php/publications/show/772, pp. 14 (October 10th, 2014).

200. Ruffner, K. C. (1995). CORONA: Americas first satellite program. Central Intelligence Agency, Washington, D. C. USA, pp. 360.

201. Rx-VRT Providers LLC, (2014). Variable rate mapping – Precision Farming for the 21st century. http://www.rxvrt.com/index.html, pp. 106 (October 26th, 2014).

202. Sankaran, S., Mishra, A., Maja, J. M., Ehsani, R. (2011). Visible near infrared spectroscopy for detection of Huanglongbing in Citrus orchards. Computers and Electronics in Agriculture 77: 127–134.

203. Santangelo, T., Lorenzo, R. D., Loggia, G. L., Maltese, A. (2013). On the relationship between some production parameters and a vegetation index in Viticulture. SPIE proceedings. 8887 http://dx.doi.org/10.1117/12.2029805, pp. 1 (October 23rd, 2014).

204. Satellite Imaging Corporation, (2014a). SPOT-5 Satellite sensor. http://www.satimagingcorp.com/satellite-sensors/spot-6/pp. 1–5 (October 5th, 2014).

205. Satellite Imaging Corporation, (2014b). IKONOS Satellite Sensor (0.82 m). http://www.satimagingcorp.com/satellite-sensors/ikonos/, pp. 1–4 (October 5th, 2014).

206. Satellite Imaging Corporation, (2014c). Quickbird Satellite Sensor (0.65 m). http://www.satimagingcorp.com/satellite-sensors/ quickbird/, pp. 1–6 (October 5th, 2014).

207. Satellite imaging Corporation (2014d). Natural Resources. http://www.satimaging.com/applications/natural-resources/, pp. 1–3 (October 1st, 2014).

208. Satellite Imaging Corporation (2014e). Agriculture. http://www.satimagingcorp.com/applications/natural-resources/agriculture/, pp. 1–5 (November 8th, 2014).

209. Satellite Imaging Corporation (2014f). AgroWatch: Tree Grading maps. http://www.satimagingcorp.com/application/natural-resources/agriculture/agriculgture-treegrading/maps.html, pp. 1–3 (October 1st, 2014).

210. Satellite Imaging Corporation 2014 g Forestry. http://www.satimagingcorp.com/applications/natural -resources/forestry/, pp. 1–3 (October 1st, 2014).

211. Satellite Imaging Corporation (2014h). Satellite Images for Environmental Monitoring. http://www.satimagingcorp.com/application/environmental-impact-studies/, pp. 1–3 October 1st 2014).

212. Satellite Imaging Corporation (2014i). Global Climate changes and satellite image data. http://www.satimaging.com/application/environmental/-impact-studies/global-warming, pp. 1–3 (October 1s, 2014).

213. Satellite Imaging Corporation (2014j). Agrowatch: Green Vegetation index. http://www.satimaging corp.com/applications/natural/natural-reasons/agriculture/agriculture-green-vegetation-index-es.html, pp. 1–3 (October 1st, 2014).

214. Scanex, (2014). Application of satellite imagery in Agriculture. http://www.scanex.ru/en/monitoring/default.asp?sub-farming&it-index, pp. 1–3 (December 3rd, 2014).

215. Schearer, S. A., Higgins, S. G., McNeill, S. G., Watkins, R. I., Barnhisel, J. C., Doyle, J. H., Fulton, J. P. (1997). Data filtering and Correction techniques for Generating Yield maps from Multiple-Combine harvesting systems. http://www.bae.uky.edu/precag/precisionAg/reports/Date-filtering/yieldsite.htm, pp. 1–7 (April 25th, 2014).

216. Schellberg, J., Hill, M. J., Gerhards, R., Rothmund, M., Braun, M. (2008). Precision agriculture on grassland: Application, perspectives and Constraints. European Journal of Agronomy 29: 59–71.

217. Schepers, J. S., Schlemmer, M. R., Ferguson, R. B. (2000). Site-specific considerations for Phosphorus. Journal of Environmental Quality 29:125–130.

218. Schepers, A. R., Shanahan, J. F., Liebig, M. K., Schepers, J. S., Johnson, S. H., Luchiari, A.Jr, 2004. Appropriateness of management zones for characterizing spatial variability of soil properties and irrigated corn yields across years. Agronomy Journal, 96:195–203.

219. Schirrmann, M., Domsch, H., 2011. Sampling procedure simulating on-the-go sensing for Soil nutrients. Journal of Plant Nutrition and Soil Science 174: 333–343.

220. Schirmann, M., Gebbers, R., Kramer, E., Seidel, J. (2011). Soil pH Mapping with an On-The-GO Sensor. Sensors 11: 573–598 doi: 10.3390/s110100573 (October 23rd, 2014).

221. Schuman, A. W. (2010). Precise placement and variable-rate fertilizer application technologies for horticultural crops. HorTechnology 20:34–40.

222. Seelan, S. K., Laguette, G. M., Casady, G. M., Seielstad, G. A. (2003). Remote sensing applications for Precision Agriculture: A learning community approach. Remote Sensing of Environment 88:157–169.

223. Seney, G. B., Lyon, J. G., Ward, A. D., Nokes, S. E. (2000). Using high Spatial resolution multispectral data to classify corn and soybean crops. Photogrammetric Engineering and Remote Sensing 66: 319–327.

224. Shanahan, J. F., Kitchen, N. R., Raun, W. R., Schepers, J. S. (2007). Responsive in-season Nitrogen Management for Cereals. Computers and Electronics in Agriculture 61:51–62.

225. Shaver, T., Ferguson, R., Hegert, G., Shapiro, C., Wortmann, C. (2012). Nutrient Management: Now and in the Future. http://www.cpc.unl.edu/includes2012/pdf/NutrientManagementNowfuture.pdf?exampleusertabel-your+Name.html, pp. 1–4 (July 22nd, 2012).

226. Shrivastava, R. J., Gebelein, J. L. (2007). Land cover classifications and economic assessment of Citrus groves using Remote sensing. ISPRS Journal of Photogrammetry and Remote Sensing 61:341–353.

227. Sibley, K. J. (2008). Development and use of an automated on-the-go soil nitrate mapping system. Wageningen University, Wageningen, the Netherlands PhD Thesis, pp. 172 http://www.edpot.wur.nl/121988/, pp. 1–87, 9 April 23rd, 2014).

228. Sibley, K. J., Adsett, F. F., Struik, P. C. 2008a. An On-the-go soil sampler for an automated soil nitrate mapping system. Transactions of the ASABE 51: 1895–1904.

229. Sibley, K. J, Astatkie, T., Brewster, G., Struik, P. C., Adsett, J. F., Pruski, K. (2009). Field scale validation of an automated soil nitrate extraction and measurement system. Precision Agriculture 10:162–174.

230. Sibley, K. J., Astatkie, T., Lada, R., Struik, P. C., Adsett, J. F., Pruski, K. 2008b. Using an automated on-the-go soil nitrate mapping system to investigate plant and soil nitrate responses in wheat and carrot production systems. http://www.edepot.wur.nl/121988, pp. 105–119 (April 25th, 2014).

231. Sinfield, J. V., Fageman, D. F., Colic, O. (2009). Evaluation of sensing technologies for on-the-go detection of micro-nutrients in cultivated soils. Computers and Electronics in Agriculture 70:1–18.

232. Singh, M., Sharma, A., Singh, and Sharma 2012. Investigations into Yield monitoring sensor installed on indigenous grain combine harvester. Proceedings of 6th International Sensing Technology, pp. 46–51.

233. Smadar, M. (2012). Israel's Agriculture. Ministry of Agriculture and Rural Development. Beit Dagan, Israel, pp. 1–11.

234. Smith, A. M., Blackshaw, R. E. (2003). Weed-Crop discrimination using Remote Sensing: A detached leaf experiment. Weed Technology 17: 811–820.

235. Song, X., Wang, J., Huang, w. 2010 Winter wheat at growth and grain protein uniformity monitoring through remotely sensed data. Proceedings of SPIE. doi: 10.1117/12.865162 (October 25th, 2014).

236. Soria-Ruiz, J., Fernandez-Ordonez, Y. (2011). Maize Crop yield map production and update using Remote sensing. Progress in Electromagnetics Research Proceedings, Marrakesh, Morocco, pp. 995–998.

237. Soria-Ruiz, J., Fernandez-Ordonez, Y., McNairn, H. (2009). Corn monitoring and crop yield using optical and microwave remote sensing. In: GeoScience and Remote Sensing. Ho, P. P. (Ed.). In-Tech Publishing, Croatia, pp. 405–419.

238. Soria-Ruiz, J., Fernandez-Ordonez, Y., McNairn, H., Bugden-Storie, J. (2007). Corn monitoring and crop yield using optical and Radarset-2 images. Geoscience and Remote Sensing symposium 2007. IGARSS., Barcelona, Spain., pp. 65–72.

239. Srivastava, A. K., Singh and Das, S. N. (2010). Nutrient optima-based productivity zonality delineation in citrus orchards of Northeast India. Proceedings of World Congress of Soil Science, Brisbane, Australia, pp. 136–138.

240. Stafford, J. V., Lark, R. M., Bolam, H. C., 1998. Using Yield Maps to Regionalize Fields into Potential Management Units. *In*: Robert, P. C., Rust, R. H., Larson, W. E. (Eds.), Proceedings of the 4th International Conference on Precision Agriculture. ASA, CSSA, SSSA, Madison, WI, USA, p. 225–237.

241. Stafford, J. A., Werner, A. (2003). Precision Agriculture Wageningen Academic Publishers, Wageningen, The Netherlands, pp. 783.

242. Stanford University 2014 Satellite data provide a new way to monitor groundwater in Agricultural regions. http://news.stanford.edu/news/2010/december/agu-water-imaging-121310.html.

243. Stiers, J. (2013). From Horses to Hands-Free with Satellite-guided tractors. http://ilfbpartners.com/farm/from-horses-to-hands-free-with-satellite-guided-tractors/, pp. 1–8 (April 15th, 2015).

244. Sudduth, K. A., Chung, S., Andrade-Sanchez, P., Upadhyaya, S. K. (2008). Field comparison of two prototype soil strength profile sensors. Computers and Electronics in Agriculture 61: 20–31.

245. Sudduth, K. A., Drummond, S. T. (2007). Yield Editors: Software for removing errors from yield maps. Agronomy Journal 99:1471–1482.

246. Sudduth, K. A., Drummond, S. T., Kitchen, N. R. (2001). Accuracy issues in Electromagnetic induction sensing of soil Electrical conductivity for Precision Agriculture. Computers and electronics in agriculture 31: 239–264.

247. Sylvester-Bradley, R., Lord, E., Sparkes, D. L., Scot, R. K., Wiltshire, J. J., Orson, J. (2006). An analysis of the potential of Precision Farming in Northern Europe. Soil use and Management 15: 1–8 (October 2nd, 2014).

248. Taylor, J. A., Short, M., McBratney, A. B., Wilson, J. (2010). Comparing the ability of multiple Soil sensors to predict soil properties in a Scottish Potato production system. Proximal Soil Sensing. Rossell, R. A. V. (Ed.). Progress in Soil Science 1. Springer Science Business Media B. V., pp. 413–422.

249. Thessler, S., Kooistra, L., Teye, F., Huiti, H., Bregi, A. K. (2011). Geosensors to support Crop production: Current applications and User requirements. Sensors 11: 6656–6684.

250. Thylen, L., Gilbertsson, M., Rosenthal, T., Wrenn, S. (2014). An on-line Protein sensor- From Research to product. Htttp://www.zeltex.com/online-proteinsensor.doc, pp. 26 (November 7th, 2014).

251. Trimble (2014a). Trimble Agriculture http://www.trimble.com/agriculture/index. aspx, pp. 1–3 (October 27th, 2014).

252. Trimble (2014b). Soil Information systems http://www.trimble.com/Agriculture/sis. aspx, pp. 1–2 (October 27th, 2014).

253. Trimble (2014c). Irrigate-IQ Precision irrigation Solution. http://www.trimble.com/Agriculture/irrigate-iq.aspx, pp. 1–4 (October 28th, 2014).

254. Trimble (2014d). Sugarcane Planting solutions. http://www.trimble.com/agriculture/sugarcane-planting.aspx., pp. 1–3 (October 27th, 2014).

255. Trimble (2014e). Field-IQ Crop Input control system. http://www.trimble.com/Agriculture/field-iq.aspx, pp. 1–3 (October 28th, 2014).

256. Trimble (2014f). Sugarcane Nutrient Management solutions. http://www.trimble.com/Agriculture/sugarcane-nutrient.aspx, pp. 1–4 (October 28th, 2014).

257. Trimble (2014g). Vine and Tree Planting. http://www.trimble.com/Agriculture/viticulture-planting.aspx, pp. 1–4 (October 28th, 2014).

258. Trimble (2014h). Canopy Management Solutions. http://www.trimble.com/Agriculture/viticulture-planting.aspx, pp. 1–3 (October 28th, 2014).

259. Trimble (2014i). Nutrient and Pest management Solutions. http://www.trimble.com/Agriculture/viticulture-nutrient-pest.aspx, pp. 1–4 (October 28th, 2014).

260. Trout, T. J., Johnson, L. F., Gortung, J. (2008). Remote sensing of canopy cover in Horticultural crops. HortScience 43: 333–337.

261. University of Nebraska-Lincoln CropWatch (2014). yield Monitoring and mapping. http://cropwatch.unl.edu, pp. 18 (November 7th, 2014).

262. UNCSAM-MARDI (2012). Background on Appropriate Precision Farming for Enhancing the Sustainability of Rice production. United Nations Centre for Sustainable Agricultural Mechanization (UNCSAM) and Malaysian Agricultural Research and Development Institute, Kaulalumpur, Malaysia, pp. 1–9 http://un-csam.org/publication/PreRiceFarm.pdf.

263. Upadhyaya, S., Texiera, A. (2011). Sensors for information gathering during Precision Farming. http://www.docstoc.com/docs/34311781/Sensors-for-Information-Gathering, pp. 1–9.

264. USDA-ARS (2005). Precision Farming and Remote Sensing. United States Department of Agriculture-Agricultural Research Service http://www.ars.usda.gov/News/docs.htm?docid=9717pp 1–8 (November 27th, 2014).

265. USDA-NASS (2009). History of Remote sensing for Crop Acreage. United States Department of Agriculture, Beltsville, USA, pp. 1–4 http://nass.usda.gov/Surveys/Remotely_Sensed_Data_Crop_Acreage/index.asp (October 6th, 2014).

266. USGA (2012). Topographic mapping United States Geological Survey, Washington, USA, http://pubs.usgs.gov/gip/topomapping/topo.html., pp. 1–12 (December 22nd, 2014).

267. USGS (2013). Deforestation Watch: First imagery from Landsat 8. United States Department of Geological survey. Washington D. C. http://www.deforestationwatch. wordpress.com/2013/06/06/first-imagery-from-landsat-8-good-news-for-the-detection-of-deforestation-but-what-are the-implications. Html p 1–10 (November 30th, 2014) p.

268. USGS (2014). Landsat and Water: Using Space to advance Resource solutions. http://www.usgs.gov/blogs/features/usgs_top_story/landsat-and-water-using-space-to-advance-resource-solutions/, pp. 1–8, 9 December 22nd, 2014).

269. Vega-Pro (2014). Vega constellation: Satellite based service for vegetation monitoring. http://pro-vega.ru/eng/.html, pp. 1–8 (December 3rd, 2014).

270. Vellidis, G., Tucker, M., Perry, C., Kvein, C., Bednarz, C. (2008). A real time wireless smart array for scheduling irrigation. Computers and Electronics in agriculture 61:44–50.

271. Veris Technologies Inc. (2014). Organic matter Mapper from Veris. http://veristech. com (October 25th, 2014).

272. Vina, A., Gitelson, A. A., Rundquist, D. C., Keydan, G., Leavitt, B., Schepers, J. (2004). Monitoring maize *(Zea mays)* Phenology with Remote sensing. Agronomy Journal 96: 1139–1147.

273. Vrindts, E., Mouazen, A. M., Reyniers, M., Maertens, K., Maleki, M. R., Ramon, H., de Baerdemaeker, J., 2005. Management zones based on correlation between soil compaction, yield and crop data. Biosystems Engineering 92 (4):419–428.

274. Wetterlind, J. (2009). Improved Farm Soil Mapping using Near Infrared Reflection Spectroscopy. Swedish University of Agricultural Sciences, Skara, Sweden Doctoral Thesis, pp. 148.

275. Whelan, B., Taylor, J. (2013). Precision Agriculture: for Grain production systems. CSIRO Publishing, Australia, pp. 199.

276. Xia, Y., Guru, S., Vankirk, J. (2009). pestMapper-A internet-based software tool for reporting and mapping biological invasions and other Geographical and Temporal events. Computers and Electronics in Agriculture 69:209–212.

277. Xiaoyu, S., Bei, C., Guijun, Y., Haikuan, F. (2014). Comparison of Winter wheat growth with multi-temporal Remote sensing imagery. Proceedings of Conference Series on Earth and Environmental Science DOI: 10.1088/1755–1315/17/1/012044, pp. 1–7 (October 25th, 2014).

278. Yang, C. (2012). Using Remote sensing and GIS for detecting and mapping invasive weeds in Riparian and Wetland Ecosystems. USDA-ARS, Weslaco, Texas, Internal Report, pp. 1–26.

279. Yang, Z., Mueller, R., Crow, W. (2014). US National Crop land soil moisture monitoring using SMAP. http://www.nass.usda.gov/research/cropland/docsyangIGARS-SI3SMAP.pdf., pp. 1–17 (December 22nd, 2014).

280. Yasrebi, J., Saffari, M., Fathi, H., Karimian, N., Emadi, M., Baghernejad, M. (2008). Spatial variability of Soil fertility properties for Precision Agriculture in Southern Iran. Journal of Applied Science 8: 1642–1650.

281. Yoshino, K., Ishioka, Y. (2005). Guidelines for soil conservation towards integrated basin management for sustainable development: a new approach based on the assessment of soil loss risk using remote sensing and GIS. Paddy and Water Environment 3: 234–247.

282. Zak, A. (2014). Kosmos Ska satellite to monitor Agricultural development from space. http://www.russianspaceweb.com/ 2015.html, pp. 1–12 (December 3rd, 2014).

283. Zhang, Q., Pierce, F. J. (2013). Agricultural automation: Fundamentals and Practices. CRC Press, Boca Raton, Florida, USA, pp. 422.

284. Zhang, J., Pu, R., Yuan, L., Wang, J., Huang, W., Yang, G. (2014). Monitoring powdery Mildew of winter wheat by using moderate resolution multi-temporal satellite imagery. PLoS ONE 9(4): e93107. doi: 10.1371/journal.pone.0093107, pp. 1–7 (November 23rd, 2014).

285. Zhang, M., Qin, Z., Liu, X., Ustin, S. L. (2003). Detection of stress in tomatoes induced by late blight disease in California, USA, using hyperspectral remote sensing. International Journal of applied Earth Observation and Geoinformation. 4: 295–310.

286. Zhong, W. X., Guo-Shun, L., Hong-Chuo, H., Zhen-Hai, W., Qing-Hua, L., Xu-Feng, L., Wei-Hong, H., Yan-Tao, L. (2009). Determination of Management zones for a Tobacco field based on Soil fertility. Computers and Electronics in Agriculture 65: 168–175.

CHAPTER 5

PUSH BUTTON AGRICULTURE: SUMMARY AND FUTURE COURSE

CONTENTS

The general perception among agricultural scientists, farm holders and to certain extent policy makers is that, techniques such as robotics, drones and satellite-guided farm operations, will creep into all farm activities beginning with land preparation, seeding, manure supply, irrigation, harvesting and until post-harvest grain processing. In other words, 'Push Button Agriculture' a concept touted here will proliferate into almost every nook and corner of agrarian zones. A revolution in farm activities is perhaps at its rudiment that promises to flourish and help human beings in reducing drudgery and make hard tasks look easier in the crop fields. It supposedly improvises on crop productivity, allows farmers to fix higher yield goals and at the same time restricts use of high amounts of fertilizers and other chemicals. It surely assists in reducing greenhouse gas emissions and climate change. Historically, we have progressively mechanized and energized our farms to replace involvement of human energy and drudgery in crop fields. Human physiological limits and other pre-occupation necessitate engineering research and development of robots, drones and computer software capable of guiding them to conduct variety of tasks. This process perhaps needs to be continued to its eventual fruition, when human labor and time in crop fields will be least. Push Button Agriculture (PBA), for now seems an apt option. It is expected to reduce human drudgery in farms to further lower level than previously known. At the same time, it may offer greater accuracy of farm operations, economize on inputs and allow us to achieve higher grain productivity.

The need to improve agricultural crop productivity stays intact and continues to demand greater attention in future too. We are now left with warning that by 2050 human population that is expected to reach 9 billion demands proportionate increase in grain harvests. The forecasts indicate a need for over 25% increase in crop productivity relative to present levels. We may adopt several techniques to achieve it either gradually in programed fashion or in spurts and lax periods. Regarding agricultural methods envisaged, currently, precision techniques that involve intensive study of soil fertility variations; variable rate application of fertilizers/ water at rates that match crop growth is spreading rapidly among farmers worldwide (Stafford, 2005; Lowenberg-DeBoer, 2006; Van Henten et al., 2009; Khosla, 2010; Krishna, 2012). Robotics, drones and satellite mediation are among prime factors that could aid better adoption of Precision Techniques. In other words, Push button techniques could be popular choice in order to induce rapid spread of Precision Farming methods and enhance productivity and economic gains.

5.1 ROBOTS: THEIR FUTURE IN GLOBAL AGRICULTURE

Robots are beginning to appear in farming zones. As time passes by, they may even dominate all sorts of farm activities related to crop production, grain/forage processing and transport. Robots as a class of gadgets are amenable to be operated in different geographical regions. They only need modifications to suit a particular terrain and agricultural practices that need to be accomplished. So, sooner or a bit later, we will trace agricultural robots all over the agrarian zones of the world. Future for Agricultural robotics is promising. They will be sought by farmers. Grift et al. (2013) opine that human urge to mechanize and automate the world to serve beneficially, be it agricultural crop production or any other aspect of life, is just unstoppable. Human effort to transfer intelligence, dexterity and ability for tasks will ultimately succeed. Almost all agricultural operations amenable for treatment by robots will be attempted. They will progress to the extent that they can reduce human drudgery and be economically efficient.

Actually, 'Agriculture was left behind by Modern Technology until now.' There has been lag even in infusing use of broad band and mobile based operations in rural areas that are predominately agrarian (Precision Farming Dealer, 2014a). Of course, currently there are villages in North America, wherein, almost all farmers adopt software for soil mapping and variable-rate methods, and use mobile based tracking systems of their vehicle. However, we are at the threshold of a new farming revolution full of robotic farmers and other autonomous vehicles, pre-programed to perform tasks accurately. Blackmore (2014) forecasts that in future, small sized robotic tillers that are less intrusive, but capable of accurate shallow tillage at low energy costs may flourish in many farming zones. Most importantly, such small robots may have less deleterious effects on soil. Such small robots do not cause soil compaction that occurs when heavy tractors with 30–40 tines/discs move on the fields during traditional farming practices (Winsor, 2014).

Robots influence various walks of human life, including agriculture that spreads as small farms and large expanses in almost 33% of global land surface. Panin (2014) opines that we have to introduce robots in large scale into farming first, then watch a while and realize its effects on human welfare and earth's environment and evolutionary processes. Of course, computer-based simulations and educated guesses are always possible.

One of the basic questions asked regarding agricultural robots is that are they good enough to replace human labor in farms? In a larger context, will they be able to deliver, in terms of higher food grain output, will they enhance farmer's profits, will they be able to perform tasks more easily and yet protect earth's environment (Crow, 2012). Blackmore (2014) opines that robots in farms should be selective, accurate and economically efficient for them to be successful in different agrarian zones. Selective and swift operation based on computer decision support is almost essential in accomplishing various tasks during precision farming. According to engineers dealing with Robotics at the Massachusetts of Institute of Technology, Cambridge, USA, despite good efforts to improve robot's abilities and dexterity in performing various tasks, it looks as though even simplest of the tasks performed with ease by humans seems difficult for a robot to accomplish. Not a good situation to report. However, with massive investment of capital, human intellect and several more engineers taking up research on robots in different aspects of human activity, we may develop robots of higher ability (Panin, 2014). In some of the developing nations and socialistic group of countries like Russia and others in Eastern Europe, plus those in Asia, it seems cost of robots and easy access is a problem that will have to be solved (Panin, 2014). As an alternative to delays due to difficulties in design of efficient robots that operate swiftly with excellent dexterity, Zhang and Pierce (2013) suggest that we should first focus robotics towards problems that could be attended with greater ease and accuracy. For example, if robotic tractors and variable nutrient management is amenable, it should be attempted first and whatever advantages possible should be accrued. Further, we may note that all of the agronomic procedures adopted during production of a particular crop species need not be conducted through robots. A step further, crop species that need robotic help need to be focused first.

5.1.1 ROBOTS AND CROP PRODUCTION STRATEGIES

Crop production occurs both under high-input intensive and at low-input subsistence levels. Robots that match the situation are needed. Agricultural enterprises in Brazil and Argentina are relatively large, at over 10,000 ha per farm. They are mostly supported by farm companies and consortia. In fact, agribusiness is a priority item in this region. Here, robots that perform hard tasks and those reducing human labor and minimizing farm drudgery are perhaps most useful. Robots connected to GPS-RTK and able to add accuracy to farm operation are apt for large farms of Cerrados and Pampas. During recent years, precision agriculture that includes variable-rate technology is becoming a useful aspect in cereal/legume fields of South America. Robotic tractors and VRT, together are expected to spread into greater areas of Brazilian Cereal/soybean intercropping zones and wheat/soybean belts of Argentina (Godoy et al., 2012). Robots with yield maps, GPS guided VRT applicators and computer support systems are said to help Brazilian farmers in terms of inputs needed, revising yield goals and enhancing economic gains. Robots along with drones and/or satellite guidance could become common in the South American agrarian regions. Small robots, used in swarms, with inter-vehicle communication and satellite controls could again revolutionize farm production strategies in this region. Such an inference holds true with various other regions that practice expansive agriculture with large farming units, for example in Northern Great Plains.

Lack of agricultural labor is not a constraint, if swarms of robots could be used to intensify the crop production zones. Timely-ness is not a factor because robots could be commissioned instantaneously, if sufficient reason is available. Robots could help farmers to achieve well-tuned and programed intensification using appropriate soil testing and computer-based decision support systems. Robots impart accuracy during application of seeds, fertilizers, water and pesticide. Sensors and computer programs located on robots provide data incessantly about plant nutrient status. Therefore, farmers can revise input schedules and yield goals appropriately. Such flexibilities are unheard of in traditional intensive farming systems. Factors such as fertile soil, weather pattern, availability of robotic tractors, small robotic weeders, inter-culture robots, and pesticide applicators play a vital role during intensification of cropping systems. Since robots impart a degree of accuracy during fertilizer supply and chemical sprays, climate change effects could be reduced enormously. Robots may actually allow intensification of cropping systems in a systematic fashion.

Robots are adaptable to variety of farming situations. Regions accustomed to low input sustainable farming too may become host to robot usage. Robots that are proportionately low cost, efficient in terms of fuel usage, mounted with easily affordable software and accessories are required. Farmers have to be selective in opting for robots and farm operations that could be conducted. In some low or moderately intensive zones, farmers could still adopt large GPS guided tractors for plowing, planting and fertilizer placement. They may have to hire usage of such large farm robots through cooperatives. Development of Robots to apply basal dosage of fertilizers and plant seeds rapidly, to take advantage of scanty precipitation is a clear possibility. In Northern plains of India, subsistence farming is fragmented and farm holdings are small. Yet, they are able to conduct initial trials using GPS-RTK connected tractors through co-operatives.

Farmers in Sahel depend on onset of rains to plant seeds. Planting of large tracts has to be accomplished rapidly to take best advantage of short spells of rains. Seed germination and establishment of seedlings is excessively dependent on moisture still left in soil after first rains. Failure of seed germination and gaps in crop stand are just too common. Droughts too affect crop stand. Farmers in Sahel will be at advantage if they can use robots for repeated sowing. Robotic planters can get into fields any time, unlike farm workers. Robots could be most useful when rapid re-seeding is necessary in Sahel immediately after rains. Utilization of rain water could be efficient. Small robotic planters can handle more area. One possible example is an 'Autonomous Micro Planter' such as *'Prospero'* that places seed plus some organic matter at each spot as required, to help the Sahelian farmers (Baichtal, 2011; Morris, 2011; Trossen Robotics

Community, 2014). Some modifications required are like broad and flat tires that do not dig into sand dunes. Robots to harvest pearl millet, sorghum and cowpea are also required in Sahelian West Africa. Robots that help in planting agroforestry seedlings should be a boon to farmers and foresters who intend planting millions of seedlings of trees each year in the dryland belts of Sahel.

5.1.2 ROBOTS AND NO-TILLAGE

We have already seen in previous chapters that autonomous tractors, or semi-autonomous versions with GPS-RTK connectivity and others make land tillage and ridge preparation easier than ever. Tillage is an easier and less costly proposition. Tillage could be attempted day and night in all-weather situations, if robotic vehicles are used. Yet, we may note that worldwide, it is 'No-tillage' systems that are getting increasingly preferred. About 23% of farm land is under No-tillage systems (Derpsch, 2004). Tillage is resorted only once in 3–4 years. However, no-tillage systems lead us to an endemic problem of weeds and volunteers that need to be eradicated, in time, before they start to dominate the crop canopy. Human labor needs to weed the field increase. Seasonal constraint of labor availability accentuates the problem and it also results in cost escalation. Hence, robots could be most useful and efficient replacers of human labor in no-tillage systems. Since timely weeding using a swarm of robots that ply in the inter-row spacing is a clear possibility in near future, they are expected to get popular. Mind you, robotic weeders could be as efficient as human labor, given that they are capable of both physical eradication of weed by uprooting and applying herbicides. It means each robot replaces human labor, often many (up to 20 workers) and reduces that much human drudgery in the field. A single farm technician can control several robots at a time. Plus, current research is focused on inter-robot connectivity and inter-phasing techniques that allow swarms of robots that could be pre-programed and coordinated (e.g., AgAnt, see Grift, 2007).

Precision farming depends to a certain extent on the intensity of soil sampling, demarcation of field into 'Management Zones' and accurate placement of inputs using robots. Robots have been touted as very useful items during precision farming. For example, there are robots that help in sampling surface and subsurface of soil profile, such as *AutoProbe* and *RapidProbe* (see Chapter 4). It therefore aids to prepare soil fertility maps and obtain digitized data for variable-rate inputs. Tillage robots such as large sized tractors are used, but at the same time many reports suggest that small robots in a swarm could accomplish the tillage without causing soil compactness and at lower cost. Robots have also been suggested to eradicate weeds in the inter-row space. Next, robots are more popular as vehicles to spray pesticides and place fertilizers either in liquid or granular form. Some of the advantages counted for rapid deployment of robots in farms are as follows:

- Robots used in agricultural farms are generally smaller than the earlier semi-autonomous tractors;
- Robots save on labor requirements. In some cases, such as Lettucebot, a single robot could replace 20 farm workers per day;
- Robots energized using electric batteries or solar power does not emit greenhouse gases;
- Robots emit low amounts of CO_2 even though energized by fossil fuel;
- Robots employed at right time could accomplish tasks such as fertilizer supply, irrigation, harvesting and improve farmer's profits;
- Robots do not cause wastage of inputs such as fertilizers or pesticides;
- Robots could be employed rapidly during precision farming.

Push Button Agriculture, as stated earlier includes a composite of ground robots and drones inter-connected to operate in an orchestrated fashion. This aspect needs careful testing and evaluation. We should note that, farmers can imbibe only standardized methods. They have no excheckr, and time for preliminary stage trials like those done in agricultural experimental stations. On the other hand, we should realize that these robots and drones are excellent instruments to keep track of field experiments, collect data and help researchers in rapidly assessing the crop genotypes, cropping systems and management methods. Routine data is needed both by Farm Scientists and Farmers, particularly regarding crop phenomics. In future, autonomous tractors, small robotic seed planter, and drones may find immense demand in farms and experimental stations, mainly to accomplish planting of multiple crop/genotype, regular phenotyping, applying pesticides at variable rates to crop genotypes, etc. (Blackmore et al., 2005; Blackmore, 2009; Gomez, 2013; Perry et al., 2012; Precision Farming Dealer, 2014c). Crop phenotyping reduces cost on skilled technicians during experimentation (see Perry et al., 2012). It offers scientists an immediate view of growth pattern in visual format not possible earlier. Farmers can sequentially observe crop as it grows through the stages and compare the performance of different cultivars sown in their farms using drone pictures. Computer monitors can allow them to regularly compare the growth rate and yield formation of different crops/genotypes. Therefore, decisions to pick a cultivar or a method would be more appropriate based on farmer's needs. In a nutshell, farm experiments done using Push Button techniques may well be economical, more accurate and show greater details to farmers/scientists. Influence about crop's performance or efficiency of a methodology could be validated more accurately.

5.1.3 ROBOTS, THEIR IMPACT ON LAND USE AND CLIMATE CHANGE FACTORS

Robots accomplish tasks programed by farmers. Farmers may try to mimic and repeat farm operations such as traditional deep plowing, turning of soil, ridging, supplying fertilizers at high rates based on blanket recommendations, and repeated pesticides sprays. Such an effort will induce greenhouse gas emissions and climate change effects, similar to that felt all along. Farmers will have to adopt Precision techniques, make imagery and collect digital data showing soil fertility variations. Robots with ability for variable-rate inputs will supply accurate quantities of nutrients in to soil phase. Robots that function taking commands from computer-decision support systems will reduce undue accumulation of nutrients/pesticides and herbicides, not otherwise. Robots are generally capable of accurate placement of fertilizers, and other chemicals. Therefore, it reduces chances of soil deterioration, seepage and contamination of ground water. However, a large tractor with GPS-RTK connectivity, if it still discs the field deep, turns the soil and exposes it to rapid oxidation and microbial activity, then CO_2 emissions are expected to be higher, immaterial whether it is a robotic tractor or not. However, reports suggest that, farmers may prefer small tractors and robots that just loosen surface soil. Robots are generally capable of planting under no-tillage systems. Such robots will automatically reduce CO_2 emission. Soil degradation and soil organic carbon loss would be reduced perceptibility. Over all, robotics will be efficient, only if precision techniques such as soil sampling, testing for nutrients, mapping, and variable rate supply based on accurate computer programs are adopted. We may have to acknowledge that robots take care of difficult tasks. Some of these aspects of farm robotics also reduce soil degradation.

Robots used in farms could be driven using petroleum fuel. In that case, CO_2 emissions are expected to be contributed by them. Robots, as they gain in popularity and usage, their number in farms will proliferate. There are farm methods envisaged, where in swarms of robots and autonomous tractors inter-linked through GPS and coordinated through control stations are expected to throng the farms. In fact, they are expected to dominate farming scenario, in due course. Hitherto, densities of auto-motives that depend on petroleum were relatively feeble in farms, excepting for tractors, a few sprayers or combines. In a way, farm machines and vehicles which emit CO_2 were few. Also, farmers may detest usage of robots in excessive number, if they clog or occupy more area in farms. As time passes by, we may also expect piles of junked robots and their wreckage across farm lands. Just like, we notice metal junk yards of passenger cars near cities. Each small robot, such as Rowbot, Hortibot, AgAnt, or Wall-Ye consigned to junk pile, may add 5–50 kg metal into farm environment. Many of them are not made of recyclable material or degradable wood-like material. In China, even subsistence farmers are getting exposed to drone technology. They are experimenting with low cost degradable wood/bamboo based drone. However, there are also robots designed and developed shrewdly, that use solar energy. They are equipped with solar panels on their body that provides for energy needed. To quote an example, Vitirover, is an autonomous weeder, a robot that has solar panels to generate energy. It operates efficiently as weeding machine in grapevines and several other plantations (see Plate 2.19; Vitirover, 2014). There are many models of robots that utilize captured electric energy that has no perceptible ill effects on environment.

5.1.4 ROBOTS AND FARM ECONOMICS

Robots to be used in farms need investment. Robots should be affordable and be economically advantageous. Therefore, tasks for which robotic machines are researched and designed should be selected carefully. Further, robots could be initially restricted to high value crops so that they are profitable. However, they will face tougher testing ground in low input farms, operated by not so well to do farmers. In fact, Cockburn-Price (2012) states that, unlike industries such as Aerospace, farm robotics is a low margin domain. Therefore, it is mandatory that new robots developed should be low cost, efficient in terms of fuel needs and yet turn out good work.

They say a Lettucebot, as it thins and /or harvests it reduces need for farm laborers. A single robot that moves at 2 mph in the lettuce fields of Southern California is equivalent to 20 human farm laborers per day. Lettucebots are sought in high numbers because yearly Lettuce production is worth 1.6 billion US$. Lettucebot ensures efficient harvest. Lettucebots operate in all weather conditions and are not affected by seasonal fluctuation noticed with human labor. Clearly, cost per unit area of lettuce field harvested is much less compared to human skilled laborers (Wiser, 2014).

Robots are also expected to throng agrarian zones that support specialty crops, such as fruit and nut plantations, flowers, nursery crops, and glass house crops. However, specialized robots need to be developed and tested before they go commercial. For example, in case of fruit trees, some of the robot designs tested still need improvement regarding sensors, ability of sensors to judge the ripeness of fruits, apt computer software that initiates signals for robot arms to pluck a fruit, etc. Often, intricate branching and canopy of fruit trees are impediments to rapid action by robots (English, 2013; Zkotala, 2013; Payne, 2013). Some of these obstacles need to be overcome. However, if standardized, we may note that robots in horticultural zones could economize on costs of labor, as well improve productivity of horticultural species (Vangioukas, 2013). Horticultural plantations too form large expanses in many regions of the world. A robot that serves to apply fertilizers automatically or one that irrigates or that which sprays pesticides/fungicides autonomously taking digitized data from satellite imagery or drones is a kind of boon to planters. Robots adaptable to each agrarian zone and crop species that flourish in the area are needed. It seems, in some farms, that produce fruits such as apples or citrus, robots built to identify, pluck and collect fruits were slightly slow and looked cumbersome in movements. A few designs meant high cost to farmers to buy the equipment. Yet, we could be optimistic that several of the plantations worldwide would eventually become host to several types of robots. Again, small robots inter-phased with drones capable of close-up imagery are perhaps still a best bet. They could use satellite imagery for coarse actions and depend on guidance from drone data for accurate actions. Overall, it should now be clear that robots would throng fruit plantations.

Lleo et al. (2009) point out that, robots could be of immense utility during identification, detachment and collection of fruits. Selection of ripe fruits using correct wave length band is the crux of the problem. The vision system of robots should be able to discriminate between green unripe regions/ fruits using chlorophyll pigment measurements and ripe yellow or orange carotenoid pigments. Robotic action seems rapid once the ripeness is judged correctly. Fruit picking robots do save immensely on human labor costs. Since they could be operated at short notice in all-weather conditions they may find acceptability, but have to be economically feasible.

5.1.5   INDUSTRIES THAT PRODUCE GROUND-BASED FARM ROBOTS

Several of the major tractor-producing industries have almost permanently shifted to production of GPS-RTK connected vehicles, which are meant for plowing and a few other farm operations. There is strong interest in using satellite controlled robotic tractors (i.e., steer-less, autonomous) in most of the intensive farming zones. There are already pilot trials and adoption of robotic tractors for plowing in Asia. Such GPS guided tractors are also popular among farm companies of Brazil and Argentina. The demand for GPS-RTK connected large autonomous tractors is still, small but in due course, it is expected to increase. We have to take note that, in addition to influence of robots on farm operations, productivity and fiscal gains to farmers; robot production industries and ancillaries are expected to proliferate. Forecasts indicate that as farmers shift towards using smart robots such as Lettucebot, strawberry fruit pickers, weeders, etc., robot production has to match the demand. Already, start-ups for farm robots are on increase. Most recent reports suggest that during 2006 to 2014, on an average, 135 farm robot companies involved in ground robots were initiated in USA alone. In four years from 2008 to 2012, it seems, over 10 billion US$ have been invested in farm robots in South-western USA to augment need for autonomous farm equipment (Precision Farm Dealer,

2014a). The general belief is that farm robot production, particularly, those with ability to accomplish field work, assess crop health and harvest grains will explode in Americas and Europe, which possess large farming enterprises. For example, utility of 'Lettucebot' attracted over 20 licenses for new start-ups in California alone (Precision Farming Dealer, 2014a).

Reports from European Community nations suggest that projects to develop sophisticated robots capable of fruit picking and sorting are being developed. However, considering economic aspects, some of the research groups are first concentrating on robots to harvest high value fruit crops (Hicks, 2012). In North Americas too start-ups for harvesting robots are traceable. They are mostly concentrating on automation of maintenance and harvesting of greenhouse crops. It is believed that each such harvester could cost 25–50,000 US$. In Japan, strawberry pickers fitted with 3D vision that can carefully pick over 60% of ripe fruits in one go in the field are being developed. It turns out highly economical because such a fruit picker uses, on an average only 9 sec fruit^{-1} to detach and store it in a basket. These are not cumbersome robots! The general opinion is that research on Agricultural Robots is now heading from predominantly academic pursuit to private industries and commercial enterprise zone. We can guess its rapid spread in commercial farms sooner than ever before.

Robot types and number in a crop field may vary, but the trend would be to use more of them, if economics allow. Robot population density in a farm could vary, but this aspect may need careful examination. Government stipulations need to be prepared based on several considerations like availability of robots, energy needs of each instrument, influence on climate, storage, re-use and servicing of robot. Agreed, that robots replace human labor needs. However, their activity in the crop fields needs energy, derived from either fossil fuel (petroleum), or electric batteries or solar energy panels. Agricultural expanses are large and found in different continents. Further, farmers may use robots in higher density as their needs manifests itself. Therefore, we should note that energy needs, particularly, petroleum and electric batteries may increase. All robots and their models are not equipped with solar panels (Vitirover, 2014).

Zemlicka (2014) has pointed out that robotics, particularly those embedded with computer based decision supports and remote control may experience rapid improvement in technology. During yester years, a large tractor meant a good buy that lasts for years. This led to obsolescence of equipment and instruments used in farms. It did not make a big dent to farm operational efficiency or profits. However, GPS-RTK connected tractors, at present, get revised in a matter of couple of months. New computer programs and electronics may have been attached to latest models, time and again. Therefore, upgrading farm robots with new advanced models is a matter to be considered by famers who adopt Push Button Techniques. To quote an example, a GPS connected tractor bought today may go obsolete in a matter of 6–7 cropping seasons (not years) (Zemliicka, 2014). Further, there are protagonists who advise more small tractors and several different robots that specialize a particular farm operation. This leads us to situation, where in, we have to upgrade the technology with more of newer models of robots; newer software; more engineering repairs if they brake-down, etc. Robots, if used in larger number need better electronic co-ordination. More fuel and cost to purchase are other points to consider, if we agree to use small robots in large numbers.

5.1.6 ROBOTS, HUMAN DRUDGERY, MIGRATION AND SETTLEMENTS

Regarding robots of utility in grape vines, forecasts are that, equipment like Wall-Ye capable of pruning, de-suckering and collecting ripe fruit bunches will definitely compete and displace, significantly large number of farm workers. Wrenn (2012) argues that "a new grape vine worker called Wall-ye that has four wheels, two arms capable of pruning and six cameras, prunes about 600 vines day^{-1}. It never falls sick and fatigue is not a problem to think about. Therefore, if farm workers and others aim at a summer time job or even a regular one in grape orchards, they may after all loose out to Wall-Ye. Robots are efficient and less problematic, since seasonal insufficiency such as one encountered with human laborers (farmer workers) does not occur with them. In addition, there are several models of robots being developed. For example, in a small agricultural nation such as New Zealand, invasion of several models of agricultural robots is said to save wine industry there with 17.6 million Euros year^{-1} through increased productivity. In addition to reduction of labor costs, robots may provide better quality pruning. In many grape producing regions, or even elsewhere, getting well trained farm labor is not easy. Farm skills may vary with each human worker. Grape farmers questioned in many regions of Europe, it seems, state that intrusion of grape vine robots is inevitable. The point to ponder is just when they will proliferate to high numbers.

Robots may have specific effects on human migration and settlement trends. Robots could of course up-root and make farm workers to abandon a location and move away or discontinue the farm jobs altogether to take up a different profession. Such events have occurred historically due to various other factors such as droughts, dust bowls, economics, war, etc. Historically, farm mechanization has reduced need for farm workers. However, in the present case, it could be robots stealing the farmer workers' jobs in entirety. For example, there are strawberry fruit picking robots, lettuce thinning and harvesting robots being rapidly developed, refined and tested. They could be swarming the vegetable production zones near the Mexican border. It is expected to displace farm worker, induce migration and affect their settlement pattern in Northern Mexico and Californian border zones. Future for robots in Californian farm land is promising. So, there could be need for skilled technicians who may be employed in robot manufacturing units. Technicians who are trained to use and service the robots will be in demand. It means introduction of specific robots will induce some jobs although in relatively small scale. We may note that depending on farm operation conducted, crops and economic advantages, farm worker displacement could be permanent, never to return. Robots are expected to induce permanent effects on farm worker settlements and migratory trends in many agrarian regions.

The forecasted invasion of robots in farms does not seem to be a transitory phenomenon. For example, in Japan, rice production involves periodic crop dusting with pesticides. These chemicals affect the farm workers' health and environment, if used rampantly. Drone usage to spray pesticides seems apt since it avoids exposure of farm worker to harmful chemicals. During recent past, robotic drones and land vehicles are being used, so that human labor needs are reduced. The effect on displacement of human labor is marked and definite. Drones are particularly popular in replacing human crop dusters (Tobe, 2014). In USA, crop dusting is known to be an important factor that contaminates environment and affects farm worker health. Also, it is forecasted, that once robot/drone usage to spray pesticides picks up; there will be rampant retrenchment of farm workers who were engaged to dust crops.

Farm labor usage is rampant in any agrarian zone. Farm workers' availability, their skills and economic costs are points to consider. In many of the developing nations, several of the farm activities needing drudgery, like hand planting, weeding, thinning seedlings, harvesting, collection of produce, separating grains and processing (winnowing, de-husking, etc.) are operations mostly conducted manually by women laborers. In the rice fields of Asia, it is a common feature to see herds of women farm workers transplanting seedlings by hand, thinning them and picking weeds. The proportion of women farm workers is high. Therefore, as we develop robots to conduct these set tasks, we should realize that it is women laborers who will be out of work. As an alternative, they have to be trained to use robots, get accustomed to computers, and study satellite imagery. They may even learn trades other than farming and that allow them to migrate.

5.2 AGRICULTURAL DRONES (UNMANNED AERIAL VEHICLES)—THEIR FUTURE

Drones are useful in aerial remote sensing and digital imagery of crop fields. They offer a series of advantages to farmers hence they are becoming popular (Yan et al., 2009). Drones might find their way into almost all agrarian regions because of the imagery obtained, particularly close-up shots from above the crop is highly useful to famers and agencies that prescribe agricultural operations. The high resolution and details about crop growth stages, chlorophyll content, NDVI, and water status (Thermal IR images) are good attractions to farmers to buy the equipment. Since drones operated periodically replace crop scouting through human labor, farmers prefer them. Crop monitoring is among the most popular and useful tasks for which drones are used in many agrarian zones. Drones are apt to be used in farms specializing in cereals such as wheat, maize or other arable crops. They are preferred to monitor the crop for canopy growth, leaf area, chlorophyll and leaf-N status (Lumpkin, 2012; Raymond Hunt et al., 2010; Torres-Sanchez et al., 2014). Drones reduce costs on labor and at the same add accuracy. They offer digital data for other ground robots to operate. Hence, drones might be expected to have greater demand in future. They have several advantages over satellite imagery. Hence they get preferred. Drones are also getting increasingly popular with farmers producing pastures and operating cattle farms. Large pastures with single or mixed species could be regularly monitored for growth and chlorophyll status using drones (Von Bueren and Yule, 2013). Drones could also be used to spray fertilizers based on digital data supplied to its computers. Interestingly enough, there are farms in Europe where cattle herding and monitoring is conducted using drones. Drone derived imagery is used to monitor animal movement and feeding habits.

There are also reports about novel methods of testing the air above the crop in farms. Drones could be used to detect disease propagules. To quote an example, Potato late blight pathogen could be monitored using drones flown above the crop. The sporangia of *Phytophthora* above the canopy of potato crop could be trapped by drones fitted with air sampling devices. Drones flown above crops periodically may be used to pick up spores of fungi commonly affecting the crop. This allows farmers to judge the movement of disease and based on it prophylactic measures could be envisaged (Aylor et al., 2011). Similarly, if the crop is situated close to industries, we can fly drones to judge the smog, dust and chemical toxicant, if any, in the atmosphere above the crop. There could be many other airborne plant disease organisms that could be studied and control measures adopted (see Schmale et al., 2012). We may note that, in addition to aerial surveillance of disease propagules, drones are also used to apply fungicides and pesticides at variable rates, based on digital data supplied to their computers or by referring to satellite imagery. Drones have the ability to reach the top of the fruit tree canopies with greater ease. Drones indeed are potentially very important farm gadgets. They could revolutionize disease control methods (Ministry of Agriculture, 2013).

Currently, there are several start-up companies producing drones of different designs and capabilities to suit crop scouting and variable-rate sprays. Report by production engineers of Trimble Corporation at Sunnyvale in California, USA states that drones have taken big leap into farming regions, individual enterprises and agriculture related activities. They are expected to flourish and play a major role (Caldwell, 2014). An educated guess about Unmanned Aerial Vehicles and their usage in farms states that 'An army of drones' is coming to agrarian zones of the world. They have the potential to change farming trends (Wiser, 2014). In California alone, they expect to create over 18,000 jobs for engineers manufacturing drones, during next 3–5 years. Drones could reduce usage of farm labor for scouting, spraying and mapping crop stand in the field. Plus, it may help by creating jobs. Farm technicians needed to operate drones and decipher maps using appropriate software. This is another avenue by which drones create jobs and affect economy of a location.

In terms of economics, reports have clearly shown that farmers have been quick to learn methods that reduce inputs and costs. There are indeed several ways to obtain advantages from drone use during crop production (Sieh, 2014; Tigue, 2014). It is said advantages that accrue from drones are often attributable to precision techniques. The imagery derived is used in creating management zones and digital data is adopted for variable-rate techniques. These aspects reduce need for inputs such as fertilizers, pesticides and irrigation. In due course, as drone based methods get refined, we may expect to consistently derive economic benefits, better than that obtained using traditional farming methods. Sieh (2014) further states that for now, farmers may use drones as private farm equipment or a flying instrument within their farms. However, in a long run, issues such as privacy, public interest and copy rights of drone imagery of farms (crops) need to be sorted out. Right now, reports by Drone-based associations suggest that investment is due to potentially immense economic advantages from drones. To quote an example, drones can cover up to 640 acres in a matter of 18 minutes. Regarding economics of drone production industries, it is said in USA about 103,776 jobs related to agricultural drone industries were created. It is not saturated yet, in fact, there is too much of a gap to cover, considering the expected demand for variety of drones all across different continents (Sieh, 2014). Agricultural drone industries may reach a turnover of 82 billion by year 2015.

Drones such as Yamaha R-Max weigh 180 kg per unit. It carries quite an amount of metallics and plastics. Drones discarded into farm junk yards do add their material into environment depending on their size. Drone copter models built so far range from 2–50 kg in weight. Flat wing drones too could weigh between 2.5–20 kg depending on model. Drones possess both metallic and plastic components. Anderson (2013, 2014) reports about drones that there are small, have light weight foam, rubber, wood or pith and cost less, around 1,000 US$ a piece. We may have to search for perishable and environmentally friendly material to prepare drones. Drones with wings or blades and bodies made of wood, hard card board and degradable material will be apt. There are actually other options while using drones in agrarian zones occupied by farmers practicing low input techniques and those preferring environmentally more friendly material in their drones. Flat-wing drones made of thick paper card board or thin wood foils fitted with light weight cameras (sensors) could perform a few sorties, after which they could be discorded to perish in the farm. The cameras (sensors) could be removed and re-used by fitting them to new card board frame of a new drone instrument. This way, cameras and computer fittings stay intact but wings and fuselage are discarded to get degraded.

Drones could crowd farm air space. They may intrude or drift into airspace of other farms. They could also be used to conduct unwanted surveillance of farm regions affecting privacy about crops, their productivity and yield levels attained by farms. Too many drones flying in different directions and erratic speeds could be problematic (Ansley, 2014). A few other concerns are when drones spray pesticides and other chemicals, depending on wind conditions spray material may drift into other crops/farms. There are possibilities that farmers who fly drones above their crops may be regularly trespassing and affecting privacy of farms situated close by. Drone intrusions may become common, if it gets popular and used too frequently to fly over other farms (Green, 2013; Hetterick, 2013; Kimberlin, 2013). Regulations for drone usage with a long-list of do's and don'ts are being prepared. In due course, perhaps each nation, will have it's rules and regulations in place (EPIC, 2014; Ehmke, 2013; Redmond, 2014). We ought to realize that drone machine should be perfect and air worthy, plus the technician(s) who operate the machine need training to use them correctly in a given field. There are nations that insist that farmers should first get trained in drone usage (Epp, 2013).

5.2.1 FUTURE OF DRONE USAGE IN DIFFERENT AGRARIAN REGIONS

Drone population in farming belts is expected to increase in future years. Let us consider an example. Soybean/maize intercrops in North American plains constitute a large well developed farming zone. The crop production strategy adopted is highly intense. Maize yield reaches 9–11 t ha^{-1} and that of soybean over 3.2 t ha^{-1}. Farmers are congenial to any new and advanced crop production system. New equipment with GPS guidance and autonomous navigation are highly welcome, since they reduce farming company's dependence on human skilled labor. As such labor resources could become erratic due to migration and other reasons. According to Wiser (2014) drones are expected to throng soybean expanses of Indiana, Illinois and Iowa. Farmers may entrust large patches of soybean to be scouted and imaged by drones, so that other procedures that utilize GPS guidance and computer-decision systems could be performed with greater ease. Digital imagery provided by drones will be essential. Hence, demand for small agricultural drones that fly close to soybean and maize crop at low altitude will be sought in future. Drones that could obtain imagery of pest attack, and provide the digital data to spray disease/pest control chemicals will be in greater demand. They after all reduce exposure of human skilled labor to harmful chemicals. Drones guided by GPS signals and digital imagery restrict chemical spray to only spots affected and in quantities that match the intensity of malady. Hence, they save chemicals and reduce costs on chemical spray. Traditional blanket sprays are costlier by many folds compared to use of drones.

Drones will eventually be used more frequently than ever because of their utility in crop monitoring, particularly in estimating biomass and leaf-N status that form the basis for fixing fertilizer-N inputs via variable rate applicators. I'nen et al. (2013) and Kaivosoja et al. (2013) believe that precise supply of fertilizer-N using hyperspectral image derived from drones is most important function of European cereal farmers. Thermal Infrared imagery by drones shall be sought while devising irrigation schedules. Hyperspectral imagery and close-up shots showing insect infestation, drought effects and disease affliction is among the most useful items related to agricultural drones. These advantages make them sought after gadgets in the near future. There are also light weight drones capable of multispectral imagery of a large farm extending over 500 acres, within 15–20 min. The digital data could be rapidly used to judge dosages of fertilizers, fungicides and pesticides using appropriate computer software. For example, drones such as AgEagle, SensFly's eBee, and several others capable of multiple functions are becoming popular in farming belts of North America and Europe (Precision Farming Dealer, 2014b; Grassi, 2013).

Mulla (2014) estimated that use of drones to survey for soil nutrient status and following it with GPS guided variable-rate applicators saves 10–30 US$ acre^{-1}. Given that farms larger than 10,000 ha are in good number in North American farming zones, such advantages amount to several thousand US$ in savings.

Reports suggest that drones with ability for sharp imagery of weed infestation in large cereal farms of North America would be very effective. Drones supplying data about density of weeds in a location of field and revealing the coordinates could become popular. Incidentally, there are several models of such light weight drones capable of rapid imagery of weeds. They are expected to spread rapidly in North America and European agrarian zones (New Scientist Tech, 2013). Certain reports related to agricultural uses of drones, suggest that these autonomous flying machines with sensors and chemical spraying ability, will be most sought after for a few reasons. They are rapid mapping of fields, their soil type and crop stand; crop scouting at any instance and exceedingly rate; locating cop patches and even single plants and identifying them by their GPS coordinates, ability to replace ground equipment during crop dusting, liquid fertilizer application and general surveillance of fields/farms (Paul, 2014).

5.3 SATELLITES AND FARMING ENTERPRISES—FUTURE

Satellites have been used to spread latest information about soil and crop management procedures, since past 4–5 decades. Satellite imagery of farming belts, natural resources and environmental effects (e.g., droughts, floods, erosion, etc.) has been made available to state agricultural agencies. Satellite imagery is currently most frequently used to study the crop productivity variations in a field or a large area, say a county or even a vast agrarian zone. Satellites possess a range of sensors and offer field imagery drawn using different wavelength bands. During recent years, farmers are prone to consult private agricultural agencies that supply satellite imagery and timely prescriptions related to fertilizer, irrigation and pest control measures. During 'precision farming,' farmers are usually asked to subdivide their farms into easily manageable areas. They are called 'management zones.'

Farmers often try to map out several aspects of the field, such as soil type, moisture fluctuations, fertility variations, crop growth variations, insect pest and disease distribution. They are at an advantage if they farm 'management zones' based on closely related aspects. At present, farm consultancies are selling services such as marking 'management zone' using multi-layered data. Such multi-layered digital data could be used in variable-rate applicators. Hence, we may forecast that satellite imagery, multi-layered field maps and satellite guided VRT equipment will take over a large portion of core farming activity. In its simplest form, farmers may frequently opt to use multi-layered field imagery/digital data that depicts soil type, soil fertility variations related to N, P and K. Yield and crop revenue maps are among popular methods that depend on satellite techniques (MZB, 2014). Satellite imagery is currently used frequently to study the climate change effects on the vast agrarian belts. Such imagery helps farming agencies to judge soil and crop moisture status, impending droughts if any, floods, soil erosion, and top soil loss if any. Satellites are used effectively in drought surveillance of dry land tracts. For example, farming agencies are provided with information about crop stand using satellite imagery rather frequently. Agrarian zones practicing rain-fed crop production strategies will be prone to go far satellite pictures as often during rainy season for reasons such as crop stand, soil moisture distribution, soil erosion, floods, etc. Satellite imagery using hyperspectral sensors allows farmers get an idea about attack by insect pests or diseases, if any. Periodic imagery suggests the rate of progression of disease/pest, so that it alerts them to adopt remedial measures before the malady gets intense. Inter-phasing drones/robots with satellite imagery may help immensely while devising large scale robotic spraying to control disease and epidemics. Satellite imagery has also been effective in informing state agencies about the extent and intensity of soil maladies such as salinity, alkalinity, erosion, acid rain effects, etc. (Feng et al., 2009; Galileo Geo Inc., 2014; Okwu-Delunzu et al., 2013; Qin et al., 2009; Qin and Zhang, 2005; Satellite Imaging Corporation, 2014; Vina et al., 2004; Xiaoyu et al., 2014; Zhang et al., 2014).

Satellite imagery and digital data accrual for use in robots is an important research aspect that needs greater attention. Almost every aspect of field operation, right up to harvest of grains by Combine harvesters, needs careful and correct navigation in the field. A robotic vehicle will have to overcome obstacles and navigate straight ahead performing the tasks such as weeding, fertilizer application, pesticide application or harvesting. Sensors that detect the surface topography are essential. Satellite or drone derived imagery could provide 3D imagery of the field (Rovira-Mas et al., 2005; Rovira-Mas, 2009; Moorehead and Bergeman, 2012). In fact, all the robots to be used in future will need 3D view of the field on which they operate using GPS coordinates. There is need to invest research time on 3D imagery and the way it is utilized with great accuracy by robots in agricultural fields (Rovira-Mas 2009). Such 3D vision is also essential while weeding a field using robotic weeder (Wooley, 2014; University of Southern Denmark, 2014). We may realize that skilled human scouts and vehicle drivers are endowed with excellent 3D perception of the field and knowledge of obstacles if any in a field. Inter-phasing satellite signals to robots on ground is essential. It is indeed a crucial aspect of 'Push Button Agriculture.' However, simultaneously, we may have to device harrows, ridgers and land preparation methods that do not create or leave large clods of soil and obstacles that may affect free movement of robots, particularly smaller versions. Satellite imagery may not provide greater details. Sensors mounted on ground robots/vehicles perform better and offer high resolution imagery. A large tractor may just trample a big clod, but small robot gets stuck. Softer tilth of field and totally clear, obstacle-free inter-row space seems essential. Tillage experts need to take note of this fact. In a horticultural plantation, detection of fruit location and its size plus color (ripening) needs perfect 3D imagery, so that picking arms could move rapidly (NREC, 2014; Carnegie robotics, 2014). Sensors mounted on ground vehicles may perform better than satellite imagery. This aspect too needs attention. It seems in some cases, selecting/developing suitable crop genotype, say those farming fruits in bunches and clusters and localizing themselves at a particular point is pertinent. However, rapid 3D detection is also essential. Further, in case of strawberry pickers, it seems ridges of slightly greater height allow the ripe fruits to hang at the edges so that 3D vision and rapid picking is accomplished quickly.

5.4 PUSH BUTTON FARMING: A REVOLUTION ON THE RISE

Push Button Agriculture as defined in this volume is a composite of few technological aspects as applicable to farms and crop production, such as robotics, unmanned aerial vehicles (drones) and satellite guidance. Push Button

Agriculture, for example, composite of recent robotic methods is forecasted to encompass most if not all farming stretches of the world, irrespective of whether they are intensive high input, profitable enterprises or low input subsistence patches. Push Button Agriculture is also expected to creep into every aspect of crop production in the field and post-harvest processing, if not quickly, perhaps in near future. Most importantly, we may note that Push Button Agriculture includes sophisticated engineering, electronically controlled gadgets and satellite guided systems that replace human drudgery in open field. It supposedly enhances farmer's profits immediately after adoption or in due course. It adds strikingly high accuracy to all operations that it engulfs and reduces ill effects of excessive or inappropriate use of fertilizers/pesticides. It avoids several inevitable errors that may otherwise occur due to human factor. A few of the robots are all weather gadgets and so add to certain advantages to farmer's abilities. Several of the robots (types, models) could be made of environmentally friendly features such as solar energy driven systems, or electric batteries, so that they could avoid fossil fuel and reduce atmospheric pollution. Over all, it seems to be a matter of time before farmers flock towards 'Push Button Agriculture.' A rapid shift will cause a definite and perceptible revolution in global farming. Push Button Agriculture seems essential and a good idea to generate food grains. At a higher productivity level, it may help us in keeping pace with higher demand for food grain production in next few decades.

Push Button Agriculture includes a range of autonomous farm equipment that could be operated using switches, buttons or touch screen systems available on computers tablets. The digitally controlled farm equipment, both small and large, accomplishes several tasks in the farm without much physical effort by human labor. Obviously, if not all, most farm vehicles and equipment would be autonomous, driver-less, and guided through satellite aided GPS-RTK or pre-programed using computers. In due course, several more autonomous farm vehicles and robots with specific abilities may find their way into farms. Push Button Agriculture, no doubt makes life easy for farmers and helps them to program their farm activity using appropriate computer programs. Some of the more relevant aspects that will need greater attention with regard to use of autonomous agricultural vehicles and farm operations are sensors, global navigation satellite systems (GNSS), machine vision, laser triangulation, geomagnetic positioning, computer decision support systems and small robots (Keicher and Suefert, 2000).

Push Button systems do consider wide range of factors including those related to natural soil fertility related factors, water resources, yield potentials and economic gains. It is not so easy for farmers to judge the variable influence of those many factors at a point of time or a stretch and arrive at accurate recommendations. Computer software can accomplish these calculations in a couple of seconds. Computers allow revisions to farm operations, their speed, intensity, timing and accuracy. We can revise input schedules and yield goals based on weather parameters, to a certain extent within the season. Some of these options are impossible in usual traditional farming methods.

Among the various aspects of Push Button Agriculture discussed in the previous three chapters, there are few techniques that are most immediately applicable and begin to offer advantages to farmers with reference to soil fertility, crop management procedures, final grain productivity and economic gains. There are others which may require persistence and need lapse of time before their advantages are felt by farming community. According to some opinions expressed by farmers in North America, Push Button techniques of greater impact to farmers, right now, are as follows:

a) The emerging Autopilot Tractors (GPS-RTK Autonomous Driverless Tractors);

b) Swarms of tractors with inter-phasing and inter-vehicle communication;

c) Irrigation using GPS connected Pivot system controlled using Smart Phones;

d) Sensing crops using drones fitted with visual and infra-red cameras;

e) Variable rate technology based-on high resolution multispectral satellite imagery or close-up shots from low-flying drones;

f) Field documentation and extensive GIS that stores details of each farm or field location regarding agronomic procedure and grain yield data (Scott, 2012);

g) Data storage and computer guided decision-support systems are indeed a boon to farmers while making decision about farm operations or revising them within season.

There are indeed innumerable advantages that could accrue to farmers adopting Push Button Agricultural systems. Some are of great value and others of lesser importance, depending on geographical area, type of farm operation conducted, crop species, value of crop produce, etc. Following are few easily recognizable features that offer advantages to farmers. Push Button Agriculture (PBA) as envisaged in this book, that involves robots, drones and satellite guidance has specific features and offers a series of advantages to farmers when they are adopted singly or as an integrated process. Following are few points that ensure that in future Push Button Agriculture will spread and flourish:

- Large farms that are over 1000s of ha in size could be covered using autonomous tractors, weeding robots, center pivot irrigation, pesticide spraying drones, autonomous ground vehicles and GPS controlled large combine harvesters. This is not easily possible with traditional farms.
- There are several rough tasks in land management that are accomplished best by semi-autonomous or totally pre-programed autonomous robots (tractors). Human physiological limits do not allow heavy drudgery to be efficiently done.
- PBA offers greater accuracy, uniformity and efficiency in most of the farm operations accomplished. However, there are some activities such as fruit picking in a tree plantation or searching for weeds in a thickly planted cereal field where in robots could be too slow. Human dexterity and rapid visual judgment is still better than machine. This aspect of engineering needs due attention and improvement.
- PBA allows easy, rapid and accurate scouting of crop fields that are excessively large and variegated, because we adopt satellite imagery and/or drones that fly quickly over the entire farm. Human scouts, even if employed in 100s, sometimes will not be able to cover the farms in entirety and to a required satisfactory accuracy. Human laborer's judgment about a patch of crop, its health, nutritional needs or ripening could be highly variable. Satellite imagery reveals comparative situation of a very large farm in one stretch that is perhaps impossible to depict if human scouts record it manually.
- PBA techniques are mostly all weather systems. Autonomous tractors with planters can seed the entire field at defined pace whether, it is day or night, drizzling or dry and sunny, without fatigue.
- PBA offers better input efficiency, particularly seed rate, fertilizers, irrigation water and pesticides.
- Since inputs are channeled to crops using variable rates, excessive supply that occurs due to blanket recommendations of fertilizer- N, P and K is avoided. The climate change effects and GHG emissions

emanating due to excessive fertilizer-N is totally avoided. Fertilizer seepage into channels gets reduced since fertilizer placement, timing and quantity are highly accurate, if PBA is adopted. *In fact, we are at a crucial juncture in agricultural history, where in PBA is among important concepts to follow during crop production.* Mainly because, it offers important advantages related to remediation and lessening of climate change effects. Such options are not available with traditional crop production techniques.

- Adoption of PBA methods reduces requirement of human skilled labor in large numbers. Drones too offer advantages related to farm labor needs, particularly those involved in scouting, mapping and spraying. However, flying drones, fixing routes, obtaining images of crops, and preparing digitized maps may need technicians with specific skills in computer science. If private Drone or Satellite companies are hired to do the imagery and arrive at recommendations, then a single set of trained technician crew will be able to perform and offer service to several farms.

- Computer controlled vehicles can be guided midway during a task to change course. Computer software allows rapid decision support that is perhaps impossible with human skills, if so many factors and algorithms are to be considered and previous data has to be compared to arrive at suggestions. PBA is impeccable on these aspects due to digitized data and computer programming.

- Data storage using computer programs and their accurate retrieval in few seconds is something restricted to domains of Push Button Agriculture.

- PBA will eventually engulf global farming zones in entirety not just for its economic advantages, accuracy of operation, but ease with which it allows humans to produce food grains. PBA lessens number of man hours of surveillance and drudgery. It just offers easier and happier life to farmers and that seems more than enough for PBA to spread in every nook and corner of global farming belts.

- Since, fertilizers, chemicals and other soil amendments are supplied accurately and to match the crop's need at different stages, accumulation of excess inputs in the soil profile is avoided. Fertilizers that are otherwise vulnerable for emission, loss via percolation and use by weeds are avoided. Ground water contamination is avoided, perhaps totally, because of accurate placement of fertilizers in matching amounts near the roots, and in split dosages. These are accomplished using computer software and decision support systems provided on fertilizer inoculators (variable-rate applicator).

Push Button Agriculture is already in vogue in a few agrarian zones. Its adoption is transitional in several other geographical locations. It is evolving from being a restricted precision farming or site-specific technique to full-fledged satellite guided farming in entirety, encompassing all activities of crop production. The extent of farm operations accomplished using PBA too varies. In some farms, only plowing, land surface preparation and ridging are contemplated using GPS-RTK connected tractors. In some others, inter-culture, earthing-up and weeding are emphasized and so PBA is adopted to address only these aspects, using autonomous robots fitted with sensors that detect and eradicate weeds. There are farms concentrating on fruit picking using PBA. In many farms, grain harvest, yield mapping and grain transport are accomplished using GPS guided autonomous combine harvesters and companion transport vehicles (see Plates 2.22 and 2.23; Kinze Manufacturers Inc., 2013, 2014).

Regarding Crop Improvement Programs, it seems there are several crop species particularly fruit plantations where robots find it difficult to negotiate, identify, sense the ripening stage and detach the fruit. In many, it is robots that look cumbersome. Similarly, while weeding intra-row spaces, it seems ground robots find it difficult and are slow to perform tasks. However, there are options to breed a certain cultivar that forms grains/pods/fruits at locations easily reachable by robots. The interiors of tree canopy are not easy to reach for robot hands. They get stuck with cris-crossing twigs and branches. If the fruits localize at the rim of canopy we may find suitable robots. Field crop breeders too may take robotics, seriously in the future and try to tailor their varieties/hybrids to suit the machine's ability. In fact, it is said short statured wheat was preferred, to certain extent, because they are amenable for easy harvesting and processing by combine harvesters.

5.4.1 SENSORS IN PUSH BUTTON AGRICULTURE

Sensors, their vantage placements on satellites, drones, farm pedestals or ground robots are forecasted to play a vital role in global agriculture in future. Sensor's distance from crops, their resolution and band width range of cameras, forms the crux of revolutionary methods to be adopted in future. Sensors have a strong say in the ability of various gadgets used during Push Button Agriculture. As stated earlier, whether it is a ground robot operated in between crop rows, an autonomous tractor performing plowing, ridging, measuring Soil EC, moisture, pH or nutrients to chart out management zones; a drone hovering over crops to assess growth and leaf chlorophyll or even a satellite offering imagery about crops or general vegetation; it is sensors that are crucial components. Sensors come in wide range of variations. The cameras fitted to satellite, drones or ground robots are usually designed to match the task they have to perform and they are connected to computers with apt programs to decipher digital data and arrive at suggestions using GIS. Agricultural ground robots currently possess sensors that pick reflections from visual, IR, NIR and thermal IR ranges. Such sensors aid in identification of crops and weeds. Sensors help in recording NDVI and mapping field crops in a location. Sensors also help in identifying and detaching individual fruits in orchards, for example in citrus groves, strawberry farms, etc. In due course, we may encounter robots with versatility and accuracy to perform several more tasks. Autonomous small robots that dibble seeds in individual spots and cover it up with soil/organic matter mix are being tested. Autonomous tractors with ability to plant large fields are already operative. They depend excessively on sensors and signals from satellites. High resolution sensors are needed if satellite imagery is to be used extensively.

Sensors are extensively used to study soil characters, map various crop production factors and provide the digital data to farmers. Sensors help in preparing maps and management blocks for easier adoption of PBA techniques. Sensors for soil-NO_3, soil-P and K are being developed, tested and standardized. They would help farmers to ascertain fertilizer-N and nutrient needs. Soil moisture sensors are being increasingly sought to be mounted on ground robots. There is recently a satellite SMAP (Dunbar, 2009; Yang et al., 2014; National Aeronautics and Space Agency, 2014) that measures soil moisture in the surface horizon and helps farmers in deciding on irrigation. In future, sensors may serve farmers rather ably by suggesting the state of crops, with visual imagery. Specialized sensors may inform on general crop productivity (e.g., soil electrical conductivity), soil moisture, crop moisture, leaf-N status, etc.

5.4.2   INTER-PHASING ROBOTS, UAVS AND SATELLITE GUIDED FARM EQUIPMENT

Push Button Agriculture in future may actually involve complex interactions between farm vehicles depending on the tasks to be eventually performed. Inter-phasing, as it is known, involves connecting ground robot with one another and drones. Inter-connected clod crushers could move first within the field, prepare a better tilth. The clod crushers that lead could signal the seed planters to follow at appropriate speed. During aerial survey sorties, drones collect digital data and provide maps of the crop. Low flying drones can really send very detailed pictures of crops. Such information is instantaneously utilized by ground robots that are interconnected through electronic signals. Electronically, it seems ground robots could be controlled and programed to conduct weeding, irrigation and fertilizer inoculation based on digital imagery derived from robots/satellites. Push Button Agriculture manifests itself prominently in such situations. For example, reports from Minnesota suggest that field soils may show deficiencies of several nutrients and at different intensities simultaneously. Drones that send pictures of soil fertility variations could be inter-phased with ground robots. In future, radio-linked ground robots are supposed to perform series of tasks sequentially and in well programed fashion, singly or in swarms of GPS-RTK tractors with variable-rate applicators. For example, a robot conducting weeding may signal an inter-connected robot that is applying fertilizer to follow using radio signals and satellite guidance (Ramstad, 2014; Grift, 2012). Some of the most recent developments relevant to robotics in farming are called inter-phasing and inter-robot communication. This may lead us to deployment of swarms of robots that could act sequentially and in groups based on signals from other vehicles. This aspect needs complex appropriate software to control swarms and accurate programing. With regard to Centre Pivot irrigation systems it is said mobile based control of the irrigation equipment is gaining acceptance.

5.4.3   'BIG DATA' FOR PUSH BUTTON AGRICULTURE

Reports suggest that agricultural operations in farms and experimental stations are offering huge data for compilation, classification and use during computer-based decision support systems. However, it is still relatively much less if other facets of science and industry are considered (Azorobotics, 2014; Harris, 2015). During precision farming, prescriptions are calculated by the decision support computers using a vast data set available in the GIS and other digital repositories. As such, it is said 'Big Data' is the driving force that provides basic data during precision farming. We may note that inputs are decided using 'Big Data' and then yield goals fixed by farmers. Right now, 'Big DATA' is a complex set of agricultural information, particularly related to inputs and crop production trends in different geographic locations. Ground Robots, drones, along with satellites add large data sets periodically into this 'Big Data' pool. In future, perhaps it is useful to categorize and concentrate data sets relevant to a particular geographic region, cropping system, crop species or a set of procedures. At this juncture, we should note that Push Button Agriculture that encompasses data storage, computer decision support systems and advanced software offer advantages hitherto unknown. Computers make rapid judgment, at speeds never possible in traditional systems. Computers arrive at decisions after consulting huge data sets in a rational fashion. Farmers own guess work, gut feelings, folklore and scantily available previous data do not find place in Push Button Agricultural systems. Push Button Agriculture is relatively more thorough while arriving at decision based on Big Data. Hence, the system will find greater acceptance with farm engineers and production managers. 'Big Data' and its usage during precision farming adds to authenticity, accuracy of data and offers most recent data to farmers to consult and arrive at decisions (Harris, 2015). However, traditional farming thrives on decisions made of relatively very few data sets.

5.4.4 PUSH BUTTON AGRICULTURE IN DIFFERENT GEOGRAPHIC REGIONS

In any agrarian zone, 'Push Button Agricultural' system could be adopted right from land clearing. Bull dozers with remote control systems or semi-autonomous are used to dig and turn the soil after clearing forest trees and shrubs. Several of the leveling vehicles too are remote controlled or autonomous. They possess GPS connectivity and can read digital data derived from satellite. Currently both, semi-autonomous and totally autonomous tractors capable of deep disking and plowing are available for use. Autonomous seed planters attached with variable-rate seeding devices are already in vogue in many of the farms in Americas and Europe. Push Button Agriculture also engulfs a series of agronomic procedures that are related to crop growth monitoring, fertilizer and irrigation supply, plant protection procedures and harvest. Combine harvesters with ability to harvest autonomously, process the grains and collect them for transport to market yard are available (e.g., see Kinze Manufacturers Inc., 2013, 2014). Practically, we may eventually reach a situation, where most, if not all field activities that are necessary for crop production could be handled using Push Button systems. This forecast will not end up a fantasy, since several of the aspects are being already practiced.

Where do we concentrate first and adopt Push Button Agriculture? Globally, farming predominately involves three staple cereals, namely maize, wheat and rice, a few millets, legumes and oilseeds. Therefore, PBA needs to focus and device robots/drones appropriately. We have to prepare software for autonomous activity, so that we could accomplish tasks in these specific aspects. Robots that offer advantages to farmers in large expanses of cereals perhaps promise, a robotics abetted agricultural revolution. Drones could add further impetus to spread of PBA. Low flying drones could turn out marvelous results in most agrarian regions. We may also experience success, if drones are deployed in high number to attend as many tasks possible in the major field crop producing zones. Robots and drones would have served mankind best, if in near future, they wean him from drudgery, irrespective of whether it is high input commercial grain producing farm or a small low input subsistence farm. Satellites are already recognized as a boon to farmers worldwide, since 1970s. They are applicable all over the agrarian zones of the world. At the bottom line, techniques that are easy, versatile and adapt well to a particular region get preferred automatically. Robotics/drone techniques that avoid strenuous procedure will be first to get selected by farming community. Crops that accrue greater fiscal advantages to farmers, upon use of robots/drones will be under focus.

5.4.5 AREAS AMENABLE FOR PUSH BUTTON AGRICULTURE

Push Button Agriculture, as a concept is versatile and should be possible for adoption in every agrarian region of the world. Agricultural Engineers have to imagine, strive, design and develop suitable robots and drones for the region and cropping systems in question. Push Button Agricultural techniques, a few of them out of the full complement, are already in vogue in many of the agrarian regions. Push Button techniques that suit the large farming belts of Americas and Europe, those particularly useful to produce major cereals and a few horticultural species are popular, in preference to ones that are still to be experimentally tested or standardized. If we consider agro-climate, most regions are amenable for standardized robotics and drone technology. Satellite based techniques are already worldwide and cover almost every possible cropping zone on earth. Push button techniques are more popular in some areas supporting valuable cereals and fruit crops. Such techniques are also common in those regions that have highly fertile soils and farmers are rich enough to support intensive farming techniques, for example in North America and Western Europe. Far-Eastern farmers (e.g., Japan) adopt Push Button techniques such as robotic tractors and drones in many locations that support intensive rice production. Drones that spray pesticides based on digital imagery are popular in the rice fields of Japan (Huang et al., 2013; Japan UAV Association, 2014; Precision Farming Dealer, 2014d). Robotics and UAVs that suits the subsistence farming belts need to be devised and popularized. Robots that suit the low fertility sandy regions of sub-Saharan regions could be devised. In this region, all weather robots that could accomplish repeated sowing immediately after rains is a need. Actually, planting on sandy soils in Sahel is comparatively easier and could become a robots domain in near future, with some research time and capital investment. Satellite techniques need to be popularized and made easy, so that Sahelian farmers could access them. Farmer cooperatives that supply aerial imagery showing progress of the crop to farmers should be possible in most subsistence farming zones. Time and again, it has been pointed out that drone/robotic technology will get cheaper as efficiency of production of these gadgets increases and markets get flooded with varieties of models. For example, recent reports suggest that drones that cost low at 170–500 US$ are possible (Anderson, 2013; Ministry of Agriculture, 2013). Farmers can then find them easily purchasable than now. Further, Push Button techniques may get refined and become efficient allowing better profits. Above all, robot and/or drone along with satellite guidance make it easy for farmers to perform their professional tasks with least drudgery and strain. These aspects serve as a great reason for farmers in any agrarian region to shift to Push Button techniques at the earliest point of time. To quote a few examples, robots that weed the inter-row space in plantations such as grape are getting popular in North America and Europe (Murray, 2012; see Plate 2.26 in Chapter 2; Vitirover; 2014; Plate 2.19 in Chapter 2). Robotic steer-less tractors with GPS-RTK connectivity are already getting prominent in cereal farms of American and Central European plains (Kinze Manufacturers Inc., 2013, 2014). Farmer's profit margins and ease of operation may outweigh other reasons for adopting Push Button Techniques.

5.4.6 ECONOMIC ASPECTS OF PUSH BUTTON AGRICULTURE (PBA)

Push Button Agriculture imparts economic advantages to farmers due to various reasons. Firstly, it reduces farm labor requirements. In many cases, it is said that cost for robots and drones get offset through need for lower levels of labor and inputs such as fertilizers, pesticides and water. PBA derives economic advantages basically through 'Precision Farming' techniques such as accurate supply through 'Management Zone' formation, variable-rate applicators, and spot application of fertilizers, pesticides and water. Generally, fertilizer and water-use efficiency is higher, if Push Button techniques involving drones to image the crop water status and variable-rate pivot irrigation gadgets are used. In case of herbicides and pesticides, spot application brings down chemical usage by 35–75% (Soegaard, 2005; Soegaard and Lund, 2005, 2007; Pedersen, 2003; Pederson et al., 2007, 2008). Aerial imagery by drones or satellites is crucial since they allow us well directed farm operations. Harvesting using autonomous combine harvesters and driver-less self-navigating grain laden vehicles is highly efficient, although reports about exact economic gains are unknown yet. *In fact, we need several experimental evaluations in different agrarian regions to quantify the net economic gains due to Push Button Agriculture.* No doubt, certain aspects such as robotic planters, weeding machines and harvesters may impart greater advantages. A few other aspects of PBA may just add to accuracy and remove drudgery, but, economic gains may be marginal.

Intensive farming of cereals in the Corn Belt and Great Plains of USA regions involves supply of proportionately large quantities of fertilize-N, as basal dose. It is applied to soil prior to seeding. This procedure is usually accomplished using semi-autonomous tractors with variable rate applicators. However, there are now several farms that use autonomous tractors to plant seeds and inject fertilizer. The precision technique followed reduces need for fertilizer-N marginally, yet allows farmer to reap expected yield goals. A step further, we should note that farmers in intensive agrarian zones, usually subdivide and space the fertilizer-N supply into 3–4 split dosages. It allows them to match plant's demand for fertilizer-N with supply at different stages. Again, variable-rate applicators drawn by large tractors are common. However, in countries such as Japan, it seems drones with liquid fertilizer storage tanks attached to their fuselage; apply 'liquid fertilizer-N' (0.2% Urea) as foliar spray. They use digital imagery and variable-rate techniques. Drones accomplish tasks swiftly flying over the crop at a close altitude. Most striking is the fact that fertilizer need is reduced by over 90–95% compared to traditional soil application of solid granules. Foliar application of fertilizer-N avoids all sorts of physico-chemical reactions that would have occurred in soil phase and that may affect fertilizer-N efficiency. Yet, there are reports suggesting that we need to concentrate our efforts on improvising fertilizer-N efficiency. After all, fertilizer-N is the key input during crop production. For example, in the Canadian Prairies that supports large wheat belt, fertilizer-N is the major input. They suggest adoption of precision farming methods and as many other procedures that enhance fertilizer-N efficiency (Country Guide-Canada, 2015). Similarly, there are reports that if ground robots or drones are used, quantity of pesticide needed to cover a unit area of crop gets reduced by 95%. This results in enormous reduction in costs on inputs, human skilled labor, plus reduces health risks to farm workers. This aspect of Push Button Agriculture is bound to be enormously popular and gain easy acceptance in most of the intensive farming zones.

5.4.7 PUSH BUTTON AGRICULTURE, FARM WORKERS' DRUDGERY AND MIGRATORY TREND

General forecasts suggest that a greater fraction of earth's human population will inhabit and thrive in cities, away from rural farms. No doubt, introduction of robots in farms will aid and perhaps hasten the process. Robots may induce farm workers to leave the rural inhabitation zones and move into cities in search of work and livelihood. Historically, aspects such as terrain, weather and agriculture cropping related factors have

induced farm worker migration. At times, it has turned out be an exo-
dus. Dustbowls in North America during 1930s, periodic dust bowls that
occur in Sahel, drought in India are a few examples that unsettle and force
farmers/farmworkers to move in search of better pastures. Often, better
wages and facilities for daily life have induced movement of farm workers.
Migrations have also been transitory and caused due to seasonal variations
and crop production trends. Quite a few of such factors induce only a
transitory migration, but others induce a permanent shift of farmers to new
environs. In case, robots are introduced in good number, each one replaces
a set of farm laborers. Migration that sets in later is surely a permanent
shift of job and dwelling for farm workers. Robots replace the farm labor
totally and for eternity. Farm worker population is expected to reduce.
Farm worker could be seen learning new skills that fit their residence in
cities. Whatever be the migratory pattern, at the bottom line, robots will get
rid of slavish drudgery of farm workers, once and for all. This is a boon to
human kind. It corrects the mistakes of excessive farm drudgery initiated
several millennia ago, accentuated highly during recent history (17th–20th
century) and perpetuated even today. *It is literally 'Robotics Engineers'
rescue act that could help a very large population of farm workers world
over, by weaning them from drudgery and slavery to crops in open fields.'*

Robots will have a drastic effect on human settlements in villages,
towns, etc. As such, we may realize that formation of large farms by agricultural companies have reduced the number of small farms significantly,
in many regions. For example, in Brazil, larger farms maintained by private companies have reduced small farms and villages based on them.
Push Button Agriculture may be apt in regions like Japan where in farmer
population is dwindling plus the average age of farmers has been increasing. Currently, average age of farmers in Japan is 65 years; therefore they
may gleefully accept any autonomous and programmable farm equipment
that allows them to accomplish tasks that are otherwise physically strenuous and difficult. Nagasaka et al. (2010) have argued that small paddy
farms are amenable to robots, drones and satellite guided techniques.
PBA is useful to farms run by single families with elderly farmers. In
future, we may see farms with autonomously tilling GPS guided tractors,
autonomous rice seedling transplanters (e.g., Kubota SPU 650, Iseki P60),
and robotic weeders with matching sensors, and combine harvesters with

GPS-RTK connectivity all over Japan. No doubt, drones are being used in large number in Japan compared to other nations. Farm vehicles with variety of sensors, mapping devices, decoding computer soft wares and remote controller will dominate the scene.

The rate of replacement of human labor from farms in different agrarian regions partly depends on rapidity with which farm robots are devised and introduced. Robots and drones could be shared among farm settlements, depending on cropping pattern and specialized activity required to be accomplished using them. For example, seeding of wheat could be accomplished in many farms around a village or county using few robots that are used sequentially in several farms, one after the other. Such an arrangement reduces need for human skilled labor. Drones could map all farms surrounding a small village.

As stated earlier, PBA primarily aims at reducing human drudgery in the open fields. We ought to realize that historically, human species began dibbling seeds using sticks and it took a real effort to cover large areas. This was followed by human drawn plows that required good physical strength and lots of energy to accomplish line sowing. During ancient period, use of draft animals such as horse or oxen helped reduce human energy needs and reduced physical strain. This shift led farmers to reach energy levels and physical capabilities, beyond their own. Animals were able to accomplish hard tasks for farmers. Each improvement must have made farmers happier since it served him excellently. The shift from animal drawn plowing to automated tractors fitted with internal combustion engine allowed him to reduce farm worker needs rather immensely. Tractors reached energy levels beyond draft animals. Farmers, for the first time, could perform farm activities without much fatigue and beyond their own physiological limits. This shift itself must have been felt as a major invention and close to Push Button Agriculture. Now, we are again, at the door step of another development/revolution, where in it is really Push Button system that operates instead of human drudgery. Now, from a simple, human or animal drawn plow to automated GPS-RTK system, replacement of human labor is highly perceptible. Aspects such as mechanization, electrification and electronic controls reduce farm labor need and offer excellent advantages to farmers.

Agricultural Robotics experts believe that these autonomous machines, whether used singly for each and every farm task or those capable of multi-function, if deployed and left to operate in mass in the fields, all that they are capable of is to replace semi-skilled farm workers. It is quite an effect on farm worker population. However, crop production as a human endeavor will need farmers (owners), farm managers, and specialist agronomists who will devise cropping systems. Matching computer programs to control the robots are needed. It means, we will need agricultural work force trained in robotics and computer applications (Bayer CropScience, 2014).

Believe it or accept it or not, a farm robot that dibbles seeds, removes weeds, applies fertilizers, irrigates or harvests is a grand invention that rids human beings of hard toil and drudgery in fields. It liberates him from perpetual slavery to crops to a certain extent. It seems, in 1830, in USA it took 250–300 human labor hours to generate 100 bushels of wheat grains from 5 acres of fertile land. Currently, it takes about 3 human labor hours to generate 100 bushels of wheat grains (Banks, 2013). Use of Push Button techniques should further lessen need for human labor hours. Krishna (2015) has opined that farmers have concentrated on only few crop species– namely three major cereal crops, a few legumes and oil seed crops. Together, these crops literally push human beings to slavery to them. A large section of global human population spends its time and ingenuity to run farms and generate food grains. This situation will get alleviated to a great extent by the introduction of robots/drones. In many developed nations, proportion of human labor pre-occupied with farms depreciated from 60–70% during early 19th century to 30% by mid-20th century and now almost to 7–10%. There are nations with just 3% human population working in farms- thanks to mechanization and electrification. Robots, as they take over the agricultural world, they would reduce human drudgery to negligible levels, if not nil. This may be true with all regions, since robots adapted to different terrains, cropping patterns and economic capacities are being designed. This robot generated agricultural revolution that lasts is an important event in larger historical timeline. Perhaps, it is as important as domestication of crops and invention of agricultural cropping some 10,000 years ago. Human beings became attached to crops and depended on their own efforts (labor) a great deal to generate food grains. A robot is a welcome invention that weans human beings from crops, to a certain extent. Robots take over a sizeable portion of farm activity that otherwise binds human beings and requires their drudgery, or sometimes slavery to crops.

In a nutshell, since the period when agriculture was invented in Fertile Crescent, the co-existence, co-evolution and inter-dependence has driven human beings into pre-occupation with crop production and stringent dependence on them for food grains and nourishment. Agriculture became a way of life and major cultural component through the ages. Human existence and farm activity is almost inseparable in most regions of the earth. *Human drudgery crept in unnoticed into farming. After 10–12 millennia, we are on the threshold of correcting this situation. We have to capitalize for the sake of future. Robots could take over the task of food grain production and nourishing the human kind plus the domesticated animals. That will leave a large portion of each day, for human beings to think and work on other useful aspects of his future-perhaps!*

As stated earlier, 'Push Button Agriculture' involves mechanization, electrification and automation of farm equipment. Push Button Farming reduces farm worker requirements and enhances efficiency of food production. It amounts to net reduction of human drudgery in farm operation. Historical data shows that during 1930s, prior to tractors and mechanization, each farmer produced food for 10 human beings. During 1960s, prior to green revolution in Asia and Fareast, each farmer produced food grains enough for 25 human beings. During 1970s, advent of fertilizer and irrigation aided intensification, made it possible for a single farmer to generate food grains enough to feed 76 people. Currently, with the advent and wide usage of tractors/GPS guided tractors, mechanization of several farm operations, and slow but sure spread of precision techniques, each farmer is said to feed 155 people with food grains (The State Journal of Illinois, 2014). It is clear, that the trend is to reduce farmers, skilled farm labor and human number involved in agricultural crop production. Introduction of Push Button Techniques into every aspect of farming, from land preparation, sowing till harvest and processing will ultimately reduce human involvement in farming to much less-perhaps negligible. *Agricultural Engineer's acumen, particularly those related to drone and/or satellite-guided robotics that reduce human scouting and hard work in the farm, perhaps, holds the key for future.*

In a nutshell, Push Button Agriculture is expected to spread rapidly into farming zones and flourish for long time, mainly because; it has the potential to enhance crop production. Since it utilizes computer/GPS guidance it could offer greater productivity. Simultaneously, it offers accuracy hitherto not possible by human skills. Most importantly, it allows ease of operation using computers, I-tablets, and touch screen systems. These are of real value to farmers worldwide. Push Button Agriculture eventually leads us close to zero or negligible levels of human labor involvement in farming. A few highly skilled technicians will control farm activity from seeding till harvest and transport of grain to selling yards. This will release a large population of farm workers, who could be utilized in various other facets of human endeavor. The fraction of farm workers released will be expectedly higher in developing nations and regions predominantly agrarian in occupation, compared to presently already highly mechanized zones in Americas and Europe. We have to make appropriate amends by training the farm workers in other skills and allowing them to migrate to more lucrative regions. No doubt, PBA should lead us to reduced drudgery, better crop yield and enhanced economic gains.

Push Button Agriculture, as stated in the 'preface' of this book, it is not a fantasy or a concept still in the realm of science fiction. *It is a reality*. It is to be considered a kind of agricultural revolution that has started but still rudimentary. At present, a few aspects such as robotic tractors, chemicals spraying drones and fruit harvesters are in use. It is a matter of time when momentum to adopt Push Button techniques picks up. PBA is a worldwide phenomenon that would definitely affect the way farmers accomplish farm procedures. Adoption of farm robotics and satellite mediation may affect several other aspects of human welfare, in addition to food grain production efficiency. Rural Administrators should try to match the changes that autonomous vehicles and gadgets may bring about. Imparting skills to use robotic tractors, satellite mediated remote control and programming of computer decision support systems is essential. Training farm workers located *in situ* will be shrewder, since it provides occupation to those who lost their jobs to robots/autonomous vehicles. There are many gadgets that are adopted, mastered and used regularly by human beings although they are not economically efficient. They are used because of reduction in drudgery, ability to overcome limitations imposed by human physiological limits and when tasks are difficult and dangerous. Therefore, some aspects of Push button Agriculture could be rampantly used, irrespective of economics. Whatever the advantages, in terms of economics and ease of operation, also the intricacies and difficulties of adopting Push Button Agriculture, we have to realize, that it is a concept for future. Generations of farmers that appear on earth in future will experience the revolution in greater intensity. It may not be long before, we see more of robots, drones and computer stations in different agrarian zones, and less of farm workers drudging in open fields. Global Agriculture is set to experience some perceptible changes and offer higher quantity of food grains, fruits and forage at a better efficiency.

Let us consider few facts that substantiate that 'Push Button Agriculture' as a phenomenon has begun. Following are few examples:

- Robotic on-the-go soil testing and mapping vehicles such as Veris Technologies' Soil EC, pH, moisture, organic matter and soil-NO_3-N measurement that adopt proximal sensor techniques (see Krishna, 2012);
- Satellite guided rice planting and fertilizer supply in Southeast Asian region; Robotic tractors (e.g., Kinze Manufacturers Inc., John Deere Inc.) with GPS-RTK connectivity that are now gaining in popularity and usage in the United States Mid-Western Region and Great Plains of North America;
- Robotic drones (e.g., Yamaha Rmax) that fly low over the rice production zones of Japan spraying liquid fertilizer-N (0.2% Urea) and pesticides at variable-rates, robotic grapevine pruners (Wall-Ye) and inter-row weeders (Vitirover) in France;
- Autonomous Combine harvesters and grain collectors being used during corn and wheat harvest in Iowa (Kinze Manufacturers Inc., 2013, 2014);
- Computer programs that decode satellite/drone imagery to prepare field/crop maps and supply digitized data to variable-rate fertilizer applicators are some examples, which show that a few aspects of Push Button Agriculture have already initiated the agricultural revolution.

The above satellite guided and/or automatic procedures plus several others that are in the process of standardization are expected to throng the farming belts of different continents. Such an agricultural revolution seems to be a perpetual phenomenon. Agricultural crop production that is believed to have been invented some 12,000 years ago has experienced both gradual changes to techniques and few that occurred in spurts and with great rapidity. 'Push Button Agriculture,' as a revolution, it is expected to manifest itself rapidly and all across world. It is expected to remove farm drudgery, displace farm workers in quite a large number, but at the same time improve food grain generation in quantity.

Push Button Agriculture with its components such as robotics, drones and satellite guidance, no doubt affects farm production strategies and economic gains. However, adoption of Push Button Agriculture may also provide ample opportunity for administrative regulation and policy decisions. Since, Push Button Agriculture is expected to raise food grain productivity, improve farm economics, and affect human skilled laborers' employment potential leading to their migrations, there is reason to believe that policy makers too will evince interest in Push Button Agriculture.

KEYWORDS

- **Big Data**
- **Farm Drudgery**
- **Farm Robots**
- **Push Button Agriculture**
- **Revolution**
- **Sensors**
- **Unmanned Aerial Vehicles**

REFERENCES

1. Anderson, C. (2013). In: Drones and Agriculture: Unmanned Aircraft my revolutionize Farming, experts say. Ghose, T. (Ed.). Huffington Post Science http://www.huffingtonpost.com/2013/05/20/drones-agriculture-unmanned-aircraft-farming_n_3308164. html?ir=India., pp. 1–4 (January 22nd, 2015).

2. Anderson C. (2014). Lower priced UAVs give Farmers a new way to improve crop yields. MIT Technology Reviews. Lessiter Publications and Farm Equipment, Brookfield, Wisconsin, USA, pp. 1–3.

3. Ansley, G. (2014). Drone filled skies an emerging headache. The New Zealand Herald. http://www.nzherald.co.nz/world/news/article.cfm?c_id=z&objectid=11231127, pp. 1–3 (September, 2014).

4. Aylor, D. E., Schmale, D. G., Shields, E. J., Newcombe, M. and Nappo, C. J. (2011). Tracking the Potato Late bight pathogen in the atmosphere using Unmanned Aerial Vehicles and Lagrangian modeling. Agriculture and Forest Meteorology 151: 251–260.

5. Azorobotics, 2014 Explore use of Drones, UAVs and Crop Models at growing Michigan Agriculture Conference. http://www.azorobotics.com/News.aspx?newsID=5166, pp. 1–4 (January 19th, 2015).

6. Baichtal, J. (2011). Prospero, A Robotic Farmer. http://www.makezine.com/2011/02/28/prospero-a-robotic-farmer, pp. 1–3 (July 3rd, 2014).

7. Banks, J. (2013). From Drones to fertilizers, farming embraces tech. Idahostatesman. com http://www.idahostatesman.com/2013/10/22/2827197/from-drones-to-fertilizers-farming.html, pp. 1–4 (January 13th, 2015).

8. Bayer CropScience 2014 Ripe for Robots: Automated Agricultural Helpers. http://www.cropscience.bayer.com/en/magazine/Ripe-for-Robots.aspx., pp. 1–3 (July 4th, 2014).

9. Blackmore, B. S. (2009). New concepts in Agricultural automation. In: Proceedings of HGCA Conference on 'Precision in arable Farming-Current Practice and Future potential. Thessaly, Greece., pp. 28–39.

10. Blackmore, B. S. (2014). In: How far off are Farm robots? Winsor, S. (Ed.). Corn and Soybean Digest, http://cornandsoybeandigest.com/how-far-are-farm-rob, pp. 1–5 (January 14th, 2015).

11. Blackmore, B. S., Stout, W., Wang, M. and Runov, B. (2005). Robotic Agriculture-the future of Agricultural Mechanization? In: Proceedings of 5th European Conference on precision Agriculture. Stafford, J. (Ed.). The Netherlands, Wageningen Academic Publishers., pp. 621–628.

12. Caldwell, J. (2014). Trimble rolls out crop-scouting drone. https://www.linkedin.com/groups/Low-Cost-UAV-Fights-Disease-53140.S.5809026007457361924, pp. 1–3 (January 15th, 2015).

13. Carnegie Robotics 2014 Custom Products. Carnegie Robotics, LLC. http://www.carnegierobotics.com/custom-products/, pp. 1–4 (July 13th, 2014).

14. Cockburn-Price, S. (2012). New Generation of robots poised to transform Global Agricultural production. http://www.robotics-platform.eu/cms/upload/Press_Room/Agri-Robotics_2_Final.pdf, pp. 1–4 (January 14th, 2015).

15. Country Guide-Canada, 2015 Improving Nitrogen Efficiency with Precision Farming. Precision Farming Dealer. http://www.precisionfarmingdealer.com/content/improving-nitrogen-efficeincy-precision-farming., pp. 1–5 (February 2nd, 2015).

16. Crow, J. M. (2012). Farmerbots: A new Industrial Revolution. New Scientist 2888:, pp. 18.

17. Derpsch, R. (2004). History of Crop Production, with and without Tillage. The Journal of No-tillage. 3: 150–154.

18. Dunbar, B. (2009). NASA uses satellite to unearth innovation in Crop forecasting. National Aeronautic and Space Agency, USA, http://www.nasa.gov./topics/earth/ features/crop_forecast.html, pp. 1–5 (December 22nd 2014).

19. Ehmke, T. (2013). Unmanned Aerial Systems for Field Scouting and Spraying. CSA News, pp. 4–9.

20. English, J. D. (2013). Robotic mass removal of Citrus. Energid Technologies Corporation, Massachusetts, USA, Research Report, pp. 1–5.

21. EPIC 2014 Domestic Unmanned Aerial Vehicles (UAVs) and Drones. Electronic and Privacy Information Centre. https://epic.org/privacy/drones.htm, pp. 1–14 (March 14th, 2014).

22. Epp, M. (2013). UAVs Taking Agronomy to New Heights. CropLife http://www. croplife.com/equipment/uavs-taking-agronomy-to-new-heights/, pp. 1–2 (March, 20th, 2014).

23. Feng, M. C., Yang, W., Cao, L. and Ding, G. (2009). Monitoring Winter wheat freeze injury using Multi-Temporal MODIS data. Agricultural Science in China DOI: 10.1016/s1671–2927(08)60313–2, pp. 1–9 (October 25th, 2014).

24. Galileo Geo Inc., 2014 Agriscience. http://galileo-gp.com/applications/agriscience/? gclid=CPnficHgocICFVgnjgodurwARQ., pp. 1–8 (November 29th, 2014).

25. Godoy, E. P., Tangerino, G. T., Andre O'Tabile, R., Inamasu, R. Y. and OvieraPorto, J. (2012). Network control system for the guidance of a four-wheel steering Agricultural Robotic platform. Journal of Control Science and Engineering 12: 1–10.

26. Gomez, K. (2013). Australian-designed robotic tractor enables precision in Planting. University of New South Wales, Sydney, Australia http://newsroom.unsw.edu.au/ news/science-technology/robotic-tractor-deliver-precision-planting, pp. 1–3 (January 18th, 2015).

27. Grassi, M. J. (2013). Rise of the Ag Drones. Kawak Technologies Inc., http://kawakaviation.com/what-people-say/ rise-of-the-ag-drones/, pp. 1–4 (August 6th, 2014).

28. Green, M. (2013). Unmanned Drones may have their greatest impact on Agriculture. http://www.thedailybeast.com/articles/2013/03/26unmnned-drones-may-have-theirgreatest-impact-on-agriculture.html#stash.c36uDpsT.dpuf., pp. 1–4 (August 3rd, 2013).

29. Grift, T. (2007). Robotics in Crop Production Encyclopedia of Agricultural, Food and Biological Engineering. Taylor and Francis Company, Boca Raton, Florida, USA, pp. 283.

30. Grift T. (2012). Robotics in Agriculture. University of Illinois Urbana Champaign, Illinois, USA http://abe-research.illinois.edu/Faculty/grift/Research/BiosystemsAutomation/AgRobots/AgRobots.html, pp. 1–8 (May 28th, 2014).

31. Grift, T. E., Kasten, M., and Nagasaka, Y. (2013). Development of Autonomous Robots for Agricultural Applications. University of Illinois. Urbana-Champaign, USA http://aberesearch.illinois.edu/Faculty/grift/Research/BiosystemsAutomation/ AgRobots/ RoboticsUIUC_CropProtectionConf.pdf 1–9 (January 6th, 2015).

32. Harris, M. (2015). USDA: Big Data implications for the production of Official statistics. Precision Farming Dealer http://www. precisionfarmingdealer.com/content/ usda-big-data-implications-prodcution-official-statistics., pp. 1–3 (January 30th, 2015).

33. Hetterick, H. (2013). Drones can Positive and Negative for the Agricultural Industry. Ohio's Country Journal. Ag Net. http://ocj.com/2013/05/drones-can-be-positive-and-negative-for-the-ag-industry/pp 1–2 (August 14th, 2014).

34. Hicks, J. (2012). Intelligent sensing Agriculture robots to harvest crops. Forbes. http://www.forbes.com/sites/jennifer/2012/ 08/06/intelligent-sensing-agriculture-robots-to-harvest-crops/, pp. 1–4 (March 18th, 2014).

35. Huang, Y., Steven, J. T, Hoffmann, C., Lan, Y. and Fritz, B. K. (2013). Development and Prospect of Unmanned Arial Vehicle Technologies for Agricultural Production Management. Journal of Agricultural and Biological Engineering 6:1–10.

36. I'nen, I. P., Saari, H., Kaivosoja, J., Honkavaara, E. and Pesonen, L. (2013). Hyperspectral imaging based biomass and Nitrogen content estimations from light-weight UAV. Proceedings of SPIE 8887 Remote Sensing for Agriculture, Ecosystems and Hydrology. DOI: 10.1117/12.2028624, pp. 1–14 (January 12th, 2015).

37. Japan UAV Association 2014 The Japan Unmanned Arial Vehicle (UAV0 Association. http://www.juav.org, pp. 1–2 (August 8th, 2014).

38. Kaivosoja, J., Pesonen, L., Kieemola, J., Inen, I. P., Salo, H., Honkavaara, E., Saari, H., Kykenen, J., Rajala, A. (2013). A case study of a precision fertilizer application task generation for wheat based on classified hyperspectral data from UAV combined with farm history data. Proceedings of SPIE 8887 Remote Sensing for Agriculture, Ecosystems and Hydrology DOI: 10.1117/12.2029165, pp. 1–8 (January 12th, 2015).

39. Keicher, R. and Suefert, H. (2000). Autonomous guidance for agricultural vehicles in Europe. Computers and Electronics in Agriculture 25: 169–194.

40. Khosla. R. (2010). Precision Agriculture: Challenges and Opportunities in Flat World. 19th World Congress of Soil Science, Soil Solutions for a Changing World, Brisbane, Australia, pp. 1–4.

41. Kimberlin, J. (2013). Virginia Beach crop drone gives peek into farming's future. Pilotline.com, pp. 1 (August 7th, 2014).

42. Kinze Manufacturers Inc., 2013 Kinze Manufacturing Continues Progress on Kinze Autonomy. http://www.kinze.com/ article.aspx?id=152, pp. 1–5 (June 18th, 2014).

43. Kinze Manufacturers Inc., 2014 Kinze autonomous Grain Cart System. Torcrobotics http://www.torcrobotics.com/case-studies/kinze.htm, pp. 1–4 (June 18th, 2014).

44. Krishna, K. R. (2012). Precision Farming: Soil fertility and Productivity Aspects. Apple Academic Press Inc., Waretown, New Jersey, USA, pp. 189.

45. Krishna, K. R. (2015). Agricultural Prairies: Natural Resources and Productivity. Academic Press Inc., Waretown, New Jersey, USA, pp. 522.

46. Lleo, L. B., Altisent, M., and Herror, A. (2009). Multispectral images of Peach related to firmness and maturity at harvest. Journal of Food Engineering 93: 229–235.

47. Lowenberg De-Boer, J. M. (2006). Trends in Adoption of Precision Farming. 8th International Conference on Precision Agriculture. Minneapolis, MN, USA, pp. 12.

48. Lumpkin, T. (2012). CGIAR Research Programs on Wheat and Maize: Addressing global Hunger. International Centre for Maize and Wheat (CIMMYT), Mexico. DG's Report, pp. 1–8.

49. Ministry of Agriculture, (2013). Beijing applies "Helicopter" in Wheat pest control. Ministry of Agriculture of the Peoples Republic of China-Report http://english.agri. gov.cn/news/dqnf/201306/t20130605_19767.htm, pp. 1–3 (August 10th, 2014).

50. Moorehead, S. J. and Bergerman, M. (2012). R Gator: an Unmanned Utility Vehicle for off-Road operations. Robotics and Automation Society: Technical committee on Agriculture and Automation http://www.researchgate.net /publication/260710461_ IEEE_Robotics_and_Automation_Society_Technical_Committee_on_Agricultural_ Robotics_and_Automation_TC_Spotlight, pp. 1–6 (July 3rd, 2014).

51. Morris, B. (2011). Future of Farming: Prospero Robotic Farmer. http://www.polize-ros.com/2011/12/27/future-of-farming-prospro-robot-farmr/, pp. 1–3 (July 3rd, 2014).

52. Mulla, D. (2014). In: At Minnesota's Farmfest, drones and data fly onto the agenda. Ramstad, A. (Ed.). Star Tribune_Business, http://startribune.com/busi-ness/269913801.htm, pp. 1–3 (September 21st, 2014).

53. Murray, P. (2012). Automation reaches French Vineyards with vine-pruning robots. http://www.singularityhub.com /2012/11/26/automation-reaches-french-vineyards-with-a-vine-pruning-robot/, pp. 1–4 (March 20th, 2014).

54. MZB, (2014). The MZB Precision Farming System. http://www.mzbtech.com/ whatisthemzbsystem.html, pp. 1–2 (January 11th, 2015).

55. Nagasaka, Y., Tamaki, K., Nishiwaki, K. Saito, M., Motobayashi, K., Kikuchi, Y and Hosokawa (2010). Autonomous rice field operation project in NARO. Proceedings of International Conference on Mechanotrics and Automation. Beijing, China, pp. 870–874.

56. National Aeronautic and Space Agency (2014). NASA to launch soil moisture active satellite. http://www.agprofessional.com /news/NASA-to-launch-soil-moisture-active-satellite-274361731.html, pp. 1–8 (November 16th, 2014).

57. New Scientist Tech, (2013). Precision Herbicide drones launch strikes on weeds. New Scientist 2924: 34–35.

58. NREC, (2014). Strawberry plant sorter. Carnegie Mellon University, Robotics program, http://www.nrec.ri.cmu.edu/projects/ strawberry/htm, pp. 1–2 (June 10th, 2014).

59. Okwu-Delunzu, V. U., Enete, C., Abubaker, A. S. and Laundi, A. S. (2013). Monitoring gully of Enugu State South-eastern Nigeria, using Remote sensing. Proceedings SPIE 8887. Remote Sensing for Agriculture, Ecosystems and Hydrology 15: 1–13 doi: 10.1117/12.2035967.

60. Panin, A. (2014). Proletarian Robots getting cheaper to exploit. The Moscow Times http://www.themoscowtimes.com /news/article/495603.html, pp. 1–3 (March 5th, 2014).

61. Paul, R. (2014). In: Agriculture gives UAVs a new purpose. AG Professional. http:// www.agprofessional.com/news/Agriculture-gives-unmanned-aerial-vehicles-a-new-purpose-2553801.html, pp. 1–3 (April 23rd, 2014).

62. Payne, J. (2013). Transformational Robotics and its application to Agriculture. Robo-hub. http://robohub.org/transformational-robotics-and-its-application-agriculture/, pp. 1–7 (March 21st, 2014).

63. Pedersen, S. M. (2003). Precision Farming – Technology assessment of Site-specific input application in Cereals. Danish Technical University, PhD thesis, pp. 139.

64. Pedersen, S. M., Fountas, S. and Blackmore S. (2007). Economic potential of Robots for high value crops and landscape treatment. Precision Agriculture Wageningen Academic Publications 08547–465, pp. 1–12.

65. Pedersen, S. M., Fountas, S. and Blackmore S. (2008). Agricultural Robots-Applications and Economic perspectives. In: Service Robot Applications. Takahashi, Y. (Ed.) I-Tech Education and Publishing, pp. 369–382.

66. Perry, E.M, Brand, J. Kant, S. and Fitzgerald, G. J. (2012). Australian Society of Agronomy Journal http://www.regional.org.au /au/asa/2012/precision-agriculture /7933_perrym.htm, pp. 1–8 (January 12th, 2015).

67. Precision Farming Dealer, (2014a). Get Ready for Robot Farmers. http://www.precisionfarmingdealer.com/content/get-ready-robot-farmers, pp. 1–7 (January 6th, 2014).

68. Precision Farming Dealer (2014b). New AgEagleRapid UAS automation system unveiled. Precision Farming Dealer http:/ /www.precisionfarmingleader.com/content/new-ageagle-rapid-uas-automation-system-unvieled., pp. 1–4 (October 21st, 2014).

69. Precision Farming Dealer (2014c). Agrigold, Kinze partner on Multi-hybrid planter. http://www.precisionfarmingdealer.com/content/agrigold-kinze-partner-multi-hybrid-planter, pp. 1–2 (June 4th, 2014).

70. Precision Farming Dealer (2014d). Unmanned Yamaha RMAX helicopter sprayer displayed at AUVSI conference. http://www.precisionfarmingdealer.com/content/content/unmanned-yamaha-rmax-helicopter-sprayer-displayed-auvsi-conference.htm, pp. 1–3 (May 16th, 2014).

71. Qin, J., Burks, T. F. Ritenour, M. A. and Bonn, W. G. (2009). Detection of Citrus Canker using Hyperspectral reflectance imaging with Special information divergence. Journal of Food Engineering 93:183–191.

72. Qin, Z. and Zhang, M. (2005). Detection of Rice sheath blight for in-season disease management using multispectral remote sensing. International Journal of Applied Earth observation using Multispectral Remote Sensing. 7: 115–126.

73. Ramstad, E. (2014). At Minnesota's Farmfest, drones and data fly onto the agenda. Star Tribune_Business, http://startribune.com/business/269913801.htm, pp. 1–3 (September 21st, 2014).

74. Raymond Hunt, E., Hively, W. D., Fujikawa, S. J., Linden, D. S., Daughtry, S. S. T. and McCarty, G. W. (2010). Acquisition of NIR-Green-Blue digital photographs from Unmanned Aircraft for Crop monitoring. Remote sensing 2: 290–305.

75. Redmond, S. (2014). The future of UAV's for Agriculture. Hensall District Cooperatives. http://www.hdc.on.ca/ grain-marketing/hdc-reports/29-grain-marketing/253-hdc-future-ofUAV-ag-steve-redmond.html, pp. 1–3 (July 21st, 2014).

76. Rovira-Mas, F., Zhang, Q. and Reid, J. R. (2005). Creation of three dimensional crop maps based on aerial stereo images. Biosystems Engineering 90: 251–259.

77. Rovira-Mas, F. (2009). 3D vision solutions for robotic vehicles navigating in common agricultural scenarios. In: Fourth IFAC International workshop on Bio-Robotics, Information Technology and Intelligent Control for Bio-production Systems, Urbana-Champaign, Illinois, USA, pp. 221–223.

78. Satellite Imaging Corporation (2014). Forestry. http://www.satimagingcorp.com/applications/natural-resources/forestry/, pp. 1–3 (October 1st, 2014).

79. Schmale, D. (2012). In: Tiny planes coming to scout crops. Drone planes take Aerial imaging to a new level. Ruen, J. (Ed.). Corn and Soybean Digest http://cornandsoybeandigest.com/corn/tiny-planes-coming-scout-crops, pp. 1–3 (August 16th, 2014).

80. Scott, B. (2012). Twelve most Advanced Agricultural Technologies. http://12most.com/2012/03/12/advanced-agricultural-technology/, pp. 1–12 (October 14th, 2014).

81. Sieh, C. (2014). Farmers learn how UAVs can be used to reduce inputs. Farming Futures http://www.farmingfutures.org.uk /blog/farmers-learn-how-drones-can-be-used-reduce-inputs, pp. 1–5 (January 17th, 2015).

82. Soegaard, H. T. (2005). Weed classification by active shape models. Biosystems Engineering 91: 271–281.
83. Soegaard, H. T. and Lund, I. (2005). Investigation of the accuracy of a machine vision based robotic micro-spray system. In: Proceedings of the 5th European Conference on Precision Farming, pp. 234–239.
84. Soegaard, H. T. and Lund, I. (2007). Application accuracy of a machine vision-controlled robotic micro-dosing system. Biosystems Engineering 96:315–137.
85. Stafford, J. V. (2005). Precision Agriculture '05 PRECISION AGRICULTURE '05. Wageningen Academic Publishers, Wageningen, The Netherlands., pp. 1005.
86. The State Journal of Illinois (2014). Farmers trying the Power of Precision. Precision Farming Dealer. http://www. precisionfarmingdealer.com/content/farmers-trying-harness-power-precision, pp. 18 (January 6th, 2015).
87. Tigue, K. (2014). University of Minnesota Research Group pushes for Ag Drones. Precision Farming Dealer. http://www.precisionfarmingdealer.com/content/university-minnesota-research-group-pushes-ag-drones.
88. Tobe, F. (2014). The Robot Report: Are Ag Robots ready? 27 companies profiled. http://www.therobotreport.com/news/ag-in-transition-from-precision-ag-to-ful-autonomy., pp. 1–22 (December 29th, 2014).
89. Torres-Sanchez, J., Pefia, J. M., De Castro, A. I. and Lopez-Granados, F. (2014). Multispectral mapping of the vegetation fraction in Early-season wheat fields using images from UAV. Computers and Electronics and Agriculture 103:104–113.
90. Trossen Robotic Community 2014 Prospero: Robotic Farmer. http://forums.trossenrobotics.com/ showthread.php?4669-Prospero-Robotic-Farmer, pp. 1–7 (July 3rd, 2014).
91. University of Southern Denmark 2014 Weeding Robot ready for Agriculture. http://www.sdu.dk/en/Om_SDU/ Fakulteterne/ Teknik/Nyt_fra_Det_Tekniske_Fakultet/ Radrenserrobottillandbruget., pp. 3 (May 19th, 2014).
92. Vangioukas, S. G. (2013). Robotics for Specialty crops: Past, Present and Prospects. Department of Biological and Agricultural Engineering, University of California, Davis, California, USA, pp. 1–5.
93. Van Henten, E. J., Grense, D. and Lockhorst, C. (2009). Precision Agriculture 2009. Wageningen Academic Publishers, Wageningen, Netherlands, pp. 992.
94. Vina, A., Gitelson, A. A., Rundquist, D. C., Keydan, G., Leavitt, B. and Schepers, J. (2004). Monitoring maize *(Zea mays)* Phenology with Remote sensing. Agronomy Journal 96: 1139–1147.
95. Vitirover, (2014). Vitirover: Replace the herbicide chemicals with Autonomous Robots. http://www.Vitirover.com /en/ FAQ.php, pp. 1–12 (June 25th, 2014).
96. Von Bueren, S and Yule, I. (2013). Multispectral aerial imaging of pasture quality and biomass using unmanned aerial vehicles. http://www.massey.ac.nz/~flrc/workshops/13/Manuscripts/Paper_vonBueren_2013.pdf, pp. 1–8 (January 15th, 2015).
97. Winsor, S. (2014). In: How far off are Farm robots? Corn and Soybean Digest, http://cornandsoybeandigest.com/how-far-are-farm-rob, pp. 1–5 (January 14th, 2015).
98. Wiser, J. (2014). The Robots are coming. Comstock's Magazine http://www.comstocksmag.com/longread/robots-are-coming., pp. 1–5 (January 6th, 2015).
99. Wooley, D. (2014). Robotic weeding leads to big labor saving. College of Engineering, Iowa State University of Science and Technology. https//engineering.iastate.edu/

research/2014/01/21/robotic-weeding-leads-to-big-labor-savings/, pp. 1–3 (May 22nd 2014).

100. Wrenn, E. (2012). Meet Wall-Ye: The French grape-picking robot which can work day and night and may well put Vine yard workers out of job. Mail One http://www. dailymail.co.uk/sciencetech/article-2209975/Meet-Wall-ye-the French-grape-grape-picking-robot-work-day-and -night-vineyard-workers-job.html., pp. 1–13 (January 3rd, 2014).

101. Xiaoyu, S., Bei, C., Guijun, Y. and Haikuan, F. (2014). Comparison of Winter wheat growth with multi-temporal Remote sensing imagery. Proceedings of Conference Series on Earth and Environmental Science DOI: 10.1088/1755–1315/17/1/012044, pp. 1–8 (October 25th, 2014).

102. Yan, L., Gou, Z. and Duan, Y. (2009). A UAV remote sensing system: Design and tests. Geospatial Technology for Earth Observation. DOI 10.1007/978–1-4419–0050–0_2, pp. 27–33 (January 12th, 2015).

103. Yang, Z., Mueller, R. and Crow, W. (2014). US National Crop land soil moisture monitoring using SMAP. http://www.nass.usda.gov/research/cropland/docsyangI-GARSSI3SMAP.pdf., pp. 1–17 (December 22nd, 2014).

104. Zemlicka, J. (2014). Relevance in the face of obsolescence. Precision Farm Dealer http://www.farm-equipment.com/pages/Spre/PFD-Editorial-Relevance-in-the-Face-of-Obsolescence-January-22, −2015.php, pp. 1–3 (January 23rd, 2015).

105. Zhang, Q and Pierce, F. J. (2013). Agricultural Automation: Fundamentals and Practices. CRC Press, Boca Raton, Florida, USA, pp. 411.

106. Zhang, M., Qin, Z., Liu, X. and Ustin, S. L. (2014). Detection of stress in tomatoes induced by late blight disease in California, USA, using hyperspectral remote sensing. International Journal of applied Earth Observation and Geoinformation. 4: 295–310.

107. Zkotala, A. (2013). Using Robotics to detect Citrus disease. Department of Engineering, University of Central Florida http://www.ucf.edu/using-robotics-to-detect-citrus-disease., pp. 1–2 (June 5th, 2014).

INDEX

T